Breeding and Biotechnology of Tea and its Wild Species

Tapan Kumar Mondal

Breeding and Biotechnology of Tea and its Wild Species

Springer

Tapan Kumar Mondal
Division of Genomic Resources
National Bureau of Plant Genetic resources
Delhi
India

ISBN 978-81-322-1703-9 ISBN 978-81-322-1704-6 (eBook)
DOI 10.1007/978-81-322-1704-6
Springer New Delhi Dordrecht Heidelberg London New York

Library of Congress Control Number: 2013958376

Printed on acid-free paper

Springer is part of Springer Science+Business Media (www.springer.com)

Dedicated to
my beloved parents

Preface

Tea is an important industrial crop that supports the life of several million plantation workers working globally. It is the morning drink of several million people worldwide. Interestingly, several wild species such as *C. japonica* are important due to its elegant flower colour. Because of its perennial nature with a life span of more than 100 years, breeding of tea and its wild species to improve the cultivars is difficult and limited to only few aspects. During the past 2 decades, as a student, teacher and humble science worker, I was, am, and surely will remain fascinated by this beautiful plant whose not only taste but also scenic beauty of plantation always refreshes my mind. While working with this plant, at various tea research institutes in the last 2 decades, I have experienced the present practices, gaps and scope of varietal improvement works and felt the need of *in vitro* culture, molecular breeding, and genomics to supplement the conventional breeding works. With the initiation of cell culture technique in 1968, a significant amount of work on various aspects of breeding and biotechnology of tea and its wild relatives has been done. Although several topical reviews and scientific articles have been published on tea and *Camellia* species, yet they are not codified in a single document.

I am deeply indebted to my teachers who blessed me to learn about this crop and plant biotechnology as a whole. Therefore, I sincerely acknowledge my thanks to my beloved teachers of Assam Agricultural University, Prof. P. S. Ahuja, Director and other Scientists of Institute of Himalayan Bioresource Technology, India, Prof. P. K. Chand of Utkal University, Scientists of UPASI, Tamil Nadu, Tocklai Experimental Station, Assam and Prof. P. C. Deka, Vice Chancellor, Sir Padampat Singhania University, Udaipur. Few people also inspired me to work further on tea breeding and they are Prof. N. K. Jain, Mr P. Haridas, and some of my planter friends of Southern India, Dooars, West Bengal as well as Assam.

I would also like to thank my wife, Dr. Bipasa Sarkar who helped me to improve the manuscript in several ways. Lastly, my son, Vaibhav, my younger sister, Tia and her family, elder brother, Prof. Swapan Kumar Mondal and his family, Kaku and his family are also gratefully acknowledged. I am also thankful to Profs. C. R. Park of USA, A. M. Vieitez of Spain, S. Matsumoto of Japan, Z. Apostolides of South Africa, Z. Chen of China, I. D. Singh of Sri Lanka, S. C. Das, T.R. Sharma, and L. M. S. Palni, India for my personal interactions with them since my student days. I apologize for those works, if any, which did not appear in this book despite a detail search worldwide.

I am also grateful to my PhD students Dr. Pranay, Olivia, Akan, Pratap, Mainaak and Showkat as I was enriched with knowledge while working with them. It is my sincere belief that this book will serve the requirement of students, scientists and industries involved in studies, teaching, research on breeding and biotechnology of tea and other *Camellia* species with an intension of serving science and society.

Tapan Kumar Mondal
New Delhi, India

Contents

About the Author

Dr. Tapan Kumar Mondal joined at Institute of Himalayan Bioresource Technology (CSIR), Himachal Pradesh, India, for his Ph.D degree which he completed in 1998. After that, he served as Deputy Manager, Research and Development Department, Tata Tatley Ltd, Munnar, Kerala till March, 2002. Since then up to 2010, he further served as Assistant Professor at North Bengal Agricultural University, Cooch Behar, West Bengal. Later in February 2010, he joined as Senior Scientist (Plant Biotechnology) at National Bureau of Plant Genetic Resource (ICAR), New Delhi. He carried out his Post Doctoral training with Prof. J. K. Zhu of University of California, Riverside, USA, on 'Regulation of small RNA under cold stress of Arabidopsis' and later worked at University of Illinois, Urbana-Champaign, USA, on 'Identification of nitrogen use efficient genes of maize by RNAseq'.

Dr. Mondal has significantly contributed in various areas of biotechnology and genetic resource management of tea. His work leads to develop the first transgenic tea plants. He has also submitted several gene sequences of tea at NCBI and also published more than 60 publications in this area. He was PI of various projects of tea biotechnology funded by DBT, DST, ICAR and Tea Board, India.

He is the recipient of University merit scholarship, scholarship from Indian Tea association, ICAR JRF, DBT fellowship, CSIR fellowship and life member of several professional societies. He also bagged 'Young scientist award' by Korean Society of Tea Science in 2003 and Japan Tea Science Society in 2004.

Abbreviations

ABA	Abscisic acid
AS	Acetosyringone
BAC	Bacterial artificial chromosome
ANN	Artificial neural network
6-BAP	6-Benzylaminopurine
cM	Centimorgan
CTAB	Cetyl trimethyl ammonium bromide
cv	Cultivar
°C	Degree celsius
CM	Coconut milk
2,4-D	2,4-Dichlorophenoxy acetic acid
DMSO	Dimethyl sulfoxide
d	Day(s)
g	Gram(s)
g/l	Gram(s) per litre
GA3	Gibberellic acid
GUS	*β*-glucuronidase
gus	*β*-glucuronidase gene
gusint	*β*-glucuronidase gene with an intron
h	Hour(s)
ha	Hectare(s)
hpt	Hygromygcin phophotransferase gene
HPLC	High performance liquid chromatography
IAA	Indole-3-acetic acid
IBA	Indole-3-butyric acid
Kn	Kinetin
KPa	Kilo pascal
kb	Kilo base pair
M	Molar
min	Minute(s)
m	Meter(s)
ml	Millilitre(s)
mM	Micromolar
MS	Murashige and Skoog's (1962) medium
µl	Microlitre
NAA	Napthalene acetic acid
NCBI	National Centre for Biotechnology Information

nM	Nano mole
npt-II	Neomycin phosphotransferase gene
O.D.	Optical density
PCR	Polymerase chain reaction
PVP	Polyvinyl pyrrolidone
pM	Pico mole
%	Percent
QTL	Quantitative trait loci
Q-PCR	Quantitative-PCR
rpm	Revolution per minute
RAPD	Random amplified polymorphic DNA
RFLP	Restriction fragment length polymorphism
ROS	Reactive oxygen species
sdH_2O	Sterile distilled water
Sec	Second(s)
SE	Somatic embryogenesis
SSR	Suppression subtractive hybridization
SSH	Simple sequence repeat
SNP	Single nucleotide polymorphism
t	Tonnes
Taq Pol.	*Taq* DNA polymerase
TBA	Tertiary butyl alcohol
TBE	Tris borate EDTA
TES	Tocklai experimental station
TDZ	Thidiazuron
TE	Tris-EDTA
UV	Ultra-violet
UPGMA	Unweighted pair group method with arithmetic mean
UPASI	United Planter Association of South-India
v/v	Volume by volume
WPM	Woody plant medium of Lloyd and McCown (1982)
w/v	Weight by volume
mg	Micro gram
ng	Nano gram
YE	Yeast extracts
YMB	Yeast mannitol broth

1 Introduction

1.1 Tea and Camellia: An Overview

Tea (*Camellia sinensis* (L.) O Kuntze) belongs to the family Theacea. It is the oldest non-alcoholic caffeine-containing beverage in the world. Chinese were the first to use tea as a medicinal drink, later as a beverage and have been doing so for the past 3,000 years (Eden 1958). The cultivated taxa comprise of three main natural hybrids. They are: *C. sinensis* (L.) O. Kuntze or China type, *C. assamica* (Masters) or Assam type and *C. assamica* subspecies *lasiocalyx* (Planchon ex Watt.) or Cambod or southern type. Tea is an evergreen, perennial, cross-pollinated plant and grows naturally as tall as 15 m. However, under cultivated condition, the bush height of 60–100 cm is maintained (Fig. 1.1) for harvesting the tender leaves to be processed for making the beverages. The flowers are white in colour and born singly or pairs at the axils. The fruits are green in colour with 2–3 seeds and start bearing within 5–6 years after planting. Leaf is the main criterion by which three types of tea are classified. Briefly, they are: (1) Assam type with biggest leaves, (2) China type with smallest leaves and (3) Cambod, leaves size are in-between of Assam and China type.

Tea thrives well within the latitudinal ranges between 45°N and 34°S that cross about 52 countries. Tea occupies about 2.7 million ha of cultivable land of the world with an annual production of about 2.2 million t. Despite occupying only 16.4% of the total tea growing areas of the world, India ranks first as the producer, consumer and exporter. Hence, tea plays a pivotal role in the national economy of India with an annual turnover of US$ 660 million.

1.2 History

Tea plants are believed to have discovered accidentally by the Chinese legendry Emperor Sheng Nong around 2737 BC. As soon as medicinal value began to be attributed to tea by Chinese, a demand for supplies of tea sprang up which results in cultivation of tea plant in Sichuan province about 3,000 years ago. Subsequently, the knowledge of tea cultivation was spread everywhere by the fine arts of Buddhism. Though, in India, wild tea plant was discovered by C. A. Bruce in Assam during 1823 but seeds were also brought by G. J. Gordon from China in 1836 for establishing a commercial garden in India. Later C. A. Bruce was appointed as the superintendent of tea plantation who took active interest to cultivate the indigenous tea plant. Soon commercial interests moved in and the world's first privately owned Tea Company, the Assam Tea Company, Assam, India was established on 12 February, 1839 with the directives from British Parliament. This was the beginning of the present day Tea Industry of India.

T. K. Mondal, *Breeding and Biotechnology of Tea and its Wild Species*,
DOI 10.1007/978-81-322-1704-6_1, © Springer India 2014

Fig. 1.1 Tea plantation at Palampur, Himachal Pradesh, India. Plants are growing under shade trees

1.3 Origin and Distribution

Southeast Asia is the original home for tea. According to Wight (1959), the primary centre of origin of tea was considered around the point of intersection of latitude 29°N and longitude 98°E near the source of the river, Irrawaddy, the point of confluence where lands of Assam, North Burma, Southwest China and Tibet met. Secondary centres of origin were considered to be located in Southeast China, Indochina, Mizoram and Meghalaya (Kingdon-Ward 1950). The above areas were, therefore, considered to be the zone of origin and dispersion of the genus *Camellia* as a whole (Sealy 1958). However, presently tea is grown within the latitudinal range of 45°N to 34°S.

Tea was introduced to Japan from China in the early part of the eighth century. From Japan, tea cultivation extended to Indonesia during the seventeenth century. In Sri Lanka, tea was first planted in 1839 when seeds were taken from Calcutta, India. In USSR, tea cultivation started when seeds were imported from China towards the end of last century. Later, from USSR, seeds were exported to Turkey in the year 1939–1940. In Europe, tea was introduced in 1740 by the East India Company's Captain Goff but those plants which were planted in the Royal Botanic Garden at Kew in England, could not survive (Sealy 1958) and the first successful introduction was achieved by a British merchant cum naturalist, John Ellis during 1768 (Aiton 1789; Booth 1830). From there, tea cultivation was extended to the African countries at the end of the nineteenth century.

1.4 Morphological Descriptions

A summary of the morphological characters of the three races of tea plants as described by Wight (1962), Barua (1963) and Bezbaruah (1971) is given below:

1. The China type {*C. sinensis* (L.) Kuntze}: It is a big shrub, 1–2 m tall with many virgate stems arising from the base of the plant near the ground with hard, thick and leathery leaf, matty surface, marginal veins indistinct and appears sunken in lamina. Blade elliptic with obtuse or broadly obtuse apex, base cuneate, margin bluntly serrulate to sinuate-serrulate with more or less incurved teeth, glabrous above and villose below when young, becoming sparsely villose as the leaf ages, ultimately becoming glabrous. Young leaves are garnet-

brown through ox-blood to purple in colour. Petiole is short, 3–7 mm long, stout, usually giving the leaf an erect pose. Flowers are borne singly or in pairs in the cataphyllary axils. Pedicel short, 6–10 mm long, clavate, glabrous with 2–3 subopposite scars little below the middle, marking the position of caducous bractioles 2–5 mm long. Sepals are 5–6 in number, imbricate, persistent, leathery, ovate or orbicular, 3–6 mm long, glabrous green. Petals are 7–8, shallowly cup-shaped, 1.5–2 cm long, broad oval to suborbicular, generally white sometimes with pale pink pigmentation. Stamen numerous, arranged in two whorls, inner ones shorter and fewer in number, outer longer and more numerous, 8–13 mm long, united at the base for a few millimetres with the corolla lobes. Ovary is white and densely hairy; three locular ovules are present 3–5 in each loculous, placentation axial. Style generally three, sometimes up to five, free for the greater part of their length, occasionally free up to the base of the ovary. Stigma is apical. The number of capsules is one, two or three coccate, containing 1–3 nearly spherical seeds with 10–15 mm in diameter. Based on leaf sizes, Sealy (1958) recognized two forms of *C. sinensis* (a) f. macrophylla (sieb.) Kitamura, with wild leaves 4–14 cm long, 2–2.5 cm wide and (b) f. parvifolia (Miq.) Sealy, with leaves 1.5–1.6 cm long and 1–1.2 cm wide.

2. The Assam type {*C. assamica* (Masters)}: It is a small tree, 10–15 m tall with a trunk sometimes up to one-thirds of its height, possesses a robust branch system. In typical plants, leaf is thin, glossy with more or less acuminate apex and distinct marginal veins. Leaf blade is usually broadly elliptic, 8–20 cm long and 3.5–7.5 cm wide, base cuneate, margin obscurely denticulate to bluntly wide serrulate, glabrous or persistently hairy on the midrib below. Flowers are single or in pairs on the cataphyllary axils, pedicels with scars of three caducous bracteoles, smooth and green. Sepals are 5–6 unequal, leathery, persistent. Petals are white 7–8 in number, occasionally with pale yellow pigmentation at the base of the petals. Stamens are numerous as in *C. sinensis*.
3. The southern form or Cambod type {*C. assamica* subspecies. lasiocalyx (Planch. MS)}: It is a small fastigiate tree, 6–10 m tall, with several upright, almost equally developed branches. Leaf is more or less erect, glossy and yellowish-green when young, light green at maturity changing to coppery-yellow or pinkish-red from autumn till the end of the season. Petioles are pinkish-red at the base. Leaf size is intermediate between China and Assam type, broadly elliptic, marginal veins not very prominent. Ovaries are 3–4 in number with five locular. Styles are 3–5 in number, free nearly up to half the length, straight with apical or linear stigma. On the other floral characters, it resembles the Assam plant, with the difference that four or more bracteoles are found on the pedicel of flowers.

1.5 Taxonomy and Nomenclature

The taxonomic position of tea is given below (Fig. 1.2). It is noteworthy to mention that the family comprises 11 genus and the genus *Camellia* has more than 325 species. Out of that, only two are commercially cultivated for producing the tea.

1.6 Economic Importance and Health Benefits

The economic importance of the genus *Camellia* is primarily due to the tea. Though tea is mainly consumed in the form of 'fermented tea' or 'black tea', yet 'non-fermented' or 'green tea' and lesser known 'semi-fermented' or 'oolong tea' is also available. They differ in their method of manufacture, chemical constituent, appearance and organoleptic taste. While black tea is widely used in India and other European countries, green tea is popular in China, Japan and Taiwan. Oolong tea is mainly consumed in some parts of China as well as Taiwan. Worldwide, 80 % black tea, 18 % green tea and 2 % oolong tea are being produced.

Kingdom	Plantae - Plants
Subkingdom	Tracheobionta - Vascular plants
Superdivision	Spermatophyta - Seed plants
Division	Magnoliophyta - Flowering plants
Class	Magnoliopsida - Dicotyledons
Subclass	Dilleniidae
Order	Theales
Family	Theaceae
Genus	*Camellia* L.
Species	*Camellia sinensis* (L.) O. Kuntze - tea
Types	Assam, China, Cambod
Cultivar	600 recognized worldwide.

Fig. 1.2 Taxonomic position of tea

For black tea, the young tender leaves are completely fermented after withering. The fermentation results in oxidation and polymerization of polyphenols, changing the nature of the chemicals constituents of tea leaves and forming theaflavin and thearubigin. These polyphenols are responsible for the briskness, strength, colour, taste, aroma and pungency associated with black tea. The infusion of black tea has a bright red or copper colour, astringent taste and characteristic aroma. On the other hand, green tea is unfermented and is the least processed among the three types. The plucked leaves are harvested and steamed immediately to inactivate the enzymes to prevent oxidation and polymerization of primary polyphenols which result in retaining of green colour in the finish product. Green tea infusion has a smell of fresh vegetables, low caffeine content. In oolong tea, primary polyphenols are allowed to oxidize partly. Oolong tea is not common and is intermediate in characteristic between green and black tea. Immediately after plucking, the tea leaves are partially fermented for about half the time of black tea. It has the colour of black tea and flavour of green tea.

Tea was used initially as a medicine, later as a beverage and now proven well as future potential of becoming an important industrial and pharmaceutical raw material. Scientific reports in the last two decades have validated many beneficial claims for tea. The majority of the beneficial effects have been attributed to the polyphenolic constituents. Several studies suggest that phenolics may be of importance in reducing the incidence of degenerative diseases such as cancer and arteriosclerosis. The most relevant compounds in dietary regime are cinamic acid derivatives and flavonoids. As natural polyphenols remain unchanged in green tea, it can be said that green tea is more beneficial than black tea. The various health benefits in relation to cancer, arthritis, cardiovascular diseases, diabetes and obesity are described below:

1. Antioxidant activity: Most beneficial effects of tea catechins were attributed to their antioxidant properties that sequester metal ions and scavenge oxygen species and free radicals (Wiseman et al. 1977). Among the different components of catechin, (-)-Epigallocatechin 3-O-gallate (EGCG), was the most potent chemical of the epicatechin derivatives tested for biological activity. It was thought to prevent tumourigenesis by protecting cellular components from oxidative damage through free radical scavenging. Indeed many of the studies had confirmed the free radicals scavenging activity of EGCG as well as its antimutagenic, antiangiogenic, antiproliferating and /or pro-apoptotic effects on mammalian cells both *in vitro* and *in vivo* (Allemain 1999). Tea catechins had been found to be better antioxidants than vitamin C, E, tocopherol and carotene. The polyphenols blocked free radicals damage to lipids (found in cell membranes and serum lipids), nucleic acids and proteins (like those found as cellular enzymes and structural proteins). Damage to these cell components can lead to tumour formation. The oxidative damage by oxygen free radicals of low density lipoproteins (LDL) in serum led to arteriosclerosis and coronary heart diseases. The oxidation of cell membranes and other cell components led to ageing. The antioxidants activity of tea polyphenols was not only due to their ability to scavenge superoxides but also due to increase the activity of some detoxifying enzymes such as glutathione peroxidase, glutathione reductase, glutathione-S-transferase, catalase and quinine reductase in the small intestine, liver and lungs which are the of the body. However, the antioxidant activity of tea is diminished by the addition of milk to the infusion due to binding of tea polyphenols to milk proteins.

2. Cardiovascular activity: Tea polyphenols and flavonoids had been reported to inhibit either enzymatic or non-enzymatic lipid peroxidation, an oxidative process implicated in several pathological conditions including atherosclerosis. Specifically, it had been suggested that tea polyphenols lower the oxidation of LDL cholesterol, with a consequent decreased risk of heart diseases. It had been observed that green tea polyphenols significantly reduced the levels of serum LDL, very low-density lipoproteins and triglycerides. At the same time, they increased the levels of high-density lipoproteins. This observation had been strengthen by the finding that in hypercholesterolemic rats, green tea polyphenols lowered blood cholesterol levels and reduced blood pressure in spontaneously hypertensive animals. Tea polyphenols also inhibited the absorption of dietary fats and cholesterols (Chen et al. 2000).
3. Anticancer activity: It is the most thoroughly studied function of tea polyphenols. It can protect the cells against cancer at several stages of carcinogenesis including cancer prevention, endogenous carcinogenic activation, DNA damage and destabilization, cell proliferation, neoplastic growth and metastasis. Tea especially green tea reduced the incidence of cancers of the stomach, small intestine, pancreas, lung, breast, skin, urinary bladder, prostrate, esophagus and mouth (Vasisht et al. 2003). Also it had been shown to reduce tumour size and growth in cancer-bearing animals. Green tea polyphenols directly inhibited the cytochrome P450 enzyme systems (phase I enzyme) that played a pivotal role in carcinogenic activation. Concurrently, they boosted the activity of phase II enzyme (e.g. glutathione transferase) that made xenobiotics hydrophilic for clearance of the body. The process was crucial for carcinogenic detoxification. Recently, it was proposed to associate the anticancer activity of EGCG with the inhibition of urokinase, one of the most frequently expressed enzymes in human cancers. Green tea showed a protective effect against damage produced by UV radiation and reactive oxygen species to the dermis through apostasies and inhibiting lipid peroxidation.
4. Antidiabetic effect: Tea drinking was shown to possess antidiabetic activity and was effective in the prevention and treatment of diabetes. Tea polyphenols lowered the serum glucose by inhibiting the activity of the starch digesting enzyme, amylase. Polyphenol inhibits both salivary and intestinal amylase. As a result, the starch was broken down more slowly and the sudden rise in serum glucose was minimized. In addition, tea polyphenols also reduced the intestinal absorption of glucose (Gomes et al. 1995).
5. Antiarthritic activity: Tea polyphenol demonstrated an exceptional protection against arthritis. The major polyphenols showing antiarthritic effect include epicatechin, epigallo-catechin, epicatechin-3-gallate and epigallocatechin-3-gallate. It has been postulated that the antioxidants property of polyphenol might be useful in the prevention and severity of arthritis (Tapiero et al. 2002).
6. Antiplaque activity: Tea polyphenols acted in two different ways to inhibit the growth and adherence of oral bacteria. Firstly, it inhibited the growth of periodontal diseases producing bacterium, *Porphyromonas gingivilis* and decay-causing bacteria such as *Streptococcus mutans*. Therefore, green tea as mouth rinse resulted in less plaque and periodontal diseases. Secondly, it inhibited the enzyme amylase present in the saliva and the starch in the mouth did not get converted into glucose and maltose. Less nutrition was thus available to decay-causing bacteria (Yu et al. 1995).
7. Antiviral activity: Tea extract had been shown to have virucidal activity against polio, influenza, and herpes simplex virus (Okubo and Juneja 1997).
8. Anti-AIDs activity: Green tea polyphenols are antimutagenic and acts as an effective adjuvant to drug therapy. It had been discovered that polyphenols from green tea and their oxidation products could inhibit the reverse transcriptase or polymerase of several types of viruses, including HIV-1 and

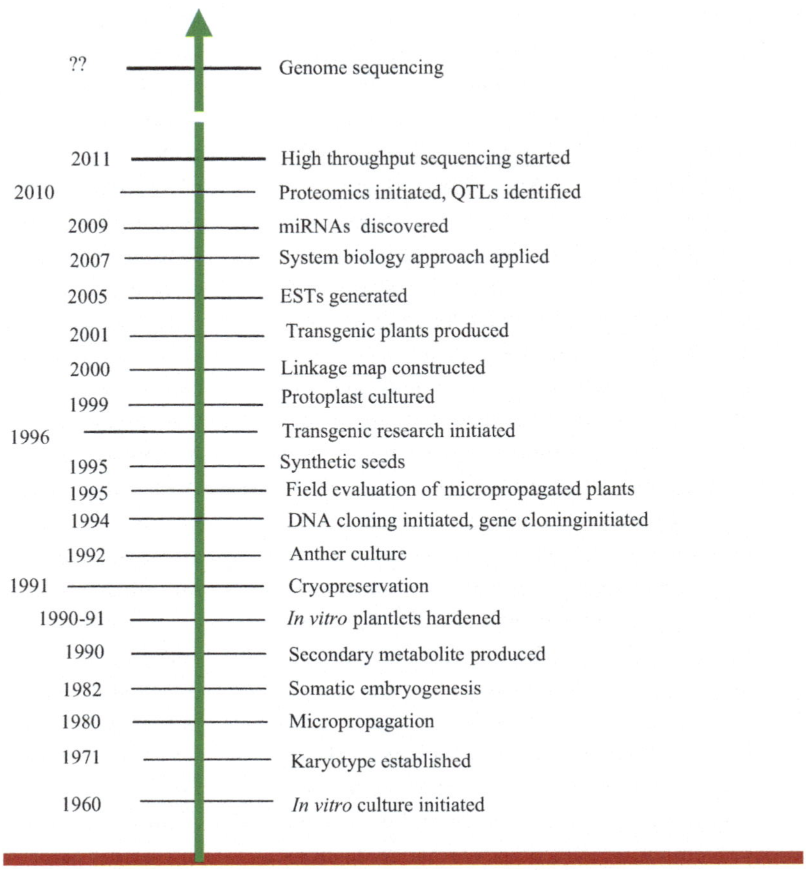

Fig. 1.3 Time-frame flow of landmark discoveries of tea

herpes simplex 1 (Hashimoto et al. 1996). However, research in this area is till in its initial stages.

9. Anorectic effect: The tea polyphenols inhibited catechol-O-methyl transferase and caffeine inhibited transcellular phosphodiesterase thus stimulated thermogensis and helped to manage the obesity. The release of glucose was slowed down by tea and thus harmful spiking of insulin was prevented (Kwanashie et al. 1989).
10. Antimicrobial activity: The crude catechins and theaflavins had been found to have an antibacterial activity. They were believed to damage bacterial cell membranes. Tea had been used in the treatment of diarrhoeal infections and cholera. Polyphenols killed the spores of *Clostridium botulinum* and thus displayed antibacterial activity against food-borne diseases and were also effective against heat-resistant bacilli such as *Bacillus subtilis, Bacillus cereus* and *Vibrio parahaemolyticus*. Green tea also had protozoacidal properties (Hamilton-Miller 1995).
11. Other biological effects: It had also been reported that green tea polyphenols exhibited neuromuscular, antiangiogenic, antihepatotoxic, antiproliferative/apoptotic and immunomodulatory effects (Sueoka et al. 2001).

1.7 Landmarks of Biotechnological Works

An overview of various landmark works of tea is depicted in Fig. 1.3. Caffeine (1, 3, 7-trimethylxanthine) is the main alkaloid for which tea is valued. It was purified from tea leaves of field grown plants during 1821 (Spedding and Wilson 1964). However, with the advancement of cell culture techniques, attentions were paid to produce in higher quantity of Caffeine from the *in vitro* callus tissue (Ogutuga and Northcote 1970a). Simultaneously, cytogenetic works were

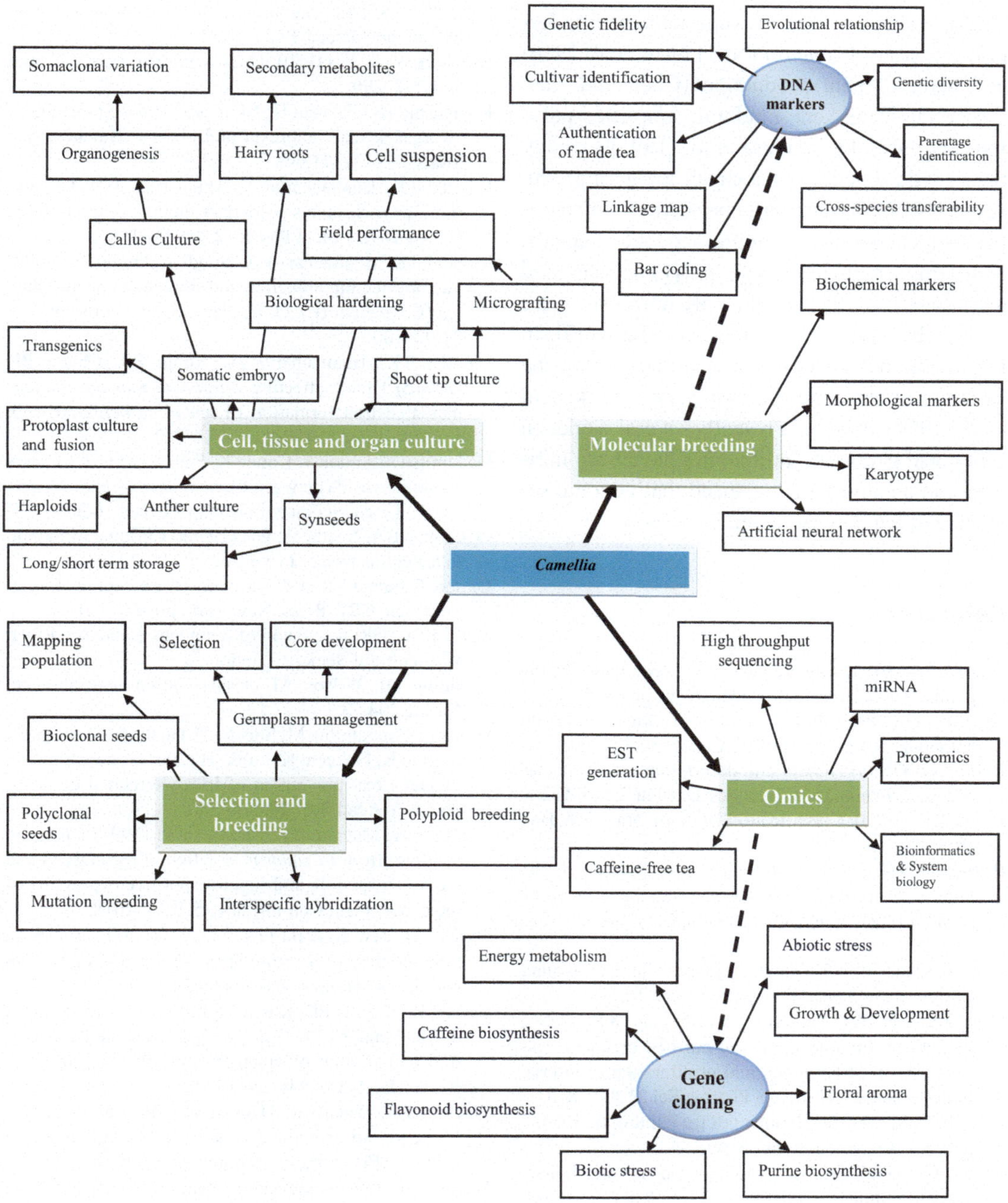

Fig. 1.4 Schematic explanation of *Camellia* improvement. The *bold arrows* are the major areas of research. *Dotted arrows* are the subareas. *Thin arrows* are the different applications with a major or submajor area

also initiated. Accordingly chromosome number had been established for the most available taxa of *Camellia* including tea (Bezbarua 1971) at TES, India and elsewhere. However, it was evident from the literature that while Forrest (1969) was pioneer to establish the *in vitro* culture of tea, Kato (1985) did a systematic study on micropropagation. Since then a significant amount of work has been done on tea and its wild relatives on various aspect of *in vitro* culture (Fig. 1.4). In tea, somatic embryogenesis had been fully exploited for clonal propagation (Mondal et al. 2001a), genetic transformation (Mondal et al. 2001b), artificial seed production (Mondal et al.

2000), embryo rescue in some interspecific hybrid crosses of *Camellia* (Nadamitsu et al. 1986) and haploid plant production (Chen and Liao 1982). DNA markers, randomly amplified polymorphic DNA by Wachira et al. (1995) to study the genetic diversity of Kenyan tea germplasm, several DNA-based markers such as intersimple sequence repeat, simple sequence repeats, restriction fragment length polymorphism and amplified fragment length polymorphism were utilized by several workers across the world. In tea, functional genomics was initiated with the isolation of chalcone synthase gene by Takeuchi et al. (1994) at Japan. Presently several works on functional genomics, proteomics have been initiated and achieved such a speed that genome sequence of tea may be unveil very soon.

References

Aiton W (1789) Hortus kewensis, or a catalogue of the plants. Royal Botanical Garden, Kew, pp 48–57

Allemain G (1999) Multiple actions of EGCG, the main component of green tea. Bull Cancer 86:721–724

Booth WB (1830) History and description of the species of *Camellia* and Thea. Hort Soc London 7:519–562

Barua PK (1963) Classification of tea plant. Two Bud 10:3–11

Bezbaruah HP (1971) Cytological investigation in the family theaceae-I. Chromosome numbers in some *Camellia* species and allied genera. Carylogia 24:421–426

Chen Z, Liao H (1982) Obtaining plantlet through another culture of tea plants. Zhongguo Chaye 4:6–7

Chen ZY, Law WI, Yao XQ, Lau CW, Ho WK, Huang Y (2000) Inhibitory effects of purified green tea epicatechins in construction and proliferation of arterial smooth muscle cells. Acta Pharma Sci 21:835–840

Eden T (1958) The development of tea culture. In: Eden T (ed) Tea. Longman, London, pp 1–4

Forrest GI (1969) Studies on the polyphenol metabolism of tissue culture derived from the tea plant (*C. sinensis* L.). Biochem J 113:765–772

Gomes A, Vedasiromoni JR, Das M, Sharma RM, Ganguly DK (1995) Antihyperglycemic effect of black tea (*C. sinensis*) in rat. J Ethnopharmacol 45:223–226

Hamilton-Miller JM (1995) Antimicrobial properties of tea (*C. sinensis* (L) kuntze). Antimicrob Ag Chemother 39:2375–2377

Hashimoto F, Kashiwada Y, Nonaka GI, Nishioka I, Nohara T, Cosentibno LM, Lee KH (1996) Evaluation of tea polyphenols as anti-HIV agents. Med Chem Lett 6960:695–700

Kato M (1985) Regeneration of plantlets from tea stem callus. Jap J Breed 35:317–322

Kingdon-Ward F (1950) Does wild tea exist? Nature 165:297–299

Kwanashie HO, Usman H, Nkim SA (1989) Screening of "Kargasok tea": anorexia and obesity. Biochem Soc Trans 17:1132–1133

Mondal TK, Bhattacharya A, Ahuja PS (2001a) Induction of synchronous secondary embryogenesis of Tea (*C. inensis*). J Plant Physiol 158:945–951

Mondal TK, Bhattacharya A, Sood A, Ahuja PS (2000) Factor effecting induction and storage of encapsulated tea (*C. sinensis* (L). O. Kuntze) somatic embryos. Tea 21:92–100

Mondal TK, Bhattacharya A, Ahuja PS, Chand PK (2001b) Factor effecting *Agrobacterium tumefaciens* mediated transformation of tea (*C. sinensis* (L). O. Kuntze). Plant Cell Rep 20:712–720

Nadamitsu S, Andoh Y, Kondo K, Segawa M (1986) Interspecific hybrids between *C. vietnamensis* and *C. chrysantha* by cotyledon culture. Jap J Breed 36:309–313

Ogutuga DBA, Northcote DH (1970a) Caffeine formation in tea callus tissue. J Exp Bot 21:258–273

Okubo T, Juneja LR (1997) Chemistry and application of green tea. CRC Press, New York, pp 109–121

Sealy JR (1958) A revision of the genus *Camellia*. Royal Horticultural Society, London

Spedding DJ, Wilson AT (1964) Caffeine metabolism. Nature 204:73

Sueoka N, Suganuma M, Sueoka E, Okabe S, Matsuyama S, Imai K, Nakachi K, Fujiki H (2001) A new function of green tea: prevention of lifestyle-related diseases. Ann N Y Acad Sci 928:274–280

Takeuchi A, Matsumoto S, Hayatsu M (1994) Chalcone synthase from *C. sinensis* isolation of the cDNAs and the organ-specific and sugar-responsive expression of the genes. Plant Cell Physiol 35:1011–1018

Tapiero H, Tew KD, Ba GN, Mathe G (2002) Polyphenols: do they play a role in the prevention of Human pathogens? Biomed Pharma 56:200–207

Vasisht K, Sharma PD, Karan M, Rakesh D, Vyas S, Sethi S, Manktala R (2003) In: Study to promote the industrial exploitation of green tea poly-phenols in India. ICSHT-UNIDO, Italy. pp. 15–22

Wachira FN, Waugh R, Hackett CA, Powell W (1995) Detection of genetic diversity in tea (*C. sinensis*) using RAPD markers. Genome 38:201–210

Wight W (1959) Nomenclature and classification of tea plant. Nature 183:1726–1728

Wight W (1962) Tea classification revised. Curr Sci 31:298–299

Wiseman SA, Balentine DA, Frei B (1977) Antioxidant in tea. Crit Rev Food Sci Nutra 37:705–718

Yu H, Oho T, Xu LX (1995) Effects of several tea components on acid resistant of human tooth enamel. J Dental Sci 23:101–105

2 Genetics and Breeding

2.1 Introduction

Although applications of breeding in tea are difficult, entire varietal development in tea and other *Camellia* species have been done through conventional breeding which started way back in 1939. Since then, several developments of genetics and breeding have taken place, which are discussed in this chapter.

2.2 Genome Size

The genome size of a tea plant was initially estimated to be 4.0 G bases (Hanson et al. 2001; Tanaka et al. 2005); however, intraspecific and interspecific variations of 2C DNA content were also observed in the genus *Camellia*. It had been found that while intrapopulation variations of 2C DNA content of tea were 5.87 to 6.4 pg of DNA, the interpopulation variations of 2C DNA (pg) content of different species varied from 2.5 to 25 (Huang et al. 2013). The higher DNA content was due to higher levels of ploidy. For example, *C. oleifera* and *C. sasanqua* had 2C DNA (pg) 17.47 and 18.79 as they were found to be octaploids (Huang et al. 2013). Tea chromosomes were found to be small and had a tendency to clump together due to 'stickiness'. Tea is diploid ($2n=30$; basic chromosome number, $x=15$) and karyotype ranges from 1.28 to 3.44 μm (Bezbaruah 1971). The r value (ratio of long arm to short arm) for all the 15 pairs of chromosomes ranged from 1.00 to 1.91. This consistency in diploid chromosome number suggested a monophyletic origin for all *Camellia* species. However, a few higher ploidy levels, such as triploids ($2n=45$), tetraploids ($2n=60$), pentaploids ($2n=75$), hexapolids ($2n=90$), octaploids ($2n=120$) and aneuploids ($2n\pm1$ to 29) had also been identified (Singh 1980; Huang et al. 2013).

2.3 Diversity of the Genus

The genus of *Camellia* had 40 species in 1920. The number of species was increased to 87 in 1958 (Sealy 1958), and more than 267 species were registered in 1982 (Chang and Bartholomew 1984). Presently, this genus is believed to comprise more than 300 species Mondal et al. 2004 with the latest discovery of *C. cherryana* in 2012 (Orel and Wilson 2012), that indicated genetically instable and high outbreeding nature of the genus. By a conservative estimation, there are more than 30,000 cultivated varieties of ornamental Camellia worldwide. The *Camellia* is the largest genus of the family 'Theaceae'. The genus is valued for tea due to the presence of caffeine, a purine alkaloid, which acts as a stimulus for the central nervous system of human beings. Nagata and Sakai (1984) reported the distribution of caffeine in 23 species of *Camellia*. The caffeine content on a dry weight basis in some of them was as follows: *C. sinensis* var. *sinensis* (3.5 %), *C. sinensis* var. assamica (4 %), *C. taliensis* (2.54 %), and *C. kissi* (0.02 %).

T. K. Mondal, *Breeding and Biotechnology of Tea and its Wild Species*,
DOI 10.1007/978-81-322-1704-6_2, © Springer India 2014

Among these, *C. kissi* belongs to the section *Paracamellia* and the other genera belong to the section *Thea*. The other three genera in the family are *Eurya* with 140 species, *Ternstroemia* with 130 species and *Adinandra* with 100 species. Apart from having caffeine in the three species, around 50 species of the genus *Camellia* were known to produce oil for industrial uses (Mondal 2011).

The classification of genus had been revisited many times by several workers (Chen et al. 2000), but Chang and Bartholomew (1984) remained the most popular who divided the whole *Camellia* genus into 4 subgenera and 20 sections in total, which are depicted below with the example of some prominent species in each section.

I Subgenus*	Protocamellia
Section I	Archecamellia
	C. granthamiana
	C. albogigas
	C. pleurocarpa
Section II	Stereocarpus
	C. krempfii
	C. dormoyana
	C. yunnanensis
	C. liberistyla
	C. liberistyloides
Section III	Piquetia
	C. piquetiana
II Subgenus	Camellia
Section IV	Olifera
	C. gauchowensis
	C. sasanqua
	C. vietnamensis
	C. oleifera
Section V	Furfuracea
	C. integerrima
	C. polypetala
	C. latipetiolata
	C. crapnelliana
	C. furfuracea
	C. oblata
	C. gaudichaudii
	C. parafurfuracea
Section VI	Paracamellia
	C. grijsii
	C. confusa

I Subgenus*	Protocamellia
	C. kissii
	C. lutescens
	C. fluviatilis
	C. brevistyla
	C. obtusifolia
	C. maliflora
	C. miyagii
	C. shensiensis
	C. brevissima
	C. puniceiflora
	C. tenii
	C. microphylla
	C. phaeoclada
	C. weiningensis
Section VII	Pseudocamellia
	C. szechuanensis
	C. chungkingensis
	C. trichocarpa
	C. ilicifolia
	C. henryana
Section VIII	Tuberculata
	C. tuberculata
	C. anlungensis
	C. obovatifolia
	C. rhytidocarpa
	C. litchii
	C. parvimuricata
Section IX	Luteoflora
	C. luteoflora
Section X	Camellia
	C. omeiensis
	C. polyodonta
	C. lapidea
	C. mairei
	C. villosa
	C. kweichowensis
	C. albovillosa
	C. albescens
	C. tunganica
	C. trichosperma
	C. phellocapsa
	C. semiserrata
	C. multiperulata
	C. lungshenensis
	C. reticulata
	C. pitardii
	C. hiemalis
	C. uraku
	C. edithae
	C. xylocarpa

I Subgenus*	Protocamellia
	C. hongkongensis
	C. cryptoneura
	C. oviformis
	C. compressa
	C. setiperulata
	C. saluenensis
	C. boreali-yunnanica
	C. lucidissima
	C. magnocarpa
	C. japonica
	C. subintegra
	C. longicaudata
III Subgenus	Thea
Section XI	Corallina
	C. coralline
	C. tonkinensis
	C. wardii
	C. pilosperma
	C. fleuryi
	C. nitidissimia
	C. paucipunctata
	C. lienshanensis
	C. pentamera
	C. scariosisepala
	C. acutiserrata
Section XII	Brachyandra
	C. muricatula
	C. szemaoensis
	C. pachyandra
	C. xanthochroma
	C. amplexifolia
	C. brachyandra
	C. nervosa
	C. nematodea
	C. gilbertii
	C. crassipetala
	C. yangkiangensis
	C. parviflora
Section XIII	Longipedicellata
	C. amplexicaulis
	C. petelotii
	C. longipedicellata
	C. indochinensis
Section XIV	Chrysantha
	C. flava
	C. aurea
	C. chrysantha
	C. flavida
	C. impressinervis
	C. euphlebia

I Subgenus*	Protocamellia
	C. chrysanthoides
	C. tunghinensis
	C. pingguoensis
	C. pubipetala
Section XV	Calpandria
	C. lanceolata
	C. connata
Section XVI	Thea
	C. quinquelocularis
	C. tachangensis
	C. crassicolumna
	C. pentastyla
	C. taliensis
	C. irrawadiensis
	C. crispula
	C. gymogyna
	C. costata
	C. yunkiangensis
	C. leptophylla
	C. pubicosta
	C. angustifolia
	C. sinensis
	C. assamica
	C. pubilimba
	C. waldenae
	C. fangchensis
	C. ptilophylla
	C. parvisepala
Section XVII	Longissima
	C. longissima
	C. gracilipes
Section XVIII	Glaberrima
	C. glaberrima
	C. kwangtungensis
IV Subgenus	Metacamellia
Section XIX	Theopsis
	C. macrosepala
	C. cuspidata
	C. grandiflora
	C. chekiangensis
	C. longicuspis
	C. crassipes
	C. longicalyx
	C. forrestii
	C. acutisepala
	C. buxifolia
	C. minutiflora
	C. parvicuspidata
	C. acutissima
	C. subacutissima

I Subgenus*	Protocamellia
	C. callidonta
	C. handelii
	C. triantha
	C. costei
	C. tsaii
	C. synaptica
	C. transnokoensis
	C. rosthorniana
	C. lutchuensis
	C. euryoides
	C. trichoclada
	C. parvilimba
	C. brevipes
	C. elongata
	C. longicarpa
	C. parvilapidea
	C. stuartiana
	C. transarisanensis
	C. fraterna
	C. dubia
	C. percuspidata
	C. membranacea
	C. rosaeflora
	C. campanisepala
	C. lancilimba
	C. tsingpienensis
	C. pubisepala
	C. parviovata
	C. viridicalyx
	C. lancicalyx
	C. parvicaudata
	C. subglabra
	C. nokoensis
	C. tsofuii
	C. trichandra
Section XX	Eriandra
	C. villicarpa
	C. cratera
	C. punctata
	C. lawaii
	C. trigonocarpa
	C. cordifolia
	C. wenshanensis
	C. melliana
	C. candida
	C. caudata
	C. assimiloides
	C. assimilis
	C. edentata
	C. salicifolia

* adopted with modification from Mondal 2011

Owing to extensive internal hybridization between different *Camellia* taxa, several intergrades, introgressants and putative hybrids had been found. They were arranged in a gradient based on morphological characters that extended from China types through intermediates to those of Assam types. Indeed, because of the extreme homogenization, existence of the pure archetypes of tea was doubtful (Visser 1969). Numerous hybrids, therefore, were referred to as China, Assam or Cambod tea depending on the morphological proximity to the main taxon (Banerjee 1992a). Tea breeds well with some of the wild relatives, and thus taxonomists had always been interested to identify such hybrids due to suspected involvements in tea genetic pool. Two particularly interesting taxa were *C. irrawadiensis* and *C. taliensis* whose morphological distributions overlapped with that of tea (Banerjee 1992a). It had also been postulated that some desirable traits such as anthocyanin pigmentation or special quality characters of Darjeeling tea might have been introduced from those wild species (Wood and Barua 1958).

Other *Camellia* species, which were suspected to have contributed to the tea genetic pool by hybridization, include *C. flava* (Pifard) Sealy, *C. petelotii* (Merrill) Sealy (Wight 1962) and possibly *C. lutescens* Dyer (Sharma and Venkataramani 1974). The role of *C. taliensis* was, however, not clear because the species itself was considered to be a hybrid between *C. sinensis* and *C. irrawadiensis* (Wood and Barua 1958; Visser 1969). Therefore, it was generally agreed that at least three taxa, i.e. *C. assamica, C. sinensis, C. assamica* sub sp. *lasiocalyx* and to an extent *C. irrawadiensis* had mainly contributed to the genome of tea. The term 'tea' should therefore cover progenies of these taxa and the hybrids thereof or between them.

Apart from this natural diversity, the different tea research institutes and dedicated planters had further developed a number of varieties with better yield, quality and traits such as tolerance to drought, diseases, etc. In my estimation, more than 1,200 such commercial cultivars of tea have been developed and released for cultivation worldwide, and many of them have special characters (Table 2.1).

Table 2.1 Different tea cultivars with special characters. (Adopted from Mondal 2009)

Special characters	Clone	Originator	Reference
Wind tolerance	UPASI-2, UPASI-10	UPASI-TRF, India	Sharma and Satyanarayana (1987)
Drought resistance	UPASI-9	UPASI-TRF, India	Sharma and Satyanarayana (1987)
Frost resistance/tolerance	BS 53	HPKV-TES, India	Deka et al. (2006)
Small leaf	CH-1, Vimtal	IHBT and TES, Assam, India	Mondal et al. (2004)
Biggest leaf	Betjan	Betjan T.E, India	Singh (1980)
Blister blight tolerance	TRI-2043, DT-1	TRI, Sri Lanka	Sivapalan et al. (1995)
High pubescence content	TRI/2043	TRI, Sri Lanka	Sivapalan et al. (1995)
High anthocyanin pigmentation	TRI/2025	TRI, Sri Lanka	Sivapalan et al. (1995)
High anthocyanin pigmentation	Cha Chuukanbohon Nou 6	NIVOT, Japan	Nesumi et al. (2012)
High tolerance to pH	TN-14–3	TRF, Kenya	Anon (1999)
Poor fermenter	12/2	TRF, Kenya	Anon (1999)
Mite tolerance	7/9	TRF, Kenya	Anon (1999)
Scale insect tolerant	TN 14–3	TRF, Kenya	Anon (1999)
High polyphenol content (53.7 %)	Luxi white tea	TRI, China	Yu and Xu (1999)
High amino acid content (6.5 %)	Anji white tea	TRI, China	Yu and Xu (1999)
Low caffeine content (0.14 %)	Guangdong tea	TRI, China	Yu and Xu (1999)
High caffeine content (6.96 %)	Wild tea at Yunnan	TRI, China	Yu and Xu (1999)
Water logged tolerant	TV-9	TES, Assam, India	Singh (1980)
High somatic embryogenesis	Makura-Ck2	NIVOT, Japan	Furukawa and Tanaka (2004)
Early germination	Tianfu 28	CAS, China	Wang et al. (2003)
Very early budding	Zhingcha 108	CAS, China	Yang et al. (2003)
Loop hopper resistance	EF	CAS, China	Hu et al. (2003)
Anthracnose	Abo	NIVOT, Japan	Yoshida and Takeda (2006)
Gray blight resistant	P11P11P12P12	NIVOT, Japan	Takeda (2002)
Trichomeless mutant	Progeny of Surugawase	NIVOT, Japan	Takyu et al. (2003)

UPASI United Planters Association of Southern India, *HPKV* Himachal Pradesh Krishi Viswavidyalaya, India, *TRI* Tea Research Institute, Sri Lanka, *NIVOT* National Institute of Vegetable, Ornamental Crops and Tea Science, Japan, *TRF* Tea Research Foundation, Kenya, *CAS* Chinese Academy of Sciences, China

2.4 Karyotype

Karyotype was considered to be the most important cytological marker for identification of the species. Karyotypes had been established for the most available taxa of *Camellia* including tea (Kondo 1975; Fukusima et al. 1966; Ackerman 1971; Datta and Agarwal 1992). However, karyotype grouping based on chromosome size was not successful in the *Camellia* taxa due to high stickiness of the chromosomes. Furthermore, even in the best preparation, homologous chromosome pairs could not appear identical in *Camellia* (Kondo 1975). Relatively little intraspecific karyotypic variation had been observed for the cultivated species of *Camellia* studied (Kondo 1975). Sat-chromosomes in karyotypes within mass accessions of certain *Camellia* species were morphologically and quantitatively variable. Thus, karyotypes including characteristics of sat-chromosomes were not of taxonomic significance for *Camellia* taxa. Among the diploid species of *Camellia* studied, *C. japonica* L. sensu lato showed the greatest karyotypic variation; many of the studied accessions indicated similar karyotypic patterns to each other (Kondo 1975). Actually, *C. japonica* L. var. macrocarpa Masamune had satellites on four submetacentric chromosomes and the other accessions had satellites on two submetacentric chromosomes (Kondo and Parks 1980). Later, it was shown by Kondo and Parks (1979) that the C-banding method could be applied to the somatic mid-metaphase chromosomes in *Camellia* taxa. These differentially stained bands in mid-metaphase chromosomes permitted the identification of 238 individual chromosomes and made it possible to match the homologous pairs of chromosomes more precisely. Karyotypic variability and divergence among the seven accessions of *C. japonica* L. sensu lato with aceto-orcein staining were revealed by C-banding method (Kondo and Parks 1980). In this way, the cytological marker was used to sort and classify the genotypes.

The karyotype characteristics of some *Camellia* species are listed in Table 2.2. The cardinal chromosome number of the 29 species was found to be stable, $2n=30$. No polyploidy were found, excluding cultivated *C. sinensis* and *C. assamica.* It indicated that the evolution of karyotype of section Thea was mainly through the gene, similar to other tropical woody plants and different from other species of genus *Camellia* in the temperate zone. The karyotype of most species in *Thea* was M (metacentric) and SM (submetacentric) chromosomes, only few species with ST (subtelocentric) chromosomes with the order of the number was M > SM > ST.

Interestingly, *C. reticulata* had a series of polyploid varying from $2n=2x=30$, $2n=4x=60$ to $2n=6x=90$, with a basic chromosome number of $x=15$. The hypothetic allopolyploid origin and parental genomes of these polyploidy types remained unknown. Genomic in situ hybridization (GISH) was used to study the genome organization and evolution of *C. reticulata.* Total genomic DNA from closely-related diploid species (*C. pitardii* and *C. saluenensis*), with the chromosome number $2n=2x=30$, were labelled and hybridized in the presence of blocking DNA onto metaphase spreads of *C. reticulata.* The *C. pitartii* probe painted part of the tetraploid and hexaploid *C. reticulata* genomes, whereas the *C. saluenensis* probe delineated part of the hexaploid *C. reticulata* genome. The results provided compelling evidence for the allopolyploid origin of *C. reticulata* genomes and demonstrated that (1) the diploid *C. reticulata, C. pitardii* and *C. saluenensis* were the progenitors of polyploid *C. reticulata*, (2) hybridization between diploid *C. reticulata* and diploid *C. pitardii* gave birth to allotetraploid *C. reticulata* and (3) subsequent hybridization between allotetraploid *C. reticulata* and diploid *C. saluenensis* formed the allohexaploid *C. reticulata* (Gu and Xiao 2003).

2.5 Propagation

Tea and its wild species are mainly propagated by three means, i.e. seeds, vegetative cuttings and nursery graftings, although budding or grafting of mature plants are also followed but very rarely.

Seeds: Conventionally tea is propagated through seeds. Seeds are generally produced in 'seed bari' (seed orchard). A fully matured healthy seed while attached to the plants or recently dehisced are collected from the ground of the seed orchards. This is primarily due to the fact that tea seeds being recalcitrant have low viability.

After eliminating the very small seeds, the remaining seeds are transferred to a tank or trough filled with water and allowed to soak for 2–3 h. The sinker seeds are taken out of water and examined for mechanical, insect and pest damage. The usual practice is to cut open a sample of 50 to 100 seeds from the batch to examine starred, cheesy, shrunken seeds or otherwise damaged seeds by pests or diseases. Floater seeds are discarded as such seeds are found to have dried cotyledons, which normally fail to germinate. Floaters are frequently the results of punctures made by Tea Seed Bug (*Poecilocoris latus*). As soon as

Table 2.2 The karyotype of some *Camellia* species. (Liang et al. 1994; Chen et al. 2000)

Species	Karyotype
C. assamica	2n=30=18m+12sm
C. sinensis	2n=30=20m+8sm+2st
C. grandibracteata	2n=30=24m+6sm
C. kwangnaica	2n=30=22m+8sm
C. quinquelocularis	2n=30=24m+6sm
C. tachangensis	2n=30=22m+8sm
C. gymnogynae	2n=30=22m+6sm+2st/20m+8sm+2st
C. ygmnogynoides	2n+30=22m+6sm+2st
C. jungkiangensis	2n=30=20m+8sm+2st
C. tetracocca	2n=30=22m+8sm
C. nanchuanica	2n=30=20m+8sm+2st/24m+6sm
C. crassicolumma	2n=30=22m+8sm/2n=30=18m+9sm+3st
C. atrothea	2n=30=20m+6sm+4st
C. taliensis	2n=30=22m+8sm
C. taliensis var. bangweicha	2n=30=22m+6sm+2st
C. irrawadiensis	2n=30=18m+12sm/22m+8sm
C. rotundata	2n=30=20m+10sm
C. makuanaica	2n=30=22m+8sm/20m+10sm
C. manglaensis	2n=30=22m+8sm+4st
C. leptophylla	2n=30=24m+4sm+2st
C. dehungensis	2n=30=20m+10sm+2st
C. gymnigyna	2n=30=20m+8sm+2st
C. costata	2n=30=20m+8sm+2st
C. parvisepaloides	2n=30=22m+8sm
C. gymnagynoides	2n=30=22m+6m+2/22m+6+2/20m+8+2st
C. purpurea	2n=30=22m+4sm+4st
C. polyneura	2n=30=22m+4sm+4st
C. sinensis	2n=30=18m+12sm+2st
C.. sinensis var. pubilimba	2n=30=20m+10sm/18m+10+2st
C. sinensis var. kucha	2n=30=22m+8sm+2st
C. ptilophylla	2n=30=22m+8sm+4st
C. assamica	2n=30=22m+8sm
C. yankiangcha	2n=30=22m+8sm+4st
C. arboewscens	2n=30=20m+10sm
C. tachangensis	2n=30=23m+7sm
C. taliensis	2n=30=1m+9sm
C. crassicolumna	2n=30=20m+9sm+1st
C. gymnigyna	2n=30=21m+8sm+1st
C. sinensis	2n=30=21m+8sm+1st
C. sinensis var. sinensis	2n=30=20m+9sm+1st
C. sinensis var. assamia	2n=30=22m+7sm+1st
C. sinensis var. pubulimba	2n=30=21m+9sm
*C. quinqueloculare*a	2n=30=22m+8sm
*C. triloculare*a	2n=30=21m+9sm
*C. triloculare*a *var. macrophyllaea*	2n=30=20m+9sm
*C. triloculare*a *var. micromidphyllaea*	2n=30=20m+10sm
*C. triloculare*a *var. kuiea*	2n=30=22m+8sm
C. cryploneura Chang	2n=90=61m+28sm+1st
C. oblate Chang ex Chang	2n=30=24m+5sm+1st

Table 2.2 (continued)

Species	Karyotype
C. meiocarpa Chang	2n=60=37m+18sm+5st
C. oleofera Abel	2n=90=60m+29sm+1st
C. grijsii Hance	2n=30=25m+5sm
C. forrestii (Diels) Cohen-Sturt	2n=60=39m+19sm+2st
C. tsaii	2n=60=38m+21sm+1st
C. tsingpienensis Hu	2n=30=21m+5sm+4st
C. yunnanensis (Pitard ex Diels) Cohen-Stuart	2n=30=19m+11sm
C. chrysantha (Hu) Tuyama	2n=30=17m+10sm+3st
C. chrysantha var. microcarpa Mo	2n=30=22m+8sm
C. impressinervis Chang	2n=30=21m+8sm+1st
C. impressinervis Chang	2n=30=20m+9sm+1st

possible, the seeds are packed after grading and sorting. For transport over long distances, packing is done in wooden boxes in units of 20 kg using moist sand, subsoil, powered charcoal or ash or a mixture of two or more of these as packing material. Moisture content of the packing material varies from 10 to 30% while in the case of powered charcoal it may vary from 25 to 30%. Seeds are spread in layers along with some packing materials, and each layer is separated from the one on the top by a thin sheet of tough paper. A kilogram of graded and sorted seed may contain 300–500 seeds depending on the size of the grader used.

After 45 days, the germinated seeds are transferred to the polythene sleeves and kept under a shaded nursery for another 12–18 months.

Vegetative cuttings: Seeds were the only commercial method of propagation till the beginning of the nineteenth century. However, due to the outbreeding nature of the plant, seedlings show a wide variability for attributes such as yield, quality, etc., and this forced people to find some alternatives. The first attempt for vegetative propagation of tea was done in Indonesia by budding as well as grafting. However, due to slow speed, this method could not serve the purpose of rapid multiplication. Hence, faster propagation by single leaf cutting was developed simultaneously in India, Sri Lanka and Indonesia (Mondal 2011). This was further fine-tuned later to fit the commercial venture that exists now. Cuttings from green and semi-hard wood are usually taken from current-year growth. Cuttings are then immediately subjected to fungicide as well as commercial grade rooting-hormone treatment and inserted in the nursery for root induction for 45–60 days depending on the location, planting material and so forth. The successful rooted cuttings are then transferred to polythene sleeves filled with good virgin soils (pH around 4.5) having adequate water-holding capacity and are kept for another 8–12 months in the nursery, by which period they become ready for field transfer. Meanwhile, propagation by cuttings was attempted in several parts of the tea-growing areas around the world (Tunstall 1931a, b; Tubbs 1932; Wellensiek 1933); however, standardization of the technique of single-leaf internode cuttings, practised today, took a long time to be successful. Following this technique, TES, Assam, India released the first lot of three clones in 1949, which revolutionized the tea industry in northeastern India and more clones have since been released from time to time.

Grafting: In recent years, nursery grafting as an alternative propagation technique has gained considerable popularity. In this technique, fresh single-leaf internode cuttings of both rootstock and scion are generally taken. Scion, commonly a quality cultivar, is grafted on rootstock, and is either drought tolerant or a high-yielding cultivar. On grafting, the scion and stock influence each other and thus composite plants combine both the characters, resulting in 100% increase of yield with better quality than either of the ungrafted cultivar. Recently, a modified improved 'second-generation' grafting had been developed, where a tender *in vitro*-derived shoot was grafted on the young seedlings of tea, which had an additional advantage over conventional grafting due to the presence of tap-root system (Prakash et al. 1999). Nevertheless, with the increasing demand for clonal tea, vegetative propagation with single-leaf internode cuttings remains the best choice in the tea industry worldwide.

2.6 Floral Biology and Pollination Mechanism

The significant differences in flower description between the China and Assam type of tea were reported by several workers (Wellensiek 1933; Bezbaruah 1975), which have been described in Sect. 1.4. Characters, such as length of the style and style arm, number and length of the outer stamens or the size of the inner petals, were able to indicate the difference in floral characters among the varieties. Importantly, anatomical study suggested that tea flower should be classified as central placental type instead of parental placental type (Bezbaruah 1975).

Tea plants showed an appreciable degree of self-sterility and invariably set a better crop of seeds with pollen from another bush nearly four times than that of selfed seeds (Wight 1938; Wu 1964). Generally, selfed seeds exhibited reduced germination. Progenies of self-pollination were inferior in vigour to those of cross pollination. Investigations on the mechanism of pollination revealed that tea pollens were heavy and sticky in nature and occurred mostly in clumps, a condition which was not favourable for carrying out by wind, rather only non-viable dried-up pollen grains can move a long distance by wind. However, insects, such as bees and wasps (*Hymenoptera*), were found to carry pollens from bush to bush. Besides, Bezbaruah (1975) observed that syrphid flies (*Diptera*) were the most common insect for natural cross-pollination. The tea flower secretes a considerable amount of honey, but they contain high phenols that cause indigestion. Perhaps, this may be the reason for not attracting the diverse insects for pollination (Barua 1989). Therefore, for higher seed setting, it may be necessary to take measures to attract insects in tea-seed orchards.

After 24–48 h of pollination, the corolla withers off and drops from the pedicel along with the anther lobes leaving the ovary exposed. The persistent calyx lobes close flatly over the ovary and the style, and the stigma gradually withers off. Although pollination takes place during flowering growth, i.e. September–January in India, the first external sign of development of fruits become evident by March and gain considerable size by May of the same year. By about August, the fruits attain full size with completely developed embryo and cotyledon. The mature embryo with two large cotyledons remains covered by a hard, deep brown testa, formed by the outer integument. The pericarp encloses 1–3 seeds inside and is made of thick, parenchymatous tissue when young but becomes sclerotic on maturity. The ripening of fruit generally takes 12 months from the time of flowering till maturity (Singh 1999).

The mechanism of self-incompatibility in tea and related species remains a challenge. A study of reproduction barrier limiting interspecific hybridization between *C. japonica* and *C. chrysantha* was undertaken in intra- and interspecies crosses. Pre-zygotic barriers were not important, pollen-type growth and penetration was good in all combination of crosses. Zygote formation and early embryo development were similar in all the crosses. While intraspecies embryos developed smoothly, interspecific crosses of embryos were aborting in various stages. Interestingly, full size but empty ovules in mature capsule resulted from embryo abortion (Hwang et al. 1992). It was reported that tea could have been considered a facultative outbreeder but with a homomorphic gametophytic self-incompatibility system. The observation of successful self-pollen tube growth in the ovary and ovule penetration clearly indicated that tea had a late self-incompatibility type of selfing control (Bezbaruah 1975).

Tea flower had also been used for value addition. Importantly, tea companies in China had begun to process fresh tea flowers for blending into specialty teas to make them flavoury. However, chemical composition of tea flowers and leaves in terms of catechins and caffeine were comparable. The flowers contained less caffeine but equivalent amount of catechins (Su et al. 2000; Lin et al. 2003). Recently, Joshi et al. (2011) did a detail characterization of various catechins and volatile compounds and found that like leaf, the unopened young flower bud contained maximum amount of flavour compounds than the fully open flower of tea.

2.7 Seed Biology

Generally, 1 to 3 seeds of 1.5–2.0 cm diameter are found in each capsule. The seeds have a hard testa outside and the embryo is covered in between two large cotyledons. Tea seeds are highly recalcitrant and lose viability within a few days after shedding from the plant (Bhattacharjee and Singh 1994). However, their viability can be maintained by surface sterilizing with mercuric chloride solution (0.01 %) for 15 min and subsequent cold storage at 4 °C. Although seeds are generally stored in moist charcoal for a few days, it is advisable to use the healthy seeds for propagation as early as possible (Singh 1999). For commercial propagation, tea seeds are produced in seed baries (orchard) planted specially for the purpose. After the release of the clonal seeds, commercial seed baries have been established for production of hybrid clonal seeds. After collection, seeds are passed through a rotary type shifter to eliminate very small seeds. Tea seeds normally vary in size from 10 to 20 mm in diameter.

2.8 Breeding Techniques

Tea breeding objectives vary from country to country, depending on the local needs (Table 2.3). However, by and large, it is aimed to improve the quality and yield. Generally, the breeding works in black tea producing countries, such as India, Kenya and Sri Lanka, are biased towards the development of high yielding and quality clones whereas the tea-producing countries near the equator, such as Japan and China, are focused on the development of cold tolerance and frost tolerance, as these countries primarily produce green tea, where the quality of made tea does not have much influence on price. Today's modern tea varieties have evolved through the sincere efforts of many years of both the tea breeders as well as experienced planters through different stages of plant introduction, selection, hybridization and physical as well as chemical breeding. The different techniques are discussed in this chapter.

2.8.1 Introduction

Introduction may involve *de novo* addition of new varieties, wild relatives or totally new species in a particular area. Often, materials are introduced from other countries or continents. In tea, the unorganized agents such as Buddhist pilgrimage or colonial soldiers made the primary introduction. However, later the secondary introduction in various tea-growing countries was done by experienced British planters or scientific communities as discussed below.

2.8.1.1 Primary Introduction

Introduction of tea into Korea was done by the troops of Emperor Wu Di during his invasion of Korea from China. In Japan, tea was introduced during the eighth century when Buddhist monk Saicho, who returned from studies of Guo Xing Si on Tian Tai Mountain of Zhe Jiang provinces, brought tea seeds to Japan and planted them at the foot of the Mountain Hi Yei in the village of Sakomoto of Omi County. Later on in 1191 A.D., Yei Sai, another Buddhist, re-introduced tea seeds from China to Japan and planted them on the hilly terrain of the Seburi Mountain, southwest of Castle Fukuoka in Chikuzen province. Yei Sai not only planted and cultivated tea plant but also regarded it as the source of a sacred remedy.

In 1690 A.D., the governor of Indonesia, J. Camphuijs brought tea seeds from China and planted the same nearby Djakarta region. Between 1828 and 1833, Mr. Jacson of East India Company

Table 2.3 Breeding objectives of tea. (Adopted from Mondal 2009 with modification)

Objectives	Importance	Regions
Improving quality	Directly linked to the profitability	Black tea producing countries such as India, East-Africa, Sri Lanka, Bangladesh and Indonesia
Increasing yield	Horizontal increase of production by extension planting is limited	Worldwide
Drought tolerance	Reduce productivity and occur all tea-growing regions of the world	Worldwide where tea grown as rain-fed crop
Reduce winter dormancy	No leaf production during winter months and occurs in North-East India, Japan and China, etc	Tea plantation near the equator
Hail/frost resistance	Causes economic loss as young leaves during rainy season are mostly affected	Hilly region of the tea-producing countries
Water log tolerance	Reduce productivity during rainy season. Generally occurs in north-eastern India	North-East India
Cold hardiness	Reduced productivity during winter due to snow. Generally occurs in China, Japan and Russia	Mainly in Japan, Russia and China
Diseases resistance, such as blister blight, stem canker, etc.	Blister blight causes severe damage as only young leaves are infected. Generally occurs in Japan, Sri Lanka, South India and Darjeeling hills of North-East India	Mainly India, Sri Lanka, Indonesia and Japan
Pest resistance, such as red spider mite, tea mosquito bug, leaf-sucking pest, etc.	Most important biotic stress as all causes severe damage to the leaves. Generally, occurs in all the tea-growing regions in the world	Worldwide
Suitability to type of manufacturing	For matching the customer's demand as well as better recovery percentage in made tea	Black-tea-producing countries such as India, East-Africa, Sri Lanka, Bangladesh and Indonesia
Low input responsive Clone	Required for organic tea farming	Organic tea

went to China and brought tea seeds again which was planted in Indonesia.

Tea was brought to Russia during 1567. Later 1735, tea was first carried overland by governmental caravans. From the Chinese border, this legendary trail lay northwestwards across 800 miles of the inhospitable Gobi Desert through Ulan Bator in Mongolia and into Russia, skirting Lake Baykal, to the town of Irkutsk.

Initially, Sri Lanka began to cultivate coffee but without success. Therefore, they decided to start tea plantation during 1841. M.B. Worms, a German who lived in Ceylon, visited China and brought tea seedlings to Pussellawa. Simultaneously, seeds were also brought from Botanical Garden of Calcutta, India during 1839 and planted in the nurseries of Royal Botanical Gardens at Peradeniya near Kandy, Sri Lanka.

2.8.1.2 Secondary Introduction

However, information regarding the secondary introduction is rather scanty due to the presence of stringent rules worldwide preventing trans-border movements of the plant propagule in later years. Additionally, the secondary introduction of tea always happened off the record by the commercial tea growers. Interestingly, the early history of the planting materials clearly indicates that being the oldest tea research Institute in the world, TES, Assam, India contributed significantly towards the secondary introduction of tea worldwide. In a conservative estimation, 60 % of the world-acreage had received its initial planting materials directly or indirectly from tea genetic resources of TES, Assam, India. Initially, the commercial tea estates had been established, and later these were used as parent source of materials for developing the new cultivars.

2.8.2 Hybridization

In natural hybridization, based on known desirable characters, such as previous performance of yield, quality or diseases tolerance capability, two parents are planted side by side in an isolated place to bear fruits. Subsequently, the seeds (F_1) are harvested, raised and planted. If average performances of these plants are found to be better than either parent, then seeds are released as hybrids or bi-clonal seeds. However, some of the outstanding performers among these progenies are marked and verified for multi-location trials and, if found suitable, are released as clones. These clones are geographically specific, and most of the tea research institutes of the world have generated the clones for their own region. Sometimes, in the above process, more than two parents are used and known as polyclonal seeds. The idea is to introduce more variability among the F_1 seeds. Since it is difficult to know about the pedigree of the cultivar (as pollen may come from any male), the chance of reproducibility of performance is low, and therefore the process is the least preferred currently.

Hand pollination or control cross, despite being an important approach, has made a limited success in tea breeding. The reasons identified could be: (1) low success rate, (2) availability (2–3 months) of tea flowers for a short period, (3) longer time taken for seed maturation (12–18 months) and (4) difference in flowering time for different clones.

2.8.3 Selection

The seeds from a particular 'seed orchard' are known by the name of that orchard or locality and are called *jats* or population, and those *jats* are the main source of planting materials of tea. There are wide variations among the offsprings for morphology, yield and quality parameters generated from a particular *jat* even though the source of the seed was the same. This is primarily due to high outbreeding nature of tea plants. Wight (1939) showed an interesting fact. About 10 % of the bushes in a commercial tea garden of northeastern India produced only 2 % of the total crops, i.e. green leaf and about 0.5 % bushes produced as much as or more than 300 % of the average crops. Thus, the planters thought that new areas planted with seeds from those 0.5 % bushes will produce more uniform and better quality tea. Those selected plants were used for future plantations. Thus, the process of selection started. The first scientific attempt to select improved tea in northeastern India was made by Stiefelhagen brothers during 1860 by establishing standard sources of tea seeds. Subsequently, scientific improvements of tea by selection were followed in many countries, such as Indonesia (Wellensiek 1934), Java (Cohen Stuart 1929), Russia (Bakhtadze 1935) and northeastern India (Wight 1939). Mother bushes were selected based on morphological characters followed by anatomical (Wight 1956) and organoleptic performance of made tea (Timoshenko 1936). Indigenous Assam tea was improved by following the technique of mass selection. The yield increased considerably, because of line breeding for desirable morphological features that are genetically linked with the characteristics of Assam tea. After that, line breeding techniques were followed to improve further by mainly evolving more uniform tea plant with better quality and adaptability. In the earliest effort, two selected *jats* were hybridized to combine desirable characters into their progenies. The 'Rajghur' *jat* was developed by combining high quality of a light leaf local *jat* with the vigour of a dark leaf 'Manipuri' *jat*.

However, the seed-grown plants were not uniform as their characters were governed by genotypes of their parents, which were diverse in some phenotypic characters in relation to environmental and soil conditions. In some cases, the yield and quality were unpredictable. It was, therefore, necessary to develop clonal cultivars in tea like many other fruit crops by multiplying the selected bushes vegetatively. Today, all tea-growing countries of the world have developed clonal materials as per their requirements.

Table 2.4 Triploid tea cultivar used under commercial cultivation along with their promising characteristics. (Gunasekara and Ranatunga 2003)

Country	Polyploid cultivar	Level of ploidy	Type of polyploidy	Promising characteristic(s)	Reference
India	Sundaram	3n	Natural	High yield and quality	Sharma and Ranganathan (1986)
India	UPASI 3	3n	Natural	High yield and overall quality	Satyanarayan and Sharma (1986)
India	UPASI 20	3n	Natural	Moderate yield, highly tolerant to drought	Satyanarayan and Sharma (1986)
India	TV 29	3n	Natural	High quality	Barbora et al. (1996)
Japan	Not-known	3n	Natural	Hardier and cold resistant	Simura and Inabe (1952)
Kenya	382/1	3n	Natural	High yield	Wachira (1994)
Sri Lanka	TRI 3069	4n	Artificial	High yield and drought tolerant	Kulasegaram (1980)
Sri Lanka	HS 10A	3n	Natural	Cold resistant	Kulasegaram (1980)
Sri Lanka	GF 5/01	3n	Natural	High yielding	Anon (1973)

2.8.4 Polyploidy Breeding

Yield is the major criteria in commercial tea cultivation which depends on the size and density of the leaves in the plucking table, i.e. upper surface area of the tea bush. A positive correlation between size of the leaf and the yield in tea was well established (Satyanarayan and Sharma 1982). Therefore, the development of tea genotypes with bigger leaves through polyploidy breeding may be useful to increase the yield of tea. Further, being vegetatively propagated plants, polyploidy breeding can be used effectively. In tea, significant amount of works on polyploidy breeding had been done (Gunasekara and Ranatunga 2003), which are reviewed below.

2.8.4.1 Naturally Occurring Polyploids

Although the cultivated species of tea was diploid (Morinago et al. 1929; Barua 1989), naturally occurring intra- and interspecies polyploids of tea (Janaki Ammal 1952; Bezbaruah 1971; Jayasuriya and Govindarajulu 1975; Wachira and Kiplangat 1991) and its wild relatives were also reported (Kondo 1977). Interestingly, natural polyploids were more common in tea populations (Table 2.4) of Japan than any other countries (Banerjee 1992b; Simura and Inabe 1952). Bezbaruah, (1968) reported that in tea, the triploids, tetraploids, pentaploids and aneuploids, resulting from open-pollinated progenies, occurred naturally but in extremely low frequency.

2.8.4.2 Artificial Induction of Polyploids

Since the discovery of the effect of colchicine in 1930s on cell division for mitotic doubling of the chromosome number, it was widely used to induce the level of polyploidy in plant (Blakeslee and Avery 1937). Colchicine inhibits mitosis in cells by interfering with the structure of the mitotic spindle, thus resulting in formation of cells with a doubled chromosome number.

Similar to other plants, colchicine had been used to induce artificial polyploids in tea (Table 2.5). In Sri Lanka, Sebasthiampillai (1976) produced five tetraploid plants, namely TRI 2023, 2024, 2025, 2026 and DT 95, by treating the meristematic tissues of the terminal bud for 2–7 days with colchicine impregnated in agar. Although he found the differential response of tea genotypes with the colchicine treatment, his ploidy plants were tetraploid as he confirmed through the cytological examination of root-tip cells. However, attempts to induce polyploids using ethyl methane sulphonate (EMS) at TES, Assam, India, though tried, were not success-

Table 2.5 Details of induction of tea colchiploids. (Gunasekara and Ranatunga 2003)

Plant part used	Treatment	% success	Reference
Axillary buds of etiolated shoots	Cotton wool moistened with colchicine (0.2%) and treatment was given in the dark	13.0%	Katsuo (1966)
Terminal buds of active shoots developing from pruned bushes	Agar impregnated with colchicines (0.2–0.5%) for 5–6 days	13.5%	Sebasthiampillai (1976)
Terminal buds	Immersion in aqueous colchicines (1–2%) for 5–7 days	6–17%	Anon (1979)
Flower buds	Colchicines (0.05%) injection and drop application for 2–6 days	30.0%	Osone (1958)

ful. Nevertheless, more than 170 and 70 polyploids were subsequently generated in the same institute through conventional hybridization and colchicine treatment, respectively (Singh 1999); yet, in tea, as high as 30% ploidy had been achieved using colchicines as mutagenic agents (Table 2.5).

2.8.4.3 Morphological, Anatomical and Cytological Markers in Polyploid Teas

Screening of polyploids has not been widely exploited in tea due to lack of reliable markers. Hence, the identification of markers related to morphology, anatomy or cytology is of great importance in screening for polyploids, whether they are artificially induced or naturally occurring. Generally, the ploidy level of tea is determined by counting chromosome numbers in meristematic tissues, i.e. root-tip cells or pollen mother cells.

Chaudhuri (1979) found wide range of phenotypic and anatomical variations, such as frequency and size of stomata and sclereids among the progenies of triploid tea (Chaudhuri and Bezbaruah 1985). Similarly, to assess the effects of the level of polyploidy on the morphogenetic attributes of the F_1 seedling population, generated from a cross between diploid and tetraploid cultivars, a clear relationship was shown between ploidy levels and morphogenetic variations (Rashid et al. 1985).

Among the different morphological markers, while leaf area was found to be higher in triploid, its expansion, i.e. growth was lower in triploids in comparison with diploid tea leaves (Ng'etich and Wachira 1992). The reason for this may be that the attributes considered were more affected by environmental factors than by ploidal status.

Anatomical markers such as stomatal density were used to differentiate the triploid than diploid as it was found to be lower in triploid than diploid cultivars (Amma 1974; Chaudhuri and Bezbaruah 1985; Wachira 1994). It was found that triploid plants had a lower stomatal density than the diploid genotypes (Wachira 1994). However, this marker could not always be used as a reliable marker for identification of polyploids in tea. In fact, Chaudhuri and Bezbaruah (1985) had indeed reported that there was a lack of correlation between the level of ploidy and stomatal density. Similarly, guard-cell sizes as well as stomata size were larger in tetraploids and triploid teas than its diploid counterparts (Amma 1974). On the contrary, Wachira (1994) found that the length to breadth ratio of the guard cells was not significantly different between diploids and triploids. Later, chloroplast number in the guard cells had been identified as a reliable ploidy marker in tea (Ahmed and Singh 1993; Koskey and Wachira 2000; Ranatunaga and Gunasekare 2002; Chen and Ye 1989). Further, Koskey and Wachira (2000) found that the ratio of the guard-cell chloroplast numbers in diploids, triploids and tetraploids was found to be 2:3:4, which was the same as the ratio of their chromosome numbers (30:45:60) (Ahmed and Singh 1993). Therefore, this finding indicated that the ploidy level of tea could be accurately and rapidly identified by the chloroplast-count method, rather than by criteria based on the size and density of stomata.

Reproductive organs such as pollen grain in most of the induced tetraploid cultivars were found to be higher than that of its diploid counterparts (Gunasekara 2000). But *in vitro* germination of pollen grain was poorer in tetraploid cultivars than in diploid cultivars (Thirukkumaran and Gunasekare 2001). Only 2% of the pollen grains of natural triploids were found to be viable (Bezbaruah 1971). It had been reported that pollen viability and fertility of triploid cultivars were unable to set seeds and fruits. In general, these morphological as well as anatomical markers were not consistent, which was why they were not accepted by tea breeder.

Therefore, the alternative cytological markers such as chromosome counts were found to be more reliable to differentiate the triploids from diploids. Chromosome counting in pollen mother cells, root-tip cells, and meristematic tissue cells at the shoot tip were subsequently standardized in tea (Gunasekara and Ranatunga 2003). Wachira and Muoki (1997) devised a new cytological technique to assess the activity of nucleoli and nucleolus-organizing regions of polyploids and diploids. Their study revealed that the mode of nucleolar number corresponded to multiples of the somatic cell number, and hence was a good marker for ploidy. Therefore, it is assumed that the attributes, related to anatomical features, are much more precise than the morphological characters which have been used to screen polyploidy genotypes in tea due to the fact that the later has larger environmental influence than the former.

2.8.4.4 Use of Polyploids in Tea Breeding

Generally, tea polyploids often lack desirable traits (Bezbaruah 1968; Sarmah and Bezbaruah 1984), and polyploidy breeding therefore requires planned hybridization, selection of promising polyploids and proper evaluation to confirm their performance as potential cultivars. High-yielding polyploids with low quality of made tea, had been improved through hybridization with a diploid cultivar of high quality traits (Sarmah and Bezbaruah 1984). Triploids had been produced by hybridizing tetraploid tea with diploid tea in Japan (Osone 1958), India (Chaudhuri 1979) and Bangladesh (Rashid et al. 1985). It had been shown that it was possible to combine good cup quality, with the superior vigour and hardiness of the polyploids, by crossing tetraploid progeny for commercialization. For example, open-pollinated tetraploid with inferior cup quality but with higher growth vigour (Bezbaruah 1976) was improved to higher cup quality by crossing with high-quality diploid clone as the male parent (Bezbaruah 1991).

The conventional method of producing triploids is through artificial induction of tetraploids, followed by hybridization with diploid cultivars. For example, out of 238 hybrids produced through hybridization between tetraploids and diploids at TES, Assam, India, only 79 hybrids were found to be triploids (Barbora et al. 1996). Since induction of diploid was time consuming, Osone (1958) used diploidized pollen of immature flowers to pollinate diploid plants for producing triploids. However, there is no evidence that this method had been widely practised in polyploid breeding programmes. Recently, to verify the quality of triploid cultivars of tea, Das et al. (2013) profiled caffeine and catechins of 97 F_1 segregating progenies of a common tetraploid and diploid parents. Catechins and caffeine level of the triploid progenies were compared against their diploid parents. Some of the progenies found to be better quality clones than their diploid parents. Most of the progenies of the diploid *C. sinensis* crossed with tetraploid, showed heterosis for caffeine and catechins. The genomic contributions of the diploid parents seem to be the main factor in the variation between the two populations. They demonstrated quantitative enhancement of some of the quality-related parameters in tea, providing a platform to refocus on this classical breeding approach for developing quality cultivars in tea.

2.8.4.5 Commercial Exploitation of Polyploids

Although emphasis was given to identify natural polyploids and develop artificial polyploids, reports on their performance and trait evaluation are scarce. After the discovery of natural polyploids of tea (Karasawa 1932; Bezbaruah 1971; Amma 1974; Katsuo 1966; Sebasthiampillai 1976), natural polyploids were included in cultivar selection programmes to identify desirable agronomic traits. Certain studies indicated that natural polyploids found in Southern India possess attributes for high yields and quality (Sharma and Ranganathan 1986). On the other hand, Banerjee (1992b) had reported that though polyploids showed high vigour and tolerance to environmental stresses, they did not always contribute towards high yields, even sometimes produced low quality of tea (Bezbaruah 1968; Sarmah and Bezbaruah 1984). The prolific growth in polyploids may be attributed to increase photosynthesis owing to the increase of chloroplast number in the guard cells.

The effects of ploidy on yield and its components had been studied in tea (Amma 1974; Banerjee 1992b; Wachira 1994; Wachira and Ng'etich 1999). It was found that triploid cultivars produced larger and heavier but fewer harvestable shoots per unit area, compared to diploids due to which triploids yield less than diploids, despite higher shoot weight (Wachira 1994). In another study by Singh (1980), it was found that out of the different type of polyploids produced in India, the dry weights of five fully formed leaves in triploids and tetraploids were higher than that of diploid leaves by 14 and 109 %, respectively. Other pentaploids and aneuploids, however, had relatively low leaf dry weights.

Although it was shown that polyploidy in tea enhanced the yield (Jayasuriya and Govindarajulu 1975; Kulasegaram 1980; Sharma and Ranganathan 1986), this was not always the case, as there were instances where increased polyploidy was demonstrated to depress the productivity (Banerjee 1992b; Wachira 1994). However, it was clear from those studies that though increased ploidy depressed yields, a significant difference in production could also be observed among the genotypes at the same ploidy level. In certain cases, triploids outyielded diploids, which indicated the potential for selecting or developing high-yielding polyploid cultivars. Nevertheless, rooting ability, leaf size and leaf dry weights of triploids and tetraploids were higher that of diploids but lower in pentaploids and aneuploids (Banerjee 1992b).

Two triploid cultivars were commercially successful in the tea industry of Sri Lanka. The first one, TRI 3069, which was an induced tetraploid of TRI 2025, had been accepted commercially and possesses many improved traits. The second cultivar, HS 10A, a natural triploid selected from a seedling-tea population on Hethersett Estate of Sri Lanka, was found to be better adapted to high elevations than diploid cultivars (Kulasegaram 1980). It had been reported that triploid forms of tea were hardier and more resistant to cold conditions than diploids, and a clone which was widely recommended for planting in southern India was reported to be a natural triploid (Jayasuriya and Govindarajulu 1975). Commercially acceptable polyploid tea cultivars developed in some tea-growing countries are detailed in Table 2.4 which show that only eight polyploids have found their way into cultivation, including one artificially induced polyploid.

Although extensive works have been done to identify precise markers for ploidy level in tea, it is clear from the above discussion that the results obtained are not consistent. However, among the criteria studied, the number of chloroplasts in the guard cells and the stomatal density can be used with some reliability for ploidy level analysis. These markers may be used for the screening of polyploids from a larger number of tea genotypes, although chromosome counts remain the most reliable and this procedure could minimize the time and resources needed for subsequent cytological studies.

2.8.5 Mutation Breeding

The work on mutation breeding in tea was initiated during 1967–1968 at TES, India with the objective of increasing genetic variability for possible use in evaluation of superior planting materials. However, except a preliminary report on irradiation with γ rays on cuttings, no progress had been achieved till now (Singh 1984). Studies done elsewhere, had shown that a wide range of variations can be created by irradiating various plant parts, such as seeds, leaf cuttings, auxiliary and apical buds of tea to induce mutations (Tavadgiridze 1979).

2.8.6 Pre-Breeding and Distance Hybridization

Tea breeds freely among the two cultivated species, i.e. *sinensis* and *assamica,* and up to a limited extent with few wild relatives. Earlier, Wight and Barua (1957) hybridized *C. irrawadiensis* with *C. sinensis*. Although the progenies were resembled later but failed to attack tester tong due to inferior quality. Later Bazbarua and Gogoi (1972) made a successful hybridization between *C. japonica* and *C. sinensis*. Morphologically, progenies were found to be intermediate but produced low yield as well as quality. However, a commercial high-yielding clone TV-24 was produced at TES, Assam, India from the cross between F_1 hybrids from *C. irrawadensis* and TV-2, an Assam–China hybrid. Six interspecific backcross progenies were generated by crossing hybrids of *C. saluensis* × *C. japonica* (*Camellia* × *williamsii*) back to *C. japonica*. Segregation data were obtained within these six families for five traits, three involving flavonoid constituents and two pubescence characters. A single major gene seemed to be primarily involved in each case, and two of the traits exhibited linkage behaviour. The taxonomic value of such marker trait was also discussed (Parks and Kondo 1974). Li et al. (2005) made a cross between high quality *C. sinensis* and *C. ptilophylla.* Out of the 62 progenies, 3 were identified to be commercially viable on the basis of the biochemical parameters or organoleptic test. At the same time, these three progenies were cold tolerant, a character which came from *C. ptilophylla,* thus making them suitable for cultivation in sub-tropical region of China.

Interspecific cross compatibility between tea and its allied 26 species in the genus *Camellia* was examined. The interspecific crossing abilities varied among the cross combinations, and the fruit-bearing rates were in the range of 0–42.6 %. The interspecific hybrids obtained from the crossings of *C. sinensis* with *C. japonica, C. pitardii, C. assimilis, C. caudata, C. salicifolia, C. irrawadiensis* and *C. taliensis* showed very low pollen fertilities. Morphological characteristics of those hybrids, including size and shape of leaves, flowers and tree performance, were generally intermediate of their parental species. The hybrids between *C. sinensis* and *C. japonica* showed a high level of tolerance to diseases such as tea grey bright, tea anthracnose and to cold damage during winter as well. It was presumed that the F_1 hybrids which were obtained from the crossing of *C. sinensis* with *C. sasanqua, C. brevistyla* and *C. oleifera* used as a male parent might be developed through parthenogenesis of a reduced gamete. The cross compatibilities between *C. sinensis* and subgeneric sections of genus *Camellia* were found to be as follows: Thea > Camelliopsis > Paracamellia = Camellia = Theopsis > Heterogenea = Corallina (Takeda 1990).

In a natural cross between *C. taliensis* and *C. sinensis,* two low-caffeine but high-theobromine containing plants were produced. Genetic analysis indicated that caffeine-less character might be controlled by one recessive locus. Thus, this caffeineless plant might be used as a genetic resource for introducing caffeine-less trait in cultivated tea through breeding (Ogino et al. 2009).

2.9 Genetic Resources of Tea

Progress and achievements of tea breeding works in certain tea-producing countries had been well reviewed (Singh 1999; Ghosh-Hajra 2001; Deka et al. 2006). Those literatures indicated that the initial emphasis was to collect and evaluate either the indigenous or exotic germplasm for befitting the local environment. However, with the increase of the region-specific need of the industry, almost all tea-producing countries have developed their past specific clones or seed stocks which are reviewed here.

As mentioned earlier, the breeding works at TES, Assam, India which started since 1949, developed a total of 31 clones (Table 2.6), 14 bioclonal seed stocks (Table 2.7) and 134 TRA/Garden series clones (Table 2.8). In Southern India, the breeding works started at UPASI, Tamil Nadu during early 1960s, which resulted in the release of 28 clones (Table 2.9) and 5 bioclonal stocks,

Table 2.6 Descriptions of different TV clones. (Deka et al. 2006)

Name	Year of release	Preference for manufacture		Remarks
		1st preference	2nd preference	Assam–China hybrid
TV-1	1949	CTC	Orthodox	Assam type
TV-2	1949	Orthodox	CTC	Assam type
TV-3	1949	Orthodox	CTC	Assam type
TV-4	1959	Orthodox	CTC	Assam type
TV-5	1959	Orthodox	CTC	Assam type
TV-6	1959	Orthodox	CTC	Assam type
TV-7	1959	Orthodox	CTC	China hybrid
TV-8	1959	Suitable for both		Assam Type
TV-9	1959	CTC	Orthodox	Cambod type
TV-10	1963	CTC	Orthodox	Assam type
TV-11	1963	Orthodox	CTC	Assam type
TV-12	1963	Orthodox	CTC	Assam type
TV-13	1965	Orthodox	CTC	Assam type
TV-14	1967	CTC	Orthodox	Assam hybrid
TV-15	1967	Orthodox	CTC	Assam type
TV-16	1968	CTC	Orthodox	Assam hybrid
TV-17	1968	CTC	Orthodox	Assam hybrid
TV-18	1970	CTC	Orthodox	Cambod type
TV-19	1973	CTC	Orthodox	Cambod type
TV-20	1974	CTC	Orthodox	Cambod type
TV-21	1976	Orthodox	CTC	Assam type
TV-22	1976	CTC	Orthodox	Cambod type
TV-23	1976	CTC	Orthodox	Cambod type
TV-24	1979	CTC	Orthodox	Cambod and species hybrid
TV-25	1982	CTC	Orthodox	Cambod type
TV-26	1982	CTC	Orthodox	Cambod type
TV-27	1985	CTC	Orthodox	Cambod type
TV-28	1985	CTC	Orthodox	Cambod type
TV-29	1990	CTC	Orthodox	Cambod, triploid
TV-30	1993	CTC	Orthodox	Cambod type
TV-31	2006	CTC	Orthodox	Assam hybrid

Table 2.7 Descriptions of bioclonal seedling developed by TES, Assam, India. (Deka et al. 2006)

Cultivar	Parent combination	Year of release	Suitable for area
TS 378	14.5.35 X 14.6.28	1968	Hills (Darjeeling) Area
TS 379	14.5.35 X 14.12.16	1989	Hills (Darjeeling) Area
TS 397	TV-1 X 19.35.2	1976	Plains
TS 449	TV-1 X 270.2.14	1970	Plains
TS 450	TV-2 X 270.2.13	1970	Hills (Darjeeling)
TS 462	TV-1 X 124.48.8	1980	Plains
TS 463	TV-1 X TV-19	1984	Plains
TS 464	TV-1 X 19.29.2	1984	Plains
TS 491	TV-1 X S_3A_3	1989	Plains
TS 506	TV-1 X 19.22.4	1994	Plains
TS 520	TV-19 X TV-20	1992	Plains
TS 557	AV-2 X Teen Ali 17	1996	Hills (Darjeeling)
TS 569	AV-2 X Tukdah-78	1996	Hills (Darjeeling)
TS 589	TV-20 X Heeleakah 22/14	1996	Plains

Table 2.8 Descriptions of different region-specific garden series clones. (Deka et al. 2006)

Area	Clone
Darjeeling	Phoobsering 312, Phoobsering 1404, Phoobsering 1258, Kopati 1/1, Happy Valley 39, Bannockburn 157, Tukdah 145, AV2, Tukdah 253, Tukdah 246, Bannockburn 777, Rungli Rungliot 4/5, Bannockburn 688, Tukdah 78, Tukdah 383, Rungli Rungloit 17/144, CP-1, Teesta Valley 1, Badamtam 15/263, Balasun 7/1A/76, Balasun 9/3/76, Thurbo 3, Thurbo 9 and Lingia 12
South India	ATK-1 (drought tolerant clone), C-17, D-12/A2, C-1, CR-6017 (quality clone), SMP-1 (resistant to blister blight), W-35, SA-6, TTL-1, TTL-2, TTl-4 and TTL-5
Kangra Valley	Kangra and Jawala
Tripura	Huplongcherra 18, Huplongcherra 26, Meghlibundh 11, Meghlibundh 20 and Meghlibundh 25
Barak Valley	Narinpore 4, Narinpore 18, Narinpore 22, Chandighat 9, Longai 17, Longai 26, Poloi 23 and Lalamookh 7
Dooars and Terai	Hantapara 12, Huldibari 19, Leesh River 9/34, Sukna 7, Sanyasithan 8, Kamalpur 6, Mohargung and Gulma 25

Table 2.9 Descriptions of different UPASI clones. (Sharma and Satyanarayana 1987)

Name	Description
UPASI-1	Resistant to drought and tolerant to wind, very upright and vigorous growth and forming as compact bush. Suitable for mid elevations
UPASI-2	Excellent spread with a dense plucking table fairly hardy tolerant to drought and wind. Suitable for all elevation
UPASI-3	A triploid clone with an excellent spread and dense plucking table flourishing at all elevation, bright infusion and liquor
UPASI-4	With profuse branching. Suitable for mid and high elevations
UPASI-5	Compact bush and dense plucking table suitable for mid and high elevation
UPASI-6	With good spread and high plucking density, fairly tolerant to drought, suitable for mid and higher altitude
UPASI-7	With profuse branching, good spread with a dense plucking table. Tolerant to drought and wind damage. Suitable for mid and high elevation
UPASI-8	Vigorous growth, with a good spread and dense plucking table. Can grow at all elevations particularly in warm humid zones
UPASI-9	Most popular standard clone in South India, fairly tolerant to drought and suitable to all altitude and can withstand the soil pH near neutral (6.8)
UPASI-10	Forms a broad dense plucking table, hardy clone, resistant to drought, fairly tolerant to wind. Thrives even in soils with pH near neutral (6.8)
UPASI-11	Vigorous orthotropic grower, tolerant to drought, suitable for mid altitude
UPASI-12	Semi-orthotropic, good grower, suitable for mid elevation
UPASI-13	Profuse branching with dense plucking table. Suitable for high altitude
UPASI-14	Spready bush with high plucking density. Suitable for high altitude and produced flavoury tea
UPASI-15	Resistant to drought and fairly tolerant to wind and mild frost. Suitable to mid altitude and produce quality tea thought out the year
UPASI-16	Tolerant to drought and mild frost. Suitable for all elevation. Flushes during winter
UPASI-17	Good spread and dense plucking table. Suitable for mid and high elevations. Produced very bright liquor
UPASI-18	Semi-orthotropic and vigorous grower with good branching, fairly tolerant to drought. Suitable mid and high altitude
UPASI-19	Hardy, resistant to drought and tolerant to mild frost. Flushes throughout the year. Tolerance to both drought as well as winter. Suitable for high altitude. Produced flavoury tea
UPASI-20	Semi-orthotropic grower, resistant to drought, Suitable for mid altitude. Produce bright liquor

Table 2.9 (continued)

Name	Description
UPASI-21	Excellent spread and dense plucking table. Suitable for mid and high altitude. Produce flavour, bright liquor
UPASI-22	Vigorous grower but with sparse branching. Makes good, tippy, orthodox tea with good flavour and quality
UPASI-23	Good rooter, excellent spread, tolerant to drought fairly well. Suitable for mid and high elevation
UPASI-24	Flushes thought out the year. Suitable for mid and high elevation. Produce scented flavour
UPASI-25	Recovery from pruning is quick, dense plucking point. Suitable for high altitude
UPASI-26	Recovery from pruning is quick. Resistant to drought, flushes throughout the year. Suitable for high altitude
UPASI-27	Compact plucking table with moderate spread. Tolerant to drought
TRF-1	High-yielding clone with profuse branching

Table 2.10 Clones developed by Tata Tea Ltd. (Haridas et al. 2006)

Clones	Parentage	Salient features
TTL-1	UPASI 9 X TRI-2025	High yielder and moderate quality, drought tolerant
TTL-2	Estate selection	Average yielder and excellent quality
TTL-3	Estate selection	Average yielder, moderate prone to drought
TTL-4	UPASI-10 X TRI-2025	High yielder, easy rooter, good quality
TTL-5	UPASI-10 X TRI-2025	High yielder, easy rooter, good quality, fairly tolerant to drought
TTL-6	UPASI-9 X TRI-2025	High yielder, broad leaved with larger shoots
TTL-7	UPASI-9 X TRI-2025	High yielder moderate quality, drought tolerant

i.e. BSS-1 (UPASI-10 X TRI-2025), BSS-2 (UPASI-2 X TRI-2025), BSS-3 (UPASI-9 X TRI-2025), BSS-4 (UPASI-15 X TRI-2025) and BSS-5 (CR6017 X UPASI-8). Further, to widen the genetic base, clones developed in Sri Lanka were introduced and experimented, and it was found that TRI-2024 and TRI-2025 were suitable and therefore gained popularity in the industry (Sharma and Satyanarayana 1987). Efforts to generate and conserve the tea genetic resources by private funding were also made. For example, Research and Development Department of Tata Tea Ltd, Kerala, India had developed seven promising cultivars, several bioclonal seeds stock such as TTSS-1, TTSS-2, besides they maintained a tea germplasm collection more than 100 popular tea genotypes and wild species (Haridas et al. 2006) and Table 2.10.

The plant improvement programmes in Sri Lanka started during 1930s when seeds of Betjan, Manipur and Rajghur from Assam were planted in Peradoniya Botanical Garden in Sri Lanka to establish the first tea garden. However, scientific works started only during 1937 when Dr. F. R. Tubbs brought few seeds of ST 4/10 from TES, Assam, India and seedlings were raised at Tea Research Institute of Sri Lanka and subsequently eight clones were released. Since then several clones were released which were popularly known as 20, 30 and 40 series clones. At present, Sri Lanka has more than 57% clonal tea area. Out of this, around 80% is composed by only three popular clones, i.e. TRI-2023, TRI-2025 and TRI-2026.

Bangladesh Tea Research Institute, Maulvibazar, Bangladesh had developed 13 clones and two bioclonal seed stocks so far. These clones are known as BT-1 to BT-13. Among the two biclonal seed stocks, BST-1 and BST-2, the former is more popular, which is a cross between BT-1 and the popular Indian tea cultivar TV-1 (Deka et al. 2006).

Initially, the Tea Research Institute of Vietnam, Vinh-Phu had started the breeding of tea by collecting the planting materials from India. Many elite clones and *jats* such as PH-1, PH-3 and IA had been evolved through selection and hybridization (Tien 1993). Later, two clones namely LDP1 and LDP 2 had been released with a yield potential of 17,500 kg green leaf/ha and 16,900 kg green leaf/ha, respectively. Two quality clones 276 and 215 (cross between PH1 and Shan tea) had also been developed (Toan and Tao 2005).

In Indonesia, the breeding of tea which was initiated during 1980s was confined to selection only. Later ten clones, namely GMB-1 to GMB-10 had been released which were developed by hybridization at Research Institute for Tea and Cinchona, Gambung, Indonesia. While GMB-1 to GMB-5 had the potential to produce 3,500 kg/ha/year of made tea, GMB-6 to GMB-10 had a productivity potential of 5,000 kg/ha/year of made tea. Interestingly, few clones were also found to be tolerant to blister blight (Arfin and Semangun 1999).

Tea improvements in Kenya started with the introduction of seeds from Assam, India which were used to establish the first tea plantation. Since these progenies had not been particularly selected for high yield and quality, the resultant seedling populations of mixed genotypes were genetically inferior, though diverse. With this population, organized tea improvement started with the formation of *Tea Research Institute of East Africa* in 1961, and later the Tea Research Foundation of Kenya in 1980 with a mandate for research on all aspects of tea. Thus, in the first phase of the tea improvement, mass selection among introduced seedling jats based on morphological characteristics was done. As a result, several cultivars such as TRFK 6/8, TRFK 7/3, TRFK 7/9, TRFK 7/14, TRFK 11/4, TRFK 12/12, TRFK 12/19, TRFK 31/8, TRFK 31/11, TRFK 31/27, TRFK 31/28, TRFK 31/29, TRFK 54/40, TRFK 55/55, TRFK 55/56, TRFK 56/89, TRFK 100/5 and TRFK 108/82 were released for the industry. However, being heterogeneous genotypes, they formed good breeding materials for the second phase of mass selection. It was basically done by hybridization of selected parental stocks, superior in certain attributes that they were selected for. Several cultivars such as TRFK 303/35, TRFK 303/152, TRFK 303/156, TRFK 303/179, TRFK 303/186, TRFK 303/199, TRFK 303/216, TRFK 303/231, TRFK 303/259, TRFK 303/248, TRFK 303/352, TRFK 303/366, TRFK 303/388, TRFK 303/577, TRFK 303/745, TRFK 303/791, TRFK 303/978, TRFK 303/999, TRFK 303/1199, TRFK 347/314, TRFK 347/326, TRFK 347/336 and TRFK 347/573 were developed. The third phase involving selections from bi-clonal full-sib progeny resulted in the release of clones TRFK 337/3, TRFK 337/138 and TRFK 338/13. Later, the limitations of phenotypic selection encouraged, spurred the search and development of superior but genetically uniform tea clones. Presently, 45 clones had been developed out of which 24 were selected from seedling populations including the most popular clone 6/8. Thus in total, 27 clones (60 %) shared the genetic pedigree of clone 6/8. Among the cultivars, few were credited with some special characters such as S 15/10, a high-yielding clone recently developed and registered for 10,000 kg made tea/ha/annum. Similarly, SFS 150 and 303/577 were accounted for drought tolerant and TN 14–3 for tolerant to high soil pH, SFS 150 and TN 14–3 for cold tolerant, 12/2 for poor fermenter, 311/287 for tetraploids and 7/9, 57/15, SC 31/27, S 15/10 for mite tolerant (Seurei 1996).

About 2,665 accessions of tea including a few species of *Camellia* were maintained at the China National Germplasm Tea Repository which had contributed to develop more than 200 improved cultivars (Wang et al. 2011). Out of which, 97 national registered cultivars, among them 17 were *jats*, 80 were clones, 30 were landraces and 67 were improved clones. They were bred by 23 different institutions including the national and local tea research institutes, Agricultural Universities, local agricultural departments, tea experimental stations, etc. There were also about 130 registered cultivars, among them 16 were *jats*, 114 were clones, 29 were landraces and 101 were improved clones, respectively. However among them, only 54 were popular clones suitable for green tea, 32 cultivars for black tea and 33 cultivars were for oolong tea. Few cultivars, such as Zhuyeqi and Fuding Dabaicha, processed stronger prune-shock. The plucking surface of Zhuyeqi and Fuding Dabaicha contained more dense leaves, branches and shoots than that of popular clone 'Xianggbolu', hence the former two were recommended for mechanical plucking (Yongming 1999; Chen and Zhou 2005).

Apart from the Chinese main land, commendable works on tea breeding had also been done at Taiwan Tea Experimental Station, Taoyuan, Taiwan. Among the 66 cultivars, Chisin Oolong, Chinsin Dapan, TTES No. 12, TTES No. 13 and Shy jih Chuen were very popular. Nevertheless, Chinsin Oolong was the most important tea cultivars in Taiwan tea industry occupying half of total tea acreage. Besides, Chinsin Dapan had a wide manufacturing adaptability, suitable for making green tea, paochung tea, white-tip oolong tea and black tea. TTES No 12 (Kinshen) was known for its light milky flavour and hence very popular. On the other hand, TTES No. 13 had satisfactory tolerance levels of environmental stress such as drought and die-back diseases as compared to Chinese Oolong tea (Toan and Tao 2005).

In Korea, tea breeding began at Tea Experimental Station, Boseong which was established in 1992 and developed seven tea cultivars till recently. These cultivars were being propagated as cuttings for distribution to tea farmers. Some of the popular cultivars were Bohyang, Myungseon, Chamnok, Seonhyang, Mihyang, Jinhyang and Oseon. Since then, massive breeding works such as introduction of green tea cultivars, hybridization using cv. Yabukita as parent had been undertaken (Jeong et al. 2005).

Tea breeding was started way back in 1920 at Tea Experimental Station, Shizuoka, Japan. Additionally, several private tea breeders started the varietal improvement of tea which resulted in the development of many clones. Among them, Hikosaburo Sugiyama (1857–1941), a noted tea breeder popularly known as 'Burbank of tea' after US plant scientist, Luther Burbank, developed the popular clones of Japan 'Yabukita', 'Koyanishi' and 'Rokuro'. At present, 'Yabukita' alone planted 76% of all Japanese tea plantations. However, clonal selection programme was more intensified during 1950 and as a result many good clonal cultivars were released in 1970. These newly developed cultivars had contributed much to the modern Japanese tea industry. There were 70 registered varieties and few of them were, Ooiwase, Yabukita, Surugawase, Sayamakaori, Yamakai, Kurasawa, Kanayamidori, Okuhikari and Sawamizuka (Takeo 1992).

2.10 Bottlenecks of Tea Breeding

Although conventional tea breeding is well established and contributed much for tea improvement over the past several decades, it is time consuming and labour intensive. The bottlenecks of conventional breeding are: (1) perennial nature, (2) long gestation periods, (3) high inbreeding depression, (4) self-incompatibility, (5) unavailability of distinct mutant of different biotic and abiotic stress, (6) lack of distinct selection criteria, (7) low success rate of hand pollination, (8) short flowering time (2–3 months), (9) long duration for seed maturation (12–18 months), (10) clonal difference of flowering time and fruit bearing capability of some clones. Similarly, although vegetative propagation is an effective method of tea propagation, it is limited by several factors, such as, (1) slower rates of propagation, (2) unavailability of suitable planting material due to winter dormancy, drought in some tea growing areas and so on, (3) poor survival rate at nursery due to poor root formation of some clones and (4) season-dependent rooting ability of the cuttings, etc. Therefore, to overcome the problems related to tea breeding, scientists across the world started finding some alternatives through biotechnological approaches which are discussed subsequently.

2.11 Conclusion

Conventional breeding in tea is well established and contributed significantly. Several region-specific clones and bi-parental hybrids have been developed across the tea growing regions of the world. However, several important aspects of tea breeding are in their initial stages which need to be strengthened. Some of them are: (1) association mapping which has tremendous potential to identify the Quantitative trait loci (QTL) particularly in tea; (2) development of large-scale molecular markers and their utilization in marker-assisted breeding; (3) although pseudo-test cross has been utilized for developing the initial linkage map, it is necessary to develop the bi-parental mapping population for various applications where progress is very much limited in comparison to other similar woody perennials such as coffee and eucalyptus.

References

Ackerman WL (1971) Genetics and cytological studies with Camellia and related genera. Technical Bull. No 1427. USDA, US Gov Print Office, Washington, DC, p 115

Ahmed N, Singh ID (1993) A technique for rapid identification of ploidy level in tea. Two Bud 40:31–33

Amma S (1974) Characteristic of tetraploid tea induced from gamma irradiated Yabukita variety. Study Tea 46:1–6

Anon (1979) Annual report. Tea Research Institute of Sri Lanka, p 64

Anon (1999) Annual report. Tea Research Foundation of Kenya. pp 45–50

Anon TI (1973) Annual report. Tea Research Institute of Sri Lanka, pp 38–39

Arfin T, Semangun P (1999) Tea industry in Indonesia. In: Jain NK (ed) Global advance in tea science. Aravali Book International Pvt Ltd, New Delhi, pp 65–72

Bakhtadze KE (1935) Methods of tea selection. Sov Sub-Tropical 2:9–15

Banerjee B (1992a) Botanical classification of tea. In: Wilson KC, Clifford MN (eds) Tea cultivation to consumption. Chapman and Hall, London, pp 25–51

Banerjee B (1992b) Selection and breeding of tea. In: Willson KC, Clifford MN (eds) Tea cultivation to consumption. Chapman and Hall, London, pp 53–86

Barbora BC, Barua DN, Bera B (1996) Tea breeding at Tocklai. Two Bud 43:3–9

Barua DN (1989) Science and practice in tea culture. Tea Research Association, Jorhat, pp 56–58

Bezbarua HP (1968) Genetic improvement of tea in North East India- its problem and possibilities. Indian J Genet 28:126–134

Bezbaruah HP (1971) Cytological investigation in the family theaceae-I. Chromosome numbers in some *Camellia* species and allied genera. Carylogia 24:421–426

Bezbaruah HP (1975) Development of flower, pollination and seed set in tea in North-East India. Two Bud 22:25–30

Bezbaruah HP (1976) Aneuploidy in tea. Nucleus 19:167–169

Bezbaruah HP (1991) Tea breeding in North East India. Proc Int Symp Tea Sci34

Bezbaruah HP, Gogoi SC (1972) An interspecific hybrid between tea (*C. sinensis* L.) and *C. Japonica* L. Proc Ind Aca Sci B76:219–220

Bhattacharjee H, Singh ID (1994) Storage of tea seed. Two Bud 41:32–34

Blakeslee AF, Avery AG (1937) Methods of inducing doubling of chromosome in plants treated with colchicines. J Hered 28:394–411

Chang HT, Bartholomew B (1984) Camellia. Timber Press, Portland

Chaudhuri TC (1979) Studies on the morphology and cytology of the progenies of triploid tea (*C. sinensis L.*). Ph.D thesis. Assam Agricultural University, Jorhat, p 176

Chaudhuri TC, Bezbaruah HP (1985) Morphology and anatomy of the aneupolid and polyploidy tea {*C. sinensis (L.)* O. Kuntze}. J Plant Crop 13:22–30

Chen L, Yu FL, Tong QQ (2000) Discussion on phylogenetic classification and evolution of Sect Thea. J Tea Sci 20:89–94

Chen L, Zhou ZX (2005) Variations of main quality components of tea genetic resources [*C. sinensis (L.) O. Kuntze*] preserved in the China National Germplasm Tea Repository. Plant Food Human Nutr 60:31–35

Chen S, Ye D (1989) Cytological studies on polyploid tea. J Tea Sci 9:117–126

Cohen Stuart CP (1929) Research on leaf yielding capacity of tea plants (Dutch). Arch Tree Cult Ned Ind 4:276–288

Das SK, Sabhapondit S, Ahmed G, Das S (2013) Biochemical evaluation of triploid progenies of diploid 3 tetraploid breeding populations of *Camellia* for genotypes rich in catechin and caffeine. Biochem Genet 51:358-376

Datta M, Agarwal B (1992) Intervarietal differences in karyotype of tea. Cytologia 57:437–441

Deka A, Deka PC, Mondal TK (2006) Tea. In: Parthasarathy VA, Chattopadhyay PK, Bose TK (eds) Plantation Crops-I. Naya Udyog, Calcutta, pp 1–148

Fukusima E, Iwasa S, Endo N, Yoshinari T (1966) Cytogenetic studies in *Camellia.* I. Chromosome survey in some *Camellia* species. Jap J Hort 35:413–421

Furukawa K, Tanaka J (2004) 'Makura-Ck2': a tea strain with a high somatic embryogenesis. Breed Res 6:109–115

Ghosh Hajra N (2001) Tea cultivation: comprehensive treatise. International Book distribution Co., Lucknow, pp 22–27
Gu Z, Xiao H (2003) Physical mapping of the 18S-26S rDNA by fluorescent in situ hybridization (FISH) in *C. reticulata* polyploid complex (Theaceae). Plant Sci 164:279–285
Gunasekara MTK (2000) Anatomical characteristics of polyploid tea cultivars. Annual Report, Tea Res Ins of Sri Lanka, p 164
Gunasekara MTK, Ranatunga MAB (2003) Polyplody in tea (*C. sinensis* L.) and its application in tea breeding: a review. Sri Lanka J Tea Sci 68:14–26
Hanson L, Mcmahon KA, Johnson MAT, Bennett MD (2001) First nuclear DNA C-values for another 25 angiosperm families. Annal Bot 88:851–858
Haridas P, Balasubramanian S, Netto LA, Ganesh Uma M, Mohan Kumar P (2006) Studies on improving planting material in tea (*C. sinensis L.*). J Plant Crop 34:243–249
Hu KM, Zhang YM, Wang JF, Xie TH, Hu KM, Zhang YM, Wang JF, Xie TH (2003) Comparison on the population dynamics and leafhopper resistance on different tea cultivars. J Tea Sci 23:57–60
Huang H, Tong Y, Zhang Q-J, Gao L-Z (2013) Genome size variation among and within *Camellia* species by using flow cytometric analysis. PLOS One 8:64981–64995
Hwang YJ, Okubo H, Fujieda K (1992) Pollen tube growth, fertilization and embryo development of *Camellia japonica L.* X *C. chrysantha* (Hu) Tyyama. J Jap Soc Hort Sci 60:955–961
Janaki Ammal EK (1952) Chromosome relationship in cultivated species of *Camellia*. Amer Camellia Year Book
Jayasuriya P, Govindarajulu V (1975) Chromosome number of some tea clones. Planters Chron LXXX:185–186
Jeong B, Song Y, Moon Y, Han S, Bang J, Kim J, Kim J, Park Y (2005) Tea tree breeding plans for the tea Industry in Korea. In International Tea Symposium-2005, Organised by Tea Res Inst, Chinese Acad Agril Sci, China Tea Science Society Nov 11–15. pp 322–332
Joshi R, Poonam, Gulati A (2011) Biochemical attributes of tea flowers (*C. sinensis*) at different developmental stages in the Kangra region of India. Scientia Hort 130:266–274
Karasawa K (1932) On triploid tea. Bot Mag 46:458–460
Katsuo K (1966) Methods of inducing the polyploidy tea plant by colchicines treatment of the axillary bud. Study Tea 33:1–4
Kondo K (1975) Cytological studies in cultivated species of *Camellia*. Ph.D thesis. Univ NC, Chapel Hill, p 260
Kondo K (1977) Choromosome number in the genus *Camellia*. Biotropica 9:86–94
Kondo K, Parks CR (1979) Giemsa C-banding and karyotype of *Camellia* C-banned karyotypes. Am Camellia Y Book 34:40–47
Kondo K, Parks CR (1980) Giemsa C-banding and karyotype of *Camellia*. Proc Internal Camellia Cong Kyoto pp 55–57
Koskey JK, Wachira FN (2000) The use of plastid chloroplast count technique to determine ploidy levels in tea. Tea 21:15–18
Kulasegaram S (1980) Technical development in tea production. Tea Q 49:157–183
Li X, Ye T, Huang Q, fu D, Zhang C, Zeng L (2005) Study on distant hybridization for commercial tea production. 2005 international symposium on innovation in tea science and sustainable development in tea Industry, TRA, CAAS, China Tea Science Society, Nov 11–15 Hangzhou, China pp 389–395
Liang GL, Zhou CQ, Lin MJ, Chen JY, Liu JS (1994) Karyotype variation and evolution of sect. Thea in Guizhou. Acta Phytotaxon Sinica 32:308–315
Lin YS, Wu SS, Lin JK (2003) Determination of tea polyphenols and caffeine in tea flowers (*C. sinensis*) and their hydroxyl radical scavenging and nitric oxide suppressing effects. J Agric Food Chem 51:975–978
Mondal TK (2009) Tea. In: Prydarsini M, Jain SM (eds) Breeding plantation tree crops tropical species. Springer pp 545–587
Mondal TK (2011) Camellia. In: Kole C (ed) Wild crop relatives: genomics and breeding resources plantation and ornamental crops. Springer, USA, pp 15–40
Mondal TK, Bhattacharya A, Laxmikumaran M, Ahuja PS (2004) Recent advance in tea Biotechnology. Plant Cell Tissue Orga Cult 75:795–856
Morinago T, Fukusima E, Kano T, Maruyama Y, Yamasaki Y (1929) Choromosome number in cultivated plants. Bot Mag 43:569–594
Nagata T, Sakai S (1984) Differences in caffeine, flavanols and amino acids contents in leaves of cultivated species of *Camellia*. Jap J Breed 34:459–467
Nesumi A, Ogino A, Yoshida K, Taniguchi F, Maeda yamamM (2012) 'Sunrouge', a new tea cultivar with high anthocyanin. JARQ 46:321–328
Ng'etich WK, Wachira FN (1992) Use of a non-destructive method of leaf area estimation in triploid and diploid tea plants (*C. sinensis*). Tea 13:11–17
Ogino A, Tanak J, Taniguchi F, Yamamoto MP, Yamada K (2009) Detection and characterization of caffeine less tea plant originated from inter-specific hybridization. Breed Sci 59:277–283
Oreal G, Wilson PG (2012) *C. cherryana* (theaceae), a new species from China. Ann Bot Fennici 49:248–254
Osone K (1958) Studies on the breeding of triploid plants by diplodising gamete cells. Jap J Breed 8:171–177
Parks CR, Kondo K (1974) Breeding studies in the genes *Camellia* (Theaceae). I. A Chemotaxonomic analysis of synthetic hybrid and backcross involving *C. japonica* and *C. saluensis*. Brittonia 26:321–332
Prakash O, Sood A, Sharma M, Ahuja PS (1999) Grafting micropropagated tea (*C. sinensis* (L.) O. Kuntze) shoots on tea seedling- a new approach to tea propagation. Plant Cell Rep 18:137–142
Ranatunaga MAB, Gunasekare MTK (2002) Identification of polyploid marker in tea (*C. sinensis* L.). Proc Annual sessions Sri Lanka Assoc for the Adv of Sci, p 38

Rashid A, Chowdhary M, Badrul Alam AFM (1985) Studies on the progenies of a cross between diploid and tetraploid tea. Sri Lanka J Tea Sci 54:54–61

Sarmah PC, Bezbaruah HP (1984) Triploid breeding in tea. Two Bud 31:55–59

Satyanarayan N, Sharma VS (1982) Biometric basis for yield prediction in tea clonal selection. Proc. PLACROSYM IV, Dec 3–5, 1981, Mysore, India pp 237–243

Satyanarayan N, Sharma VS (1986) Tea (*Camellia L.* spp) germplasm in south India. In: Srivastava HC, Vatsya B, Menon KKG (eds) Plantation crops: opportunity and constraints. Oxford IBH Publishing Co., New Delhi, pp 173–179

Sealy JR (1958) A revision of the genus *Camellia*. R. Hortic Soc., London, pp 58–60

Sebasthiampillai AR (1976) A simple technique for the polyploids in tea. Tea Q 46:12–15

Seurei P (1996) Tea improvement in Kenya: a review. Tea 17:76–81

Sharma VS, Ranganathan V (1986) Present status and future need of tea research. In: Srivastava HC (ed) Plantation crops, vol II. Oxford and IBH Publishing Co., New Delhi, pp 37–50

Sharma VS, Satyanarayana N (1987) UPASI clones. Planter Chronic 81:28–33

Sharma VS, Venkataramani KS (1974) The tea complex. I. Taxonomy of tea clones. Proc Ind Aca Sci 53:178–187

Simura T, Inabe T (1952) Studies on polyplody of tea plants. Tokai-Kinki National Agricultural Experimental Station. Res Prog Rep 1:1–14

Singh ID (1980) Non-conventional approaches in the breeding of tea in North East India. Two Bud 27:3–6

Singh ID (1984) Advances in tea breeding in North-East India. Proc Placrosym IV:88–106

Singh ID (1999) Plant Improvement. In: Jain NK (ed) Global advances in Tea. Aravali Book International (P) Ltd., New Delhi, pp 427–448

Sivapalan P, Gnanapragasam NC, Kathiravetpillai A (1995) Field guide book. Tea Research Institute of Sri Lanka, Sri Lanka, pp 5–12

Su SK, Chen SL, Lin XZ, Hu FL, Shao M (2000) The determination of ingredient of tea (*C. sinensis*) pollen. Apicult China 51:3–5 (in Chinese)

Takeda Y (1990) Cross compatibility of tea (*C. sinensis*) and its allied species in the genus *Camellia*. JARQ 24:111–116

Takeda Y (2002) Genetic analysis of tea gray blight resistant in tea plants. JARQ 26:143–150

Takeo T (1992) Chemistry of tea. In: Willson KC, Cliford MN (eds) Tea: cultivation to consumption. Chapman and Hall, London, pp 413–457

Takyu T, Takeda Y, Nagatomi S (2003) Trichomeless mutant in tea. Tech News Int Rad Breed 67:2

Tanaka T, Mizutani T, Shibata M, Tanikawa N, Parks CR (2005) Cytogenetic studies on the origin of *Camellia × vernalis*. V. Estimation of the seed parent of *C. × vernalis* that evolved about 400 years ago by cpDNA analysis. J Jap Soc Hort Sci 74:464–468

Tavadgiridze SK (1979) Biology of growth and development in some polyploid forms of tea obtained by colchicines treatment and of irradiation. Subtropicheska Lenltry 3:137–139

Thirukkumaran G, Gunasekare MTK (2001) Use of pollen morphology and physiology to different ploidy level of tea (*C. sinensis*) clones. Proc Jaffna Sci Assoc 9:6–7

Tien DM (1993) Tea industry in Vietnam. Proc Intl. Symp. Tea Sci. Human health. January 11–14, 1993, Calcutta, Tea Research Association, India, pp 103–106

Timoshenko MT (1936) The selection of tea for its chemical composition. Sov Sub-Tropical 1:25–31

Toan NV, Tao NV (2005) Tea breeding selection by hybridization method in Vietnam. International Tea Symposium 2005. Organised by Tea Research Institute, Chinease Academy Agricul Sci. China Tea Sci Soc Nov 11–15, Hangzhou, China

Tubbs FR (1932) A note on vegetative propagation of tea by green shoot cuttings. Tea Q 5:154–156

Tunstall AC (1931a) A note on the propagation of tea by green shoot cuttings. Quart J Indian Tea Assoc 4:49–51

Tunstall AC (1931b) Experiment on vegetative propagation of tea by green shoot cuttings. Bull, Tocklai Experimental Station, pp 113–114

Visser T (1969) Tea *C. sinensis* (L.) O. Kuntze. In: Ferwerdu EP, Wit F (eds) Outlines of perennial crop breeding in the Tropics. Veenaran and Zonen, Wageningen, pp 459–493

Wachira FN (1994) Triploidy in tea (*C. sinensis*): effect on yield and yield attributes. J Hort Sci 69:53–60

Wachira FN, Ng'etich WK (1999) Dry-matter production and partition in diploid, triploid and tetraploid tea. J Hort Sci Biotech 74:507–512

Wachira FN, Kiplangat JK (1991) Newly identified Kenyan Polyploid tea strains. Tea 12:10–13

Wachira FN, Muoki RC (1997) Nucleolar and nucleolus organizer regions in tea as visualized by silver staining. Afr Crop Sci J 5:253–258

Wang X, Chen L, Yang Y (2011) Establishment of core collection for Chinese tea germplasm based on cultivated region grouping and phenotypic data. Front Agric China 5:344–350

Wang Y, Luo F, Li CH, Wang YC, Tang XB, Wang Y, Luo F, Li CH, Wang YC, Tang XB (2003) Selection of a tea accession Tianfu 28 with high quality and resistance. South-West China J Agril Sci 16:61–64

Wellensiek SJ (1933) Floral biology and technique of crossing with tea. Arch Thea Cult 12:27–40

Wellensiek SJ (1934) Research on quantitative tea selection. I. The Pajoeng reform see garden in Tjihirocan (Dutch). Arch Theecult Ned Ind 8:9–37

Wight W (1938) Recent advance in the classification and selection of tea plant. In: Proc 2nd Tocklai Annual Conference Tockali, Assam India, p 38

Wight W (1939) Report Indian tea association. Sci Dept Tocklai Assam pp 22–24

Wight W (1956) Genetic basis of yield. Proc 13th Tocklai Ann Conf., Tea Res Assoc., Assam

Wight W (1962) Tea classification revised. Curr Sci 31:298–299
Wight W, Barua PK (1957) What is tea? Nat 179:506–507
Wood DJ, Barua DN (1958) Species hybrids of tea. Nat 181:1674–1675
Wu CT (1964) Studies on hereditary, variation and morphology of pubisence on the young shoots of tea plants (China). Bull Pinchen Tea Exp Stn 20:1–23
Yang YJ, Yang SJ, Wang YS, Zeng JM, Yang YJ, Yang SJ, Wang YS, Zeng JM (2003) Selection of early budding and high quality green tea cultivar. J Tea Sci 23:9–15
Yongming Y (1999) Agrotechnology of tea in China. In: Jain NK (ed) Global advances in tea science, Aravali Books International (P) Ltd., New Delhi, pp 481–500
Yoshida K, Takeda Y (2006) Evaluation of anthracnose resistance among tea genetic resources by wound-inoculation assay. JARQ 40:379–386
Yu F, Xu N (1999) Tea germplasm resources of China. In: Jain NK (ed) Global advances in tea science. Aravali Books International (P) Ltd., New Delhi pp 393–412

3 Micropropagation

3.1 Introduction

The importances of *in vitro* multiplication are well known. It is especially important for woody perennial plants such as tea and its wild species. Several aspects of micropropagation are discussed in this chapter.

3.2 Need for Micropropagation

Although vegetative propagation is an effective method of propagation, it is limited by several factors in tea and other related species, such as (1) slower rates of propagation; (2) unavailability of suitable planting material due to winter dormancy and drought in some tea-growing areas; (3) poor survival rate at nursery due to poor root formation of some clones and (4) season-dependent rooting ability of the cuttings.

Therefore, micropropagation technique appears to be an ideal choice for circumvention of the problems related to conventional propagation. Additionally, due to faster rate of multiplication, it is highly apt for a newly developed tea cultivar which owes high demand in the industry and hence needs to be supplied in large quantities within a short span of time. Additionally, central to any successful transgenic technology is an efficient *in vitro* regeneration protocol. While an efficient regeneration protocol is essential for introduction of a foreign gene into plant tissues, micropropagation is important for the transfer of a large number of genetically modified plants to the field within a short span of time.

3.3 Tea

Several reviews on micropropagation of tea and related species have been published in the past (Kato 1989; Vieitez et al. 1992; Dood 1994; Das 2001; Mondal 2003). It is evident from the literature that till the late 1980s, the thrust of tea micropropagation was on increasing rate of multiplication. However, the emphasis on enhancing survivability of micropropagated tea at the hardening stage was given only during the early 1990s. Studies on the field performance of tea and commercial exploitation of micropropagation started only at the beginning of the new millennium, which is covered in the forthcoming discussion. Several factors from tissue type to media composition influence the success of micropropagation.

3.3.1 Explants

Fundamentals to the establishment of *in vitro* culture are the type, origin and availability of explants throughout the year. Generally, shoot tips and nodal segments with dormant axillary buds of either juvenile or adult origin of current year growth are commonly used as explants for tea micropropagation (Vieitez et al. 1992).

T. K. Mondal, *Breeding and Biotechnology of Tea and its Wild Species*,
DOI 10.1007/978-81-322-1704-6_3, © Springer India 2014

While Iddagoda et al. (1988) as well as Jha and Sen (1992) had used zygotic embryos and immature and mature cotyledons for the induction of adventitious buds, Kato (1985) employed epidermal layers of stem segments, stem segments without epidermal layer and intact stem segments for shoot regeneration. Among these, callus from the epidermal layers produced buds more rapidly than the callus from other origins. Flower stalks, stem pieces and leaf petioles had been attempted for the induction of adventitious buds without success barring callus formation (Sarwar 1985).

3.3.2 Initiation and Multiplication

Maximum efforts had been made to standardize the media formulation for increasing the shoot multiplication rate. The most common basal medium had been either full or half-strength Murashige and Skoog (MS) salts (Murashige and Skoog 1962). However, there were also a few reports on the use of other media such as woody plant medium (WPM; Lloyd and McCown 1980) and Heller's (1953) medium (Table 3.1). Nakamura (1987a) compared various basal media, namely MS, B_5 medium (Gamborg et al. 1968) and Nitsch and Nitsch (1969), and concluded that MS was the best for tea shoot multiplication. Tahardi and Shu (1992) achieved axillary shoot proliferation on WPM with various concentration of thidiazuron (TDZ) within 10–12 weeks. Although several researchers had reported that MS medium was most suitable for initiation of multiple shoots, half-strength MS salts were also reported for multiplication and shoot proliferation in tea (Phukan and Mitra 1984; Banerjee and Agarwal 1990; Agarwal et al. 1992). Manipulation of vitamin compositions along with organic and inorganic salts of MS was found to be useful for initiation and multiplication of axillary shoots of tea (Arulpragasam and Latiff 1986). Among the different plant growth regulators (PGR), addition of 6-benzyladenine (BAP, 1–6 mg/l) and indole-3-butyric acid (IBA, 0.01–2.0 mg/l) in the culture medium was suitable for both shoot initiation and subsequent multiplication. Apart from tea, BAP (1–4 mg/l) was also found to be the best cytokinin in different related species, such as *C. japonica* (Vieitez et al. 1989a), *C. oleifera* (Tian-Ling 1982), *C. reticulata* (San-Jose and Vieitez 1990; San-Jose et al. 1991) and *C. sasanqua* (Torres and Carlisi 1986). Kato (1985) succeeded in achieving indirect organogenesis via callus phase from three different types of explants, namely epidermal layer of stem segment, intact stem segment and stem segment without epidermal layer, using BAP (10 mg/l) augmented medium. Although 2,4-dichlorophenoxyacetic acid (2,4-D) and α-naphthaleneacetic acid (NAA) were found to induce callus, these were ineffective for the growth and development of tea shoots (Nakamura 1988). Auxins such as NAA in combinations with BAP produced either callus or induced 4–5 shoot buds per explant within 8–12 weeks (Phukan and Mitra 1984; Bag et al. 1997). Picloram and 2,4,5-trichlorophenoxyacetic acid (2,4,5-T) had been used successfully for the elongation of tea shoots by several researchers (Arulpragasam and Latiff 1986; Nakamura 1987 a, b; Iddagoda et al. 1988; Nakamura 1989; Jain et al. 1991). Higher concentrations of these auxins (10 mg/l) were found to be critical for shoot elongation of 30 different cultivars of tea (Kuranuki and Shibata 1993). Importance of indole-3-acetic acid (IAA, 0.1–2.0 mg/l) and kinetin (Kn, 0.21–8.0 mg/l) for both induction and multiplication of axillary shoots was emphasized by several researchers in tea (Phukan and Mitra 1984; Sarwar 1985; Das and Barman 1988) as well as in *C. sasanqua* (Torres and Carlisi 1986), *C. japonica* (Creze and Beauchesne 1980) and in some of their hybrids (Creze and Beauchesne 1980; Lammerts 1958). It was clearly evident that MS was superior to WPM in all combinations.

The effect of TDZ on micropropagation of tea was studied in detail. Mondal et al. (1998) had shown that extremely low concentrations of TDZ (0.02 mg/l) alone were effective in inducing shoot bud proliferation and maintaining high rates of shoot multiplication on hormone-free media up to 24 subcultures. On the other hand, higher concentrations of BAP (0.22–2.2 mg/l) and its continued presence were required to initiate and sustain shoot proliferation. Although all explants produced callus at higher concentration

Table 3.1 Micropropagation of *Camellia* spp. (Mondal et al. 2004)

Species/cultivar	Explant	Medium				Response	Reference
		Bud break and shoot initiation	Multiplication	Regenera-tion via shoot organo-genesis	Rooting		
C. sinensis, TV-1	Term bud and nodal stem segment of field grown plant	MS + IAA (0.25) + BA (4) + CW (10 %) + YE (200)	½ MS + IAA (0.5) + BA (4.0) + CW (10 %) + YE (200)	–	½ MS + AA (2) + IBA (7) + Sucrose (4 %) IBA (7)	Shoot proliferation and rooting (60 %)	Agarwal et al. (1992)
C. sinensis (L.) O. Kuntze cv.TRI-2025, CY9 and PK2	Shoot tip (4–5 cm) and nodal segment of mature plant	MS + modified vit + IBA (0.01) + GA_3 (0.1) + BA (1) for TRI- 2025 and MS + modified vit + IAA (0.1) + BA (1) for others	MS + modified vit + IBA (0.01) + GA_3 (0.1) + BA (2) for TRI- 2025 and MS + modified vit + IAA (0.1) + BA (2)	–	–	Axillary shoot proliferation (2–4/explant)	Arulpragasam and Latiff (1986)
C.saluenensis × *C. japonica*	Shoot tip	–	–	–	½ MS	Root induction and plantlet regeneration	Beretta et al. (1987)
C. sinensis (L.) O. Kuntze	Nodal segment	MS + BA (1.12) + NAA (0.2)	MS + BA (2.25) + IBA (0.2)	–	½ MS + NAA (3.6)	Axillary shoot multi-plication and rooting	Bag et al. (1997)
C. sinensis, TV-1	Shoot tip and nodal segment (0.7–1.0 cm) of mature field grown plant	MS + YE (200) + CM (10 %) + IAA (0.5) + BA (4)	½ MS + CM (10 %) + IAA (0.25) + BA (4)	–	½ MS + modified vit + IBA (7)	Rooting (60 %) of *in vitro* grown shoot	Banerjee and Agarwal (1990)
C. saluenensis × *C. chrysantha*	Seedling shoot tip	MS + Kin (1) + 2iP (1) + GA_3 (1) + IAA (1) + PVP (10 g/l)	MS + Kin (1) + 2iP (1) + GA_3 (1) + IAA (1) + PVP (10 g/l)	–	–	Axillary shoot development	Creze and Beauchesne (1980)
C. japonica, Purple Dawn	Shoot tip and nodal segments of 10-year-old tree	½ MS + BA (1)	½ MS + BA (1) + GA_3 (5)	–	–	Axillary shoot development	Carlisi and Torres (1986)
C. japonica	Shoot tip and nodal segment of 3–4-year-old seedling	MS + Kin (1) + 2iP (1) + GA_3 (1) + BA (1) + IAA (1) + A (20) + PVP (10 g/l)	MS + Kin (1) + 2iP (1) + GA_3 (1) + BA (1) + IAA (1) + A (20) + PVP (10 g/l)	–	–	Axillary shoot development	Creze and Beauchesn (1980)

Table 3.1 (continued)

Species/cultivar	Explant	Medium				Response	Reference
		Bud break and shoot initiation	Multiplication	Regeneration via shoot organo-genesis	Rooting		
C. sinensis L. cv. Kolkhida, kymyn, Tun-zui.	Zygotic embryo (1–2 mm)	½ MS + modified vit + GA_3 (0.1)	½ MS + modified vit + BA (2) + IAA (0.2) + AA (1) + 2 % sucrose	–	½ MS + IBA (0.5–2.0)	Axillary bud proliferation (6–8), and rooting of shoots	Iddagoda et al. (1988)
C. sinensis L.	*In vitro* micro shoot	MS + BA (6) + NAA (0.5) + GA_3 (2)	MS + BA (6) + NAA (0.5) + GA_3 (2)	–	(1) Shoots were treated with IBA (50) before putting in soil mixture (2) Liquid medium ½ MS + IBA (5) in dark (3) Filter paper bridge ½ MS + IBA (0.5) (4) Solidified agar with MS + . IBA (0.5)	Rooting: 97 % for medium 1, 50 % for medium 2 and 2.44 % for medium 3	Jain et al. (1991, 1993)
C. sinensis (L.) O. Kuntze, T-78	Shoot tip and cotyledonary node from germinated seedling; Nodal segment of field-grown plant	½ MS + CW (10 %) + CH (1 g/l) + BA (5) + IBA (0.1)	½MS + CW (10 %) + CH (1 g/l) + BA (1) + IBA (0.1)	–	MS + IBA (100) for 10 days and then transferred to MS	Axillary bud proliferation (8 and 35 per shoot tip and cotyledonary node respectively), *in vitro* rooting (80–90 %)	Jha and Sen (1992)
C. sinensis cv. Yabukita	Epidermal layer, Intact stem segment, stem segment without epidermal layer	MS + IBA (4) + BA (2) from all three explants	–	MS + IBA (0.5) + BA (10)	½ MS + IBA (0.5)	Shoot organo-genesis (20 % from epidermal layer)	Kato (1985)
C. sinensis with 30 different clones	Shoot apex	MS + BA (1) + IBA (1) + GA_3 (0.01)	–	–	–	Shoot proliferation	Kuranuki and Shibata (1992)
C. sinensis with 30 different clones	*In vitro* cutting	–	½ MS + BA (1) + IBA (0.01) + GA_3 (10)	–	–	Shoot proliferation	Kuranuki and Shibata (1993)

Table 3.1 (continued)

Species/cultivar	Explant	Medium				Response	Reference
		Bud break and shoot initiation	Multiplication	Regeneration via shoot organo-genesis	Rooting		
C. sinensis cv. Ch-hybrid	Nodal segment from *in vitro* grown seedling	MS+TDZ (1.1)+NAA (1.86)	MS	–	Treatment with IBA (500) for 30 min	Axillary shoot development	Mondal et al. (1998)
C. sinensis cv. CH 14 INTA and CH 318 INTA	Uninodal segments and axillary buds	½ MS+BAP (1)	½ MS+Kn (1)+GA_3 (1)	–	¼ MS+IBA (6)	Shoot multiplication, rooting	Molina et al. (2013)
C. sinensis cv. Yabukita	Axillary bud	MS+BA (1)+GA_3 (1)	MS+BA (1)+GA_3 (1)	–	½ MS+IBA (3) or MS+NAA (1 and 10)	Axillary shoot proliferation, rooting	Nakamura (1987a)
C. sinensis (L.) cv. Yabukita	Shoot tip of mature tea bush	MS, B_5, NN (1969)	MS+BA (1)+GA_3 (5)	–	½ MS+IBA (3)	Shoot multiplication, rooting	Nakamura (1987b)
C. sinensis with 23 cultivars.	Stem segment	MS+IAA (0.01–1.0)	MS+IAA (0.01–1.0)+GA_3 (1–5)	–	–	Adventitious bud formation (21.2 %) from callus	Nakamura (1989)
C. sinensis	Axillary buds	–	½ MS+IBA (0.1)+BA (1)+GA_3 (5)	½ MS+IAA (1)+ GA_3 (5)	–	Callus, differentiation of roots, differentiation of adventitious buds and axillary shoot proliferation	Nakamura (1990)
C. sinensis cv. Yabukita	Stem segment of *in vitro* shoot	½ MS+IBA (0.1) or ½ MS+2iP (30)	½ MS+IBA (0.1)+BA (1)+GA_3 (5)	–	–	Proliferation of shoot	Nakamura (1990)
C. sinensis UPASI-3	Nodal explant	½ MS	½ MS+BA (3)	–	¼ MS (liquid)+IBA (3)	Shoot multiplication and rooting (60 %)	Pandidurai et al. (1996)
C. sinensis (L.) O. Kuntze	Shoot tip, nodal segment of field grown plant	MS+YE (0.1 %)+CM (10 %)+NAA (2)+BA (6) for indirect regeneration (via callus) MS+IAA (2)+Kin (8) for direct regeneration.	For indirect regeneration (via callus) ½ MS+IAA (1)+Kin (3)	–	–	4–5 shoot buds from NAA+BA treated callus and 3–4 shoot buds from IAA+Kin treated explants	Phukan and Mitra (1984)

Table 3.1 (continued)

Species/cultivar	Explant	Medium				Response	Reference
		Bud break and shoot initiation	Multiplication	Regeneration via shoot organo-genesis	Rooting		
C. sinensis	Nodal segment of *in vitro* shoots	½ MS+modified vit+IAA (1)+Kin (3)+YE (0.15 %)+CW (10 %)	½ MS+modified vit+IAA (1)+Kin (3)+YE (0.15 %)+CW (10 %)	–	½ MS+IBA (8)	Shoot multiplication and rooting (60–70 %)	Phukan and Mitra (1990)
C. sinensis cv. Banuri-96	Nodal segment from field-grown plants	½ MS	½ MS+IAA (0.2)+BA (2.5)	–	Grafted on seedlings of tea plants	83.3 % graft union	Prakash et al. (1999)
C. sinensis	Embryo of TRI 3013 (female parent) × DT 95 (male parent)	MS+BAP(3)+IBA(0.5)	MS+BAP(3)+IBA(0.5)	–	IBA (50) for 3 h	*Ex vitro* rooting of tea microshoots with simultaneous acclimatization reduced the cost of production of micropropagated plant by 71 % compared to that of micropropagation with an *in vitro* rooting step	Ranaweera et al. (2013)
C. sinensis UPASI-3, UPASI-9	Nodal segment	MS+BA (5)+CM (10 %)	MS+GA_3 (1)+BA (0.5)	–	MS+IAA (1)	Shoot proliferation and rooting (18 %)	Rajkumar and Ayyappan (1992)
C. sinensis UPASI-9	Nodal explant (1.0–1.5 cm)	MS+BA (0.5)	MS+BA (5)	–	–	Axillary shoot multiplication	Rajshekaran and Mohankumar (1992)
C. japonica	Shoot apex and nodal segment	MS+modified vit+BA (1)+IAA (0.1)	MS+modified vit+BA (1)	–	½ MS+BA (1)	Shoot proliferation, rooting (91 %)	Samartin et al. (1984)
C. japonica	Shoot tip from 2–3 month old seedling	MS+BA (1)	–	–	½ MS with modified vit+IBA (1 mg/l)	–	Samartin et al. (1986)
C. sasanqua, Onigoromo Thunb	Shoot tip and nodal segment	–	–	–	Gamborg's (B_5) medium	Shoot proliferation and rooting (100 %) for *in vitro* rooting. 95 % for ex vitro rooting	Samartin (1991)

Table 3.1 (continued)

Species/cultivar	Explant	Medium				Response	Reference
		Bud break and shoot initiation	Multiplication	Regeneration via shoot organo-genesis	Rooting		
C. sinensis (TV-1, T-78, UPASI-9 and Kangra Jat)	Nodal segment from *in vitro* raised cultures	MS (liquid static cultures) + TDZ (1.1)	MS (liquid static cultures) + TDZ (1.1)	–	IBA treatment (30 min) before placing onto sand in closed jars under *in vitro* conditions	88 % rooting	Sandal et al. (2001)
C. reticulata, 'Captain Rawes'	Term-l shoot tip and node	Heller's (1953) macro + $(NH_4)_2SO_4$ (132) + MS micro + MS vit + BA (2) + Z (2) + IBA (0.01) + 2iP (2)	WPM + BA (2) + Z (2) + 2iP (2) + IBA (0.01)	–	½ WPM + dipping in IBA solution (1 g/l) for 30 min	Axillary shoot proliferation and rooting	San-Jose and Vieitez (1990)
C. reticulata cv. Captain Rawes	Shoot tip and stem node of adult trees	Heller's (1953) macro + $(NH_4)_2SO_4$ (132) + MS micro + MS vit + BA (2) + Z (2) + IBA (0.01) + 2iP (2)	WPM + BA (2) + Z (2) + 2iP (2) + IBA (0.01)	–	½ macro WPM + full micro + vit + 6 % sucrose	Axillary shoot development and plantlet regeneration (Horizontal position of explants was better than vertical position for shoot multiplication)	San-Jose et al. (1991)
C. reticulata, Captain Rawes	*In vitro* leaf	Heller's (1953 macro + $NH_4)_2SO_4$ (132) + vit + BA (2) + Z (2) + IBA (0.01) + 2iP (2)	WPM + BA (2) + Z (2) + 2iP (2) + IBA (0.01)	WPM + BA (2) + IBA (1)	½ WPM + 6 % sucrose	Adventitious shoot regeneration and rooting (61.5 %) from shoots of axillary origin	San-Jose and Vieitez (1992)
C. sinensis L.	Flower stalk, leaf piece, petiole, immature cotyledon, stem piece and shoot tip	MS + 2,4-D (4.42), NAA (3.72) + BA (2.25), Kin (2.15)	MS + 2,4-D (10) + YE (0.2 %)	Buds with 2–3 leaves were initiated on MS + BA (2.25)	–	Callus and shoot-bud regeneration (50 %) Explant browning were checked by lower concentration of MS salts ($\frac{1}{20}$)	Sarwar (1985)
C. sinensis cv. Banuri-96	Single node cutting of *in vitro* shoot	½ MS + 8 % agar	½ MS + IAA (0.2) + BA (2.5)	–	–	71.6 % root induction and 73 % field survival	Sharma et al. (1999)

Table 3.1 (continued)

Species/cultivar	Explant	Medium				Response	Reference
		Bud break and shoot initiation	Multiplication	Regeneration via shoot organo-genesis	Rooting		
C. sinensis	*In vitro* cutting	MS+BA (1)+IBA (0.01)+GA_3 (5)	–	–	–	Shoot proliferation	Shibata and Kuranuki (1993)
C. oleifera	Immature zygotic embryo and cotyledon	–	–	MS+BA (4)+NAA (2)	–	Adventitious buds and plantlet regeneration	Tian-Ling (1982)
C. sinensis, TRI-2025	Axillary bud	MS+BAP (2.2)	WPM+TDZ (1.1)	WPM+GA_3 (0.5–2.5)	NAA (5) for 7 days in the dark, followed by transfer of the shoots to an auxin-free medium in the light	–	Tahardi (1994)
C. sinensis, TRI-2025	Axillary bud	Modified WPM+BA (2.25)+IBA (1.01)+TDZ (0.01)	Modified WPM+BA (2.25)+IBA (1.01)+TDZ (0.01)	–	–	Shoot proliferation	Tahardi and Shu (1992)
C. sasanqua, Day Dream	Shoot tip, stem segment	MS+BA (1)+NAA (0.1) for juvenile plant MS+BA (1) for adult material	½ MS+modified MS vit+NAA (0.1)+BA (2)+GA_3 (5)	–	½ MS+modified vit	Axillary shoot proliferation and plantlet regeneration	Torres and Carlisi (1986)
C. x williamsii, Debbie	Internode segment	WPM macro+MS micro+MS vit+BA (0.5)+IBA (0.01)+Glucose (30 g/l)	–	MS+IBA (0.1)+ TDZ (0.67)	–	–	Tosca et al. (1996)
C. japonica, Alba Plena	Shoot tip and nodal segment	–	–	–	WPM macro after dipping in IBA (1 g/l) for 15 min followed by 12 days darkness	Rooting of micropropagated shoot (87 %)	Vieitez et al. (1989a)

Table 3.1 (continued)

Species/cultivar	Explant	Medium				Response	Reference
		Bud break and shoot initiation	Multiplication	Regeneration via shoot organo-genesis	Rooting		
C. japonica cv. Alba Plena	Shoot tip (2–4 cm), nodal segment, and whole shoot of field-grown plant	Heller's (1953) macronutrient increased by factor 1.25 + $(NH_4)2SO_4$ (132) + MS micronutrient + BA (1) + IBA (0.01) + myoinositol (100) + Jacquiots vit (Gautheret 1959)	WPM + BA (2) + Z (2) + 2iP (2) + IBA (0.01)	–	WPM after giving IBA (1 g/l) treatment for 15 min	Shoot multiplication and rooting (76 %)	Vieitez et al. (1989b)
C. oleifera	Lateral bud of adult tree	–	–	–	–	Axillary shoot development	Yan et al. (1984)

Figures in parenthesis without assigned units denote concentrations in mg/l

A adenine, *AA* ascorbic acid, B_5 Gamborg et al. (1968) medium, *CH* casein acid hydrolysate, *CW* coconut water, *macro* macronutrients, *micro* micronutrients, *NN* Nitsch and Nitsch (1969) medium, *PVP* polyvinyl pyrrolidone, *vit* vitamins, *WPM* woody plant medium (Lloyd and McCown 1980), *YE* yeast extract, *CM* coconut milk

of TDZ (1.1, 2.2 and 3.3 mg/l) in combination with either 2,4-D, NAA or IBA at concentrations ranging from 1 to 3 mg/l in MS, maximum response (98 %) of shoot proliferation was observed with a combination of TDZ (1.1 mg/l) and NAA (2 mg/l). The number of shoots formed was higher in explants initiated on medium containing TDZ when compared to medium supplemented with BAP, but multiplication rates were more or less similar (i.e. 2–3 times) after each subculture. Since very low concentrations of TDZ are used only at the initial phase, the higher cost of TDZ was compensated in overall cost. TDZ thus appeared to be a potent cytokinin-like growth factor for tea micropropagation with high proliferation rates. Despite the fact that inclusion of liquid culture step for bulking microshoots may be cost effective, it is generally limited by a low-oxygen environment coupled with the production of hyperhydric structures. Barring these limitations, establishment of cultures in liquid medium is an important step towards automation (Aitken-Christie et al. 1995). An efficient liquid culture system for tea shoot proliferation was standardized (Sandal et al. 2001). It was found that MS along with TDZ (0.55–1.1 mg/l) was the best for shoot proliferation in unagitated liquid medium. Of the different liquid volumes compared in 250 ml Erlenmeyer flasks, 20 ml was the most effective. However, with greater volume, hyperhydricity was induced. Therefore, use of 20 ml liquid medium with subculture periods at an interval of 6–8 weeks was found to be a cost- and labour-effective process in combination with the existing protocols of tea micropropagation involving solid media with subculture periods at 4-week interval. Similarly, earlier, Carlisi and Torres (1986) found that liquid medium was better than solid medium for *C. japonica* when cultured on 1/2 MS fortified with BAP (1 mg/l) and GA_3 (5 mg/l). The significant role of a growth adjuvant for micropropagation of tea was well documented. These included coconut milk (Phukan and Mitra 1984; Sarathchandra et al. 1988; Banerjee and Agarwal 1990; Nakamura and Shibita 1990; Agarwal et al. 1992; Jha and Sen 1992; Rajkumar and Ayyapan 1992), yeast extract (Phukan and Mitra 1984; Sarwar 1985; Banerjee and Agarwal 1990; Phukan and Mitra 1990), casein acid hydrolysate (Chen and Liao 1983; Jha and Sen 1992), serine and glutamine as nitrogen sources (Chen and Liao 1982), etc. Among the different carbon source, sucrose with the concentration between 15 and 60 g/l remains a unanimous choice for tea micropropagation. This was confirmed by Nakamura (1990), who found that 3–6 % sucrose was the best for adventitious bud formation as compared to four other sugars, viz. lactose, galactose, glucose, or maltose, for adventitious bud formation. Therefore, it can be concluded that MS with low cytokinin, specially BAP, seems to be the best for micropropagation of tea.

Interestingly, the effect of caffeine on *in vitro* growth of tea was studied in detail. It was found that caffeine, which is abundantly present in tea leaves, significantly retards the growth/development of tea shoots, stem and roots *in vitro*. This may be the reason why the replanting area of new plants is less and the soil requires rehabilitation as over the years tea plants may secrete caffeine into the soils (Owuor et al. 2007).

3.3.3 Rhizogenesis

Establishment of microshoots in hardening phase depends upon the efficiency of rooting. In tea, both *in vitro* and *ex vitro* root induction of microshoots were reported. While *in vitro* rooting of tea depends upon concentration and duration of auxin treatment, salt strength of basal medium or physical condition of the cultures, *ex vitro* rooting was influenced by pH of the hardening media and relative humidity of the hardening chamber. Reduction of MS salt concentrations to half strength favoured not only root induction but also root elongation in tea (Kato 1985; Banerjee and Agarwal 1990). Generally, IBA (0.5–8 mg/l) was preferred over NAA in tea (Gunasekare and Evans 2000; Bidarigh and Azarpour 2011; Bidarigh et al. 2012) as well as in *C. japonica* (Samartin et al. 1984; Samartin et al. 1986; Vieitez et al. 1989b; Bidarigh and Azarpour, 2013), *C. sasanqua* (Samartin 1991) and *C. reticu-*

lata (San-Jose and Vieitez 1990, 1992) as roots induced by NAA were shorter and thicker, with accompanying calli making subsequent transplanting difficult. Although rooting occurred much later when treated with IBA, roots were long and fibrous (Nakamura 1987a; Banerjee and Agarwal 1990). Liquid medium with filter paper bridge was successfully used for *in vitro* rooting of tea (Tian-Ling 1982; Kato 1985; Nakamura 1987b). Jain et al. (1993) found that *ex vitro* rooting was better than all types of *in vitro* rooting methods such as liquid shake culture, agar-solidified medium or filter paper bridges. They achieved 97% rooting in shoots whose cut ends were dipped in IBA (50 mg/l) for 2 h before transplanting into potting mixtures. Another important observation made by Banerjee and Agarwal (1990) was that low light with a low pH (4.5–4.6) was favourable for root induction in tea. This finding was in agreement with Nakamura (1987a), indicating that dark treatment after dipping the shoots in auxin promoted rooting in tea. *In vitro* rooting of tea also depends upon genotype. Murali et al. (1996) demonstrated that among the four different cultivars, the highest rooting was achieved with both UPASI-26 and UPASI-27 cultivars compared to two other cultivars, namely UPASI-3 and BSB-1, with an IBA treatment (1 mg/l) for 30 min. They concluded that genotypic differences strongly influence rooting ability of tea, which requires a fine tuning of the media to suit each individual cultivar.

In general, *in vitro* rooting of tea microshoots was achieved either by culturing on media containing low auxin concentration for a longer time or with an 'auxin-shock' treatment by initially culturing in high auxin concentration followed by the transfer to an auxin-free medium.

3.3.4 Hardening and Field Transfer

Perhaps, the most critical phase of micropropagation is the establishment of *in vitro* plantlets to the greenhouse. Achievement of uniform plant growth and high survival rate not only demand good greenhouse conditions but also require modification of the internal microclimate to match the local environmental need.

3.3.4.1 Conventional Hardening

Conventionally, *in vitro*-raised tea microshoots are hardened for 6 months on soil mixture containing various additives such as cow dung, soil rite, etc. and then kept inside the indigenously developed polytunnel of various shapes and sizes. Arulpragasam et al. (1988) transferred 5–8 cm rooted plantlets to small plastic pots containing fumigated soil and kept in a humid chamber for a period of 10 days. Plants were gradually acclimatized and planted in larger concrete pots. Das and Barman (1988) achieved better hardening on preconditioning plants at low temperature (22 °C) and low light intensity (250 µmol m^{-2} s^{-1}) after transferring to soil sleeves. The plants were covered with poly bags and kept in a thatch house for few weeks and were gradually exposed to sunshine before transferring to field. Generally, the standard procedure is to transfer the rooted plantlets of tea to potting mixtures containing peat and soil (1:1) under high humidity created by using misting or fogging units (Banerjee and Agarwal 1990; Agarwal et al. 1992; Jha and Sen 1992; Jain et al. 1993), but use of vermiculite and soil in equal ratio also works very well (Tian-Ling 1982; Kato 1985). A detailed study on various parameters such as time of harvesting of microshoots, shoot size, soil pH, PGRs, CO_2 enrichment and light condition was carried out by Sharma et al. (1999). They found that soil pH coupled with CO_2 enrichment was the most critical factor for hardening to achieve high percentage of survival. Following this protocol, 300 plants were transferred to the field (Sharma et al. 1999). For large-scale multiplication, micropropagated tea shoots of 4–5 cm height were harvested and treated with IBA (500 mg/l) for 30 min before transferring to Hiko tray containing pre-sterilized sand and cow dung (1:1). These Hiko trays were then kept in polytunnel with intermittent watering for 90 days inside an indigenously developed polyhouse, resulting in 90% survival. Thereafter, they were transferred to polythene sleeves containing black virgin soil for a year in the same polyhouse (Rajshekaran and Mohan Kumar 1992; Mondal et al. 1998). Following these techniques at Research and Development Department of Tata Tea Ltd, India, more than 45,000 plants of eight different cultivars were transferred to the field, and leaves had been harvested regularly to manufacture tea for past several years (Mondal et al. 2004).

Physiological and biochemical responses were compared between field-grown plants of vegetatively propagated (VP) and tissue culture raised (TC) plants. No significant variation was observed between them in terms of photosynthetic carbon assimilation rate. Carbon assimilation studies carried out with a radiotracer technique revealed that 'Assam' cultivar UPASI-27 assimilated a higher amount of labelled CO_2 followed by UPASI-3. However, UPASI-27 was marginally better than UPASI-3 in terms of mobilization of assimilates to the growing sinks. Both UPASI-3 and UPASI-27 reassimilated higher quantities of photosynthates followed by BSB-1 and UPASI-26. Although there was a marginal variation in photosynthetic pigments of TC and VP plants, it was not statistically significant. Similarly, no significant variations were observed in certain substrates (polyphenols, catechins and amino acids) and enzymes (polyphenol oxidase, peroxidase and phenylalanine ammonia-lyase) except protease involved in the formation of quality constituents of made tea. However, clonal variation was evident with respect to photosynthetic pigments, substrates/enzymes. Under soil moisture stress, no significant variation was observed between VP and TC plants in terms of proline accumulation (Marimuthu and Rajkumar 2001).

3.3.4.2 Biological Hardening

The biological hardening is an alternative concept for hardening the micropropagated plantlets. Like in any other plants, micropropagated tea plants often experience high mortality during or following laboratory-to-land transfer. Apart from various abiotic factors, one major cause of mortality of such 'aseptically' raised plants is their sudden exposure to the soil microbial communities present in the rooting media. Tissue culture raised plants, at least initially, were unable to resist sufficiently against the soil microbial attack. Therefore, microbial culture can be used to overcome the 'transient transplant shock' and better development of the plants on the transfer to the soil. Although a number of organism antagonistic fungi such as *Trichoderma,* vesicular arbuscular mycorrhiza and *Piriformospora indica* were found to be suitable for micropropagated raised woody plants (Singh et al. 2000), Pandey et al. (2000) found that two bacteria, namely *Bacillus subtilis* and *Pseudomonus corrugata,* were suitable as microbial inoculants for hardening of micropropagated tea plants prior to transfer in open land. They found that bacterial inoculation enhanced survival rate of 100, 96 and 88 % as against 50, 52 and 36 % in corresponding control plants in rainy, winter and summer seasons, respectively. Rhizoplane and rhizosphere soil analysis showed that the major biotic factor responsible for mortality of tea was the fungus *Fusarium oxyporum*. Thus, biological hardening holds a greater promise for hardening of micropropagated tea shoots.

3.3.4.3 Micrografting as a Hardening Tool

Micrografting, a hardening tool of *in vitro*-raised shoots have been used in a wide range of woody plants such as citrus, cherry, kiwifruit, pistachio, stone fruits, apple, grape, *Larix decidua* and *Picea* spp., where *in vitro*-raised scions were directly grafted either onto *in vitro* raised-rootstocks under sterile conditions or *in vivo*-raised stocks (Banerjee et al. 2000). In tea, Prakash et al. (1999) reported the grafting of *in vitro*-raised tea shoots on seedlings at the greenhouse. Microshoots of tea cultivar Banuri-96 were either grafted on seedlings of the same cultivar (autograft) or on UPASI-9 seedlings (heterograft). Four important factors were investigated: (1) effect of PGRs, (2) assessment of compatibility, (3) effect of the age of rootstock and (4) season.

A higher amount of foliar development was observed when a pulse treatment of liquid formulation consisting of BAP (5 mg/l) and NAA (5 mg/l) was given at the graft union for 10 min. Also, a significantly higher percentage of survival was observed in autografts as compared to heterografts. Age of the rootstock was found to be the most critical factor. The success rate was much higher in 4-month-old seedlings than in either 2-month- or 12-month-old seedlings. However, using the same technique, Mondal et al. (2005) reduced the hardening period up to 1 year of *in vitro*-raised tea shoots when grafted on 3-month-old tea seedlings. However, it was observed that the result of Mondal et al. (2005) differs in two aspects from the report of Prakash et al. (1999). Firstly, Mondal et al. (2005) achieved a higher rate of survival (98 %) without any PGR treatment. Therefore, it was concluded that PGR plays an insignificant role for the graft compatibility, which is an absolute necessity for commercial use of this technique to eliminate the cost of PGR. Secondly, Mondal et al. (2005) found that the growth of the grafted plants was much higher than the ungrafted *in vitro*-raised scion or seedling. One of the reasons for better growth of microprogated plants than of the vegetative counterpart was due to the higher root volume, which helped to absorb more water and nutrients from the soil. It was due to the pre-existing tap root system of the seedling. This was not only of great importance to reduce the hardening period of the tissue culture plant but also for the better root system in the plant to resist the subsequent drought periods in the field. Micropropagated tea plants of the same cultivar as well as age required 12–18 months time in the hardening phase through conventional hardening approach whereas the same cultivar of micrografted one, demanded only 6–8 months, which vitally reduced the hardening time by one planting season. This plays an important role in tea breeding where progress is slow due to the slow rate of growth.

3.3.5 Field Performance of Micropropagated Raised Plants

Several reports on micropropagation of tea stated that hardened plantlets were transplanted to potting mixture but only a few details of field performance are available. To date, no systematic study has been conducted to assess the growth performance of micropropagated tea plants grown in the field. Sharma et al. (1999) analysed the nursery performance of 17-month-old micropropagated and VP tea plants of cultivar Banuari-96. No remarkable difference was noticed at nursery level except root induction time, which was earlier by 1 month in case of micropropagated shoots. They found that while average height and stem thickness at collar region were higher in VP plants, the average number of leaves was twice (16 leaves/plant) in micropropagated plants in comparison to their *in vivo* counterparts (6 leaves/plant). However, reasons for such difference were not mentioned but it seems that 17-month time was perhaps too early to detect the actual performance for this woody perennial.

In a comparative study of 8- and 4-year-old, field-grown micropropagated and VP tea cultivars, namely UPASI-9 and TTL-1, Mondal et al. (2004) reported that overall yields were comparable. While the yield of micropropagated UPASI-9 recorded 3,656 kg made tea/ha/annum, VP plants of UPASI-9 yielded 3,596 kg made tea/ha/annum after second cycle of pruning (8 years) under the environmental conditions of South India. On the other hand, micropropagated TTL-1 plants registered 2,321 kg made tea/ha/annum and their vegetative counterparts yielded 2,383 kg made tea/ha/annum after first cycle of pruning (4 years) with 10,000 plants/ha as well as 22 % out turn (a ratio which indicates that 100 kg green leaf produce 22 kg made tea in black tea manufacturing process) in both. Chemical analysis of made tea for thearubigin, theaflavin, total soluble solids and total colour evidenced no significant difference between the two types for both the cultivars, which is obvious as both are genetically similar and planted in the same environment. Although the different physiological parameters such as photosynthetic rate and chlorophyll content etc. remained the same, two different morphological variations were detected: (1) the number of lateral shoots produced after 'centering' was significantly higher in micropropagated raised plants as compared to VP plants. This is probably due to the effects of various PGR treatments that the micropropagated raised plants experienced under *in vitro* conditions. (2) The root volumes of micropropagated plants were also higher than VP plants. This may perhaps be due to the fact that micropropagated shoots were treated with IBA before transplanting to induce rooting, which may be responsible for better root development in the field.

3.3.6 Cold Storage and Cryopreservation

Low temperature storage is an important technique for preserving the planting material. In tea long-term storage for various explants such as cut root, seed, pollen were tested to reveal suitable storage techniques. It was found that in case of vegetative organ, cut roots were most suitable for long-term storage (Amma and Watanabe 1985).

Ballester et al. (1997) reported almost 100 % survival frequencies in seven of the eight clonal shoot cultures of *C. japonica* and *C. reticulata* assayed, when stored at 4 °C for up to 12 months. Shoot tips of *C. japonica* encapsulated in alginate beads and stored at 4 °C, survived for a shorter period of time than uncapsulated ones. Encapsulated material had survival rates of 75, 50 and 10 % on 30, 60 and 75 days, respectively.

Cryopreservation of tea shoot tips was standardized (Kuranki and Sakai 1995) using both vitrification and alginate-encapsulation dehydration techniques. When the vitrification method was applied, shoot tips of 2 mm long from cold-hardened *in vitro*-grown plantlets were pre-cultured for 2 days at 5 °C on 1/2 MS medium containing sucrose (68 g/l). Following pre-culturing, tips were dehydrated for 90 min at 0 °C in a highly concentrated vitrification solution consisting of 30 % (w/v) glycerol, 15 % (w/v) ethylene-glycol and 15 % (w/v) DMSO in 1/2 MS medium supplemented with sucrose (137 g/l). Then, these vitrified shoot tips were plunged directly in liquid nitrogen and kept there for 1 h. After rapid warming in a water bath at 30 °C, successfully vitrified shoot tips resumed growth in about 5 days and developed shoots without intermediary callus formation. The average rate of normal shoot formation was 60 % about 1 month after plating on 1/2 MS culture medium. The assays performed by using the encapsulation dehydration technique (Kuranuki and Sakai 1995), indicated that shoot formation capacity of encapsulated dried shoot tips cooled to −196 °C was lower by as much as 20 % than vitrified material (60 vs. 40 %). Further, Kuranuki (2006) reported that 60 % shoot formation rate was achieved after cryopreservation of tea multiple shoot tips using alginate-encapsulation dehydration techniques. She concluded that valuable cryogenic procedure of tea explants would be vitrification method for terminal buds and alginate-encapsulation dehydration technique for multiple shoot tips.

3.4 *Camellia* Species

Among the related species, *C. japonica* had been studied extensively due to its high demand of large number of cultivars. Vieitez et al. (1991) did a systematic study of micropropagation of *C. japonica,* which elaborately highlighted various factors affecting multiplication rate in *in vitro* and subsequently hardening processes. Depending on the species-specific requirements among the wild camellias, various factors that influence the micropropagation are briefly reviewed below.

3.4.1 C. japonica

In the late 1970s, the use of *in vitro* culture methods was suggested as a means to solve the constraints in propagation mainly due to shy rooting in vegetative cuttings (Bennett and Scheibert 1982). Since then, several protocols have been described for the micropropagation of *C. japonica*. The first elaborate report to regenerate plants from shoot tips and axillary buds was made by Creze and Beauchesne (1980), who took meristems with one or two leaf primordial of 0.5 mm long from 1-year-old rooted cuttings or from 3- to 4-year-old seedlings and cultured them on a MS medium supplemented with adenine (20 mg/l), IAA (0.1 mg/l) and 1 mg/l each of Kn, BAP, GA_3 as well as polyvinyl pyrrolidone (PVP) (10 g/l). Though cultures were established and elongated to produce shoots more rapidly than shoot tips, no rooting and transfer to soil were described.

In *C. japonica,* buds of juvenile origin gave consistently better results in terms of both growth and vigour on MS as compared to other macronutrient formulae such as Lepoivre (Quoirin and Lepoivre 1977), Knop (Tabachnik and Kester 1977), Schenk and Hildebrandt (1972), and modified Heller (1953). However, regeneration from adult material of *C. japonica* cv. Alba Plena was poor on MS (Vieitez et al. 1989a). In a series of shoot multiplication experiments, Vieitez et al. (1989a) found that WPM was the best among the six macronutrient formula tested, modified Heller (1953), MS, half-strength MS, WPM, Gresshoff and Doy (1972) and Anderson (1984). In contrast to Carlisi and Torres (1986), who found that full- and half-strength MS were the best for culturing of *C. japonica,* the observations recorded by Vieitez et al. (1989a) were poor in these media. The different responses observed by Carlisi and Torres (1986) were probably genotype-dependent. For *C. japonica,* the most widely used cytokinin was BAP. However, Creze and Beauchesne (1980) reported the importance of 2-iP (1 mg/l) as an essential component for shoot proliferation, Kn was also found to have no effect on shoot multiplication when used either alone or in combination with BAP (Samartin et al. 1984). The GA_3 (5–10 mg/l) was also used for proliferation of shoots in cv. Purple Dawn (Carlisi and Torres 1986; Torres and Carlisi 1986). Among the auxins, IAA and IBA were used for shoot proliferation, but there were no reports on the use of NAA and 2,4-D for culturing of *C. japonica* (Creze and Beauchesne 1980; Vieitez et al. 1989b). Apart from PGR, another factor, which was found to be important, is the physical condition of the media. In general, liquid medium was more suitable than solid medium for shoot proliferation in *C. japonica* (Carlisi and Torres 1986; Vieitez et al. 1989a).

3.4.2 C. oleifera

Very little work has been done on this important oil-yielding species. Tian-Ling (1982) used MS medium supplemented with BAP (4 mg/l) and NAA (2 mg/l) for induction of adventitious buds leading to plantlet regeneration. In another study, lateral buds of adult trees were also used by Yan et al. (1984) for induction of axillary bud proliferation.

3.4.3 C. reticulata

Heller's (1953) macro elements with the addition of $(NH_4)_2SO_4$ (0.13 mg/l) in combination with MS vitamins were found to be the best for induction of axillary buds. WPM was also found to be superior than modified Heller (1953) as well as the recipes of Anderson (1984). A combination of BAP and zeatin had also been successfully used for promoting the growth and proliferation of axillary shoots (San-Jose and Vieitez 1990; San-Jose et al. 1991). Multiplication rates in terms of both the number of axillary buds and the length of shoots could further be improved by horizontal placement of the explants (San-Jose and Vieitez 1990).

3.4.4 C. sasanqua

While Torres and Carlisi (1986) preferred MS medium, Samartin (1991) found B_5 (Gamborg et al. 1968) macronutrients supplemented with micronutrients of MS to be suitable for the growth and proliferation of axillary shoots. A combination of BAP and NAA was found to be the most suitable for shoot multiplication in both of these studies.

3.4.5 *Camellia* Hybrids

Despite the availability of limited information, the medium of Tukey (1934) was found to be the best for *in vitro* seedling growth of three different interspecific hybrids including *C. japonica*× *C. cuspidata, C. japonica*× *C. reticulata* and *C. japonica*× *C. saluenensis* (Lammerts 1958). Creze and Beauchesne (1980) made the first attempt to regenerate plants from shoot tips and axillary buds on *C. saluenensis* × *C. chrysantha.*

3.4.6 Rooting and Hardening

Like other woody plants, rooting is a major limitation in micropropagation of *Camellia* species. Rooting of *in vitro*-raised shoots was achieved either upon continuous exposure to a low concentration of auxin or initially to a less exposure to a high auxin concentration followed by their transfer to an auxin-free medium. *In vitro* rooting of *Camellia* species had been reported by a number of workers (Table 3.1). Reduction of MS salt concentrations to half strength favoured both induction and elongation of rooting in *Camellia* species (Samartin et al. 1984, 1986; Kato 1985). However, Vieitez et al. (1989b) did not find any significant difference in rooting of *in vitro*-raised shoots of *C. japonica* cv. Alba Plena using half-strength MS medium. In camellia, IBA (0.5–8 mg/l) had been shown to give better results than NAA for *in vitro* root initiation. Roots induced by NAA were shorter, thicker, and with accompanying calli, which were undesirable features for the subsequent transplanting. On the other hand, with IBA treatments, rooting occurred much later but were long and fibrous (Samartin et al. 1986). Liquid medium with filter paper bridge was beneficial for rooting in *C. olerifera* (Tian-Ling 1982; Kato 1985; Nakamura 1987b). Torres and Carlisi (1986) reported that a pulse treatment of shoots with IBA (500 mg/l) for 30 min before placing on a root induction medium gave best results in *C. sasanqua*. Beretta et al. (1987) obtained increased rooting in *C. saluenous* × *C. japonica* hybrids with IBA (1–2 g/l) treatment for 15 min. In comparison to other woody species, the *Camellia* species requires higher IBA concentration and longer immersion time and such high treatments were not as deleterious for camellias as in other woody species like *Prunus avium* (Riffaud and Cornu 1981). Dark treatment after dipping the micropropagated shoots in auxin was reported to favour rooting in *C. japonica* (Samartin et al. 1986) and *C. reticulata* (San-Jose and Vieitez 1990). However, Samartin (1991), who also worked with *C. japonica,* did not find any significant difference between the effects of light and dark treatments on rooting. Rooting mixture alone influences the survival rate at the nursery. While 75 % survival of *C. japonica* was obtained in petrite:soil (1:1) (Samartin et al. 1984), a higher survival rate of 70–90 % of the same species was achieved in soil:quartz (1:1) mixture (Samartin et al. 1986; Vieitez et al. 1989b). In *C. reticulate,* rooted shoots were transferred to pot containing 1:1 mixture of peat and quartz and placed in a plastic tunnel with mist/fog system in lamps to give a 16 h photoperiod, which recorded 80 % survival (San-Jose et al. 1991). However, there is no report on either any nursery performance or any field performance of micropropagated camellias presently.

3.5 Problems of Micropropagation

Like other woody perennials, major problems encountered in tea micropropagation are phenolic exudation from explants and microbial contamination in tissue culture medium, which are explained below.

3.5.1 Explant Browning

High phenol contents, for which tea is valued, are also exuded in the *in vitro* culture medium from the cut ends of explants, which undergoes enzymatic oxidation to form some toxic compounds. These toxic compounds create problems by lowering down the pH of the tissue culture medium. Sarwar (1985) tried various chemicals such as ascorbic acid, catechol, l-cysteine, phloroglucinol, phenyl-thiourea, PVP-10, sodium diethyl dithio-carbonate, sodium fluoride and thiourea along with different strengths of MS inorganic salts. He found that highly reduced strengths of MS (1/20) prevented explants' browning. While ascorbic acid was used by several researchers (Iddagoda et al. 1988; Agarwal et al. 1992), PVP was recommended by Creze and Beauchesne (1980) to prevent explant browning of *in vitro* cultures. Magnitude of polyphenols exudation varied between different cultivars of tea. Murali et al. (1996) categorized that polyphenol leaching *in vitro* was highest in Cambod type followed by China type and was least with Assam type.

Pandidurai et al. (1996) reported that successive transfer of the explants into a new culture medium at periodic intervals worked well to prevent explant browning of tea. Interestingly, it had been observed that quality tea cultivars were difficult to establish *in vitro* as in the case of TTL-2 due to high phenolic content (Anon 1999). Being quality clone, TTL-2 must have had a high amount of polyphenol, which made it difficult to establish under *in vitro* culture.

3.5.2 Microbial Contamination

In tea, explants are often taken from field-grown plants that are heavily contaminated by various epiphytic and endophytic organisms, causing severe losses at every stage of micropropagation (Debergh and Vanderschaeghe 1988). Generally, cultures are pre-screened when explants from field-grown plants are used. However, in tea, several strategies had also been employed to overcome these problems. Ogutuga and Northcote (1970) used 5–15 mm stem segments from the greenhouse-grown tea plant and surface-sterilized with 70% ethanol followed by 7% sodium hypochloride solution. Kato (1985) used stems with 3–4 leaves from the greenhouse-grown plants (cv. Yabukita) and sterilized with 7% calcium hypochloride solution for 20 min. Arulpragasam and Latiff (1986) snapped off the leaf carefully with petioles as close as possible to the stem without any damage to the axillary buds. The shoots were surface-sterilized in 10–15% Clorox solution for 15 min under constant agitation and inoculated in culture medium after rinsing with distilled water. Kuranuki and Shibata (1993) minimized the problems of endogenous bacteria by reducing the size of the explant to an apical meristem and a leaf primordium less than 0.5 mm. Rate of contamination varied upon the season of explant collection (Nakamura 1989a), which was minimized by collecting the explants at the beginning of first seasonal flush when air temperature was relatively low and the plants were juvenile and vigorous. Haldeman et al. (1987) had shown significant reduction of bacterial as well as fungal contaminations with different concentrations of benomyl (1, 2 or 4 g/l) and rifampicin at 10, 25 or 50 mg/l without any phytotoxic effects. Das and Barman (1988) treated the explants with streptomycin sulphate (1%) for 1 min prior to inoculation in order to prevent contamination. However, (0.05–1%) mercuric chloride solution (0.05–1%) was also used for surface sterilization of the explants (Rajakumar and Ayyappan 1992; Rajashekaran and Mohankumar 1992; Jha and Sen 1992; Agarwal et al. 1992).

3.6 Conclusion

With available information, it can be concluded that, while extensive works have been done on micropropagation of tea, challenges remain on commercial application of micropropagation of tea. A wide range of media composition, explants, PGR, varietal difference along with physiological stage of the explants has been studied. While MS is the most preferred basal medium, BAP is the most widely used PGR. Important aspects which remain uninvestigated are: (1) field behaviour of micropropagated tea specially after 2nd round of pruning which is indicative for stabilized yield, (2) low temperature long-term storage of tea shoot, and (3) lowering down the cost of production of micropropagation techniques including hardening so that it becomes comparable with vegetative propagation.

It is noteworthy to mention here that micropropagation of tea has not been commercially exploited. Very few micropropagated plants are in the field. Akula and Akula (1999) have rightly pointed out excellent reasons. The reasons are loss of juvenility of *in vitro* culture that hinder long-term production, lack of taproot system that makes micropropagated tea more vulnerable to drought. But the major point is the higher cost of production of the micropropagated tea plantlets which is seven times higher in comparison to VP plant, although it has been reduced much in recent years with the advancement of hardening protocol such as carbon enrichment, biological hardening, micrografting and specially designed hardening chamber that leads to higher survival rate. However, when a cultivar is developed that generates high demands of large planting material of superior clone, including transgenic tea to supply within a short period of time to the growers, micropropagation is a better alternative practically for the commercial exploitation.

References

Agarwal B, Singh U, Banerjee M (1992) *In vitro* clonal propagation of tea (*Camellia sinensis* (L.) O. Kuntze). Plant Cell Tiss Organ Cult 30:1–5

Aitken-Christie J, Kozai T, Takayama S (1995) Automation in plant tissue culture. General introduction and overview. In Aitken-Christie J, Kozai T, Smith MAL (eds) Automation and environmental control in plant tissue culture. Kluwer Academic Publishers, The Netherlands, pp 1–15

Akula A, Akula C (1999) Somatic embryogenesis in tea (*Camellia sinensis* (L) O Kuntze. In Jain SM, Gupta PK, Newton RJ (eds) Somatic embryogenesis in woody plants, vol 5. Kluwer Academic Publishers, The Netherlands, pp 239–259

Amma S, Watanabe A (1985) Long term storage of germplasm of tea (*Camellia sinensis (L)* O Kuntz). JARQ 19:196–201

Anderson WC (1984) A revised tissue culture medium for shoot multiplication of rhododendron. J Amer Soc Hort Sci 109:343–347

Anon (1999) Annual report. Research and Development Department, Tata Tea Ltd, Munnar

Arulpragasam PV, Latiff R (1986) Studies on the tissue culture on tea (*Camellia sinensis* (L.) O. Kuntze). 1. Development of a culture method for the multiplication of shoots. Sri Lank J Tea Sci 55:44–47

Arulpragasam PV, Latiff R, Seneviratne P (1988) Studies on tissue culture of tea (*Camellia sinensis* (L.) O. Kuntze). 3. Regeneration of plants from cotyledon callus. Sri Lank J Tea Sci 57:20–23

Bag N, Palni LMS, Nandi SK (1997) Mass propagation of tea using tissue culture methods. Physiol Mol Biol Plants 3:99–103

Banerjee AK, Agrawal DC, Nalawade SM, Krishnamurty KV (2000) Recovery of in vitro cotton shoots through micrografting. Curr Sci 78:623–626

Banerjee M, Agarwal B (1990) *In vitro* rooting of tea, *Camellia sinensis* (L.) O. Kuntze. Ind J Exp Biol 28:936–939

Ballester A, Janeiro LV, Vieitez AM (1997) Cold storage of shoot cultures and alginate encapsulation of shoot tips of *Camellia japonica* and *Camellia reticulate* Lindley. Sci Hort 71:67–78

Bennett WY, Scheibert P (1982) *In vitro* generation of callus and plantlets from cotyledons of *Camellia japonica*. Camellia J 37:12–15

Beretta D, Vanoli M, Eccher T (1987) The influence of glucose, vitamins and IBA on rooting of *Camellia* shoots *in vitro*. In: Abstracts of symposium on vegetative propagation of woody species. Inter Soc Hort Sci Italy, p 105

Bidarigh S, Azarpour E (2011) The study effect of cytokinin hormone types on length shoot *in vitro* culture of tea (*Camellia sinensis* L.). World Appl Sci J 13:1726–1729

Bidarigh S, Hatamzadeh A, Azarpour E (2012) The Study Effect of IBA Hormone Levels on Rooting in Micro Cuttings of Tea (*Camellia sinensis* L.). World Appl Sci J 20:1051–1054

Bidarigh S, Azarpour E (2013) Study effect of IBA hormone levels on rooting in micro cuttings of tea (*Camellia sinensis* L.). ARPN J Agri Biol Sci 8:1–5

Carlisi JC, Torres KC (1986) *In vitro* shoot proliferation of *Camellia* 'Purple Dawn'. Hort Sci 21:314

Chen Z, Liao H (1982) Obtaining plantlet through anther culture of tea plants. Zhongguo Chaye 4:6–7

Chen Z, Liao H (1983) A success in bringing out tea plants from the anthers. China Tea 5:6–7

Creze J, Beauchesne MG (1980) *Camellia* cultivation *in vitro*. Int Camellia J 12:31–34

Das SC, Barman TS (1988) Current state and future potential of tissue culture in tea. Proc. 30th Tocklai Conf. TRA Jorhat, pp 90–94

Das SC (2001) Tea. In: Parthasarathy VA, Bose TK, Deka PC, Das P, Mitra SK and Mohandas S (eds) Biotechnology of Horticultural Crops, vol 1. Naya Prokash, India, pp 526–546

Debergh PC, Vanderschaeghe AM (1988) Some symptoms indicating the presence of bacterial contaminants in plant tissue culture. Acta Hort 255:77–81

Dood A W (1994) Tissue culture of tea (*Camellia sinensis* (L.) O. Kuntze)-A review. Inter J Trop Agri 12:212–247

Gautheret (1959) La culture des tissus végétaux: techniques et réalisations. Masson Edit

Gamborg O, Miller R, Ojima K (1968) Nutrient requirements of suspension cultures of soyabean root cells. Exp Cell Res 50:157–158

Gresshoff PM, Doy CH (1972) Development and differentiation of haploid *Lycopersicon esculentum*. Planta 107:161–170

Gunasekare MTK, Evans PK (2000) *In vitro* rooting of microshoots of tea (*Camellia sinensis* L.). Sri Lanka J Tea Sci 66: 5–15

Haldeman JH, Thomas RL, Mckamy DL (1987) Use of benomyl and rifampicin for *in vitro* shoot tip culture of *Camellia sinensis* and *Camellia japonica*. Hort Sci 22:306–307

Heller R (1953) Recherches sur la nutrition minerale des tissus vegetaux cultives *in vitro*. Annales des Sciences Naturelles (Botanique) Biologie Vegetale. 14:1–223

Iddagoda N, Kataeva NN, Butenko RG (1988) In vitro clonal micropropagation of tea (*Camellia sinensis* L.) 1. Defining the optimum condition for culturing by means of a mathematical design technique. Ind J Plant Physiol 31:1–10

Jain SM, Das SC, Barman TS (1991) Induction of roots from regenerated shoots of tea (*Camellia sinensis* L.). Acta Hort 289:339–340

Jain SM, Das SC, Barman TS (1993) Enhancement of root induction from *in vitro* regenerated shoots of tea (*Camellia sinensis* L.). Proc Ind Natl Sci Acad 59:623–628

Jha TB, Sen SK (1992) Micropropagation of an elite Darjeeling tea clone. Plant Cell Rep 11:101–104

Kato M (1985) Regeneration of plantlets from tea stem callus. Jap J Breed 35:317–322

Kato M (1989) *Camellia sinensis* L. (Tea): *In vitro* regeneration. In Bajaj YSP (ed) Biotechnology in agriculture and forestry, vol 7. Medicinal and aromatic plants II. Springer-Verlag, Berlin, Heidelberg, pp 83–98

Kuranuki K, Sakai A (1995) Crypreservation of *in vitro* shoot tips of tea (*Camellia sinensis*) by vitrification. Cryoletters 16:345–352

Kuranuki K (2006) Developments of cryopreservation techniques for genetic resources of tea {*Camellia sinensis, (L.)* O. Kuntze}. Tech Bull Shizuoka Tea Exp Stat 3:1–63

Kuranuki Y, Shibata M (1992) Effect of concentration of plant growth regulators on the shoot apex culture of tea plant. Bull Shizuoka Tea Expt Sta 16:1–6

Kuranuki Y, Shibata M (1993) Improvement of medium components for *in vitro* cuttings of tea plant. 2. Optimum concentration of plant growth regulators. J Tea Sci 77:39–45

Lammerts WE (1958) Embryo culture in *Camellia* seed germination. In Tourje EC (ed) *Camellia* culture. Southern California Camellia Society, Pasadena, pp 171–174

Llyod G, McCown B (1980) Commercially feasible micropropagation of mountain laurel, *Kalmia lalifolia* by use of shoot tip culture. Comb Proc Int Pl Prop Soc 30:421–427

Marimuthu S, Raj Kumar R (2001) Physiological and biochemical responses of micropropagated tea plants. *In Vitro* Cell Dev Biol Plant 37:618–621

Molina SP, Rey HY, Pérez ML, Mroginski LA (2013) Plant regeneration of tea (*Camellia sinensis*) by *in vitro* culture of meristems, axillary buds and uninodal segments. Rev FCA UNCUYO 45:127–134

Mondal TK (2003) Micropropagation of tea (*Camellia sinensis*). In: Jain SM, Ishii K (eds) Micropropagation of woody trees and fruits. Kluwer Academic Publishers, The Netherland, pp 671–720

Mondal TK, Bhattacharya A, Sood A, Ahuja PS (1998) Micropropagation of tea using thidiazuran. Plant Growth Reg 26:57–61

Mondal TK, Bhattacharya A, Laxmikumaran M, Ahuja PS (2004) Recent advance in tea Biotechnology. Plant Cell Tissue Orga Cult 75:795–856

Mondal TK, Parathiraj S, Mohan Kumar P (2005) Micrografting-A technique to shorten the hardening time of micropropagated shoots of tea {*Camellia sinensis (L)* O. Kuntze}. Sri Lank J Tea Sci 70:5–9

Murali KS, Pandidurai V, Manivel L, Rajkumar R (1996) Clonal variation in multiplication of tea through tissue culture. J Plant Crop 24:517–522

Murashige T, Skoog F (1962) A revised medium for rapid growth and bioassay with tobacco tissue cultures. Physiol Plant 15:473–497

Nakamura Y (1987a) Shoot tip culture of tea cultivar Yabukita. Tea Res J 65:1–7

Nakamura Y (1987b) *In vitro* rapid plantlet culture from axillary buds of tea plant (*C. sinensis* (L.) O. Kuntze). Bull Shizuoka Tea Expt Stat 13:23–27

Nakamura Y (1988) Effects of the kinds of auxins on callus induction and root differentiation from stem segment culture of *Camellia sinensis* (L.) O. Kuntze. Tea Res J 68:1–7

Nakamura Y (1989) Differentiation of adventitious buds and its varietal difference in stem segment culture of *Camellia sinensis* (L.) O. Kuntze. Tea Res J 70:41–49

Nakamura Y (1990) Effect of sugar on formation of adventitious buds and growth of axillary buds in tissue culture of tea. Bull Shizuoka Tea Expt. Stat 15:1–5

Nakamura Y, Shibita M (1990) Micropropagation of tea plant (*Camellia sinensis L. O Kuntze*) through *in vitro* cuttings- effects of various hormones on growth of shoots from axillary buds. Jap J Agri Q 72:9–17

Nitsch JP, Nitsch C (1969) Haploid plants from pollen grains. Science 163:185

Ogutuga DBA, Northcote DH (1970) Caffeine formation in tea callus tissue. J Exp Bot 21:258–273

Owuor PO, Wachira FN, Obanda M, Waller GR (2007) Effects on caffeine on *in vitro* tea growth. J Plant Crop 35: 36–38

Pandey A, Palni LMS, Bag N (2000) Biological hardening of tissue culture raised tea plants through rhizosphere bacteria. Biotech Lett 22:1087–1091

Pandidurai V, Murali KS, Manivel L, Rajkumar R (1996) Factors affecting *in vitro* shoot multiplication and root regeneration in tea. J Plant Crop 24:603–609

Phukan MK, Mitra GC (1984) Regeneration of tea shoots from nodal explants in tissue culture. Curr Sci 53:874–876

Phukan MK, Mitra GC (1990) Nutrient requirements for growth and multiplication of tea plants *in vitro*. Bangladesh J Bot 19:65–71

Prakash O, Sood A, Sharma M, Ahuja PS (1999) Grafting micropropagated tea (*Camellia sinensis* (L.) O. Kuntze) shoots on tea seedling-a new approach to tea propagation. Plant Cell Rep 18:137–142

Quoirin M, Lepoivre P (1977) Improved media for *in vitro* culture of *Prunus* sp. Acta Horti 78:437–442

Ranaweera KK, Gunasekara MTK, Eeswara JP (2013) *Ex vitro* rooting: A low cost micropropagation technique for tea (*Camellia sinensis* (L.) O. Kuntz) hybrids. Scientia Hort 155:8–14

Rajasekaran P, Mohankumar P (1992) Rapid micropropagation of tea (*Camellia* spp). J Plant Crop 20:248–251

Rajkumar R, Ayyappan P (1992) Micropropagation of *Camellia sinensis* (L.) O kuntze. J Plant Crop 20:252–256

Riffaud JL, Cornu D (1981) Utilization de la culture *in vitro* pour la multiplication de merisiers adultes (*Prunus avium* L.) selectionnes en foret. Agronomie 1:633–640

Samartin A (1991) Potential for large scale *in vitro* propagation of *Camellia sasanqua* Thunb. J Hort Sci 67:211–217

Samartin A, Vieitez AM, Vieitez E (1984) *In vitro* propagation of *Camellia japonica* seedlings. J Hort Sci 19:225–226

Samartin A, Vieitez AM, Vieitez E (1986) Rooting of tissue cultured camellias. J Hort Sci 61:113–120

Sandal I, Bhattacharya A, Ahuja PS (2001) An efficient liquid culture system for tea shoot proliferation. Plant Cell Tiss Orga Cult 65:75–80

San-Jose MC, Vieitez AM (1990) *In vitro* regeneration of *Camellia reticulata* cultivar 'Captain Rawes' from adult material. Sci Hort 43: 155–162

San-Jose MC, Vidal N, Vieitez AM (1991) Improved efficiency of *in vitro* propagation of *Camellia reticulata* cv. captain leaves. J Hort Sci 66:755–762

San-Jose MC, Vieitez AM (1992) Adventitious shoot regeneration from *in vitro* leaves of adult *Camellia reticulata*. J Hort Sci 67:677–683

Sarathchandra TM, Upali PD, Wijeweardena RGA (1988) Studies on the tissue culture of tea {*Camellia sinensis* (L.) O. Kuntze}. Somatic embryogenesis in stem and leaf callus cultures. Sri Lanka J Tea Sci 52:50–54

Sarwar M (1985) Callus formation from explanted organs of tea (*Camellia sinensis* L.). J Tea Sci 54:18–22

Schenk RU, Hildebrandt A (1972) Medium and techniques for induction and growth of monocotyledonous and dicotyledonous plant cell cultures. Can J Bot 50:199–204

Sharma M, Sood A, Nagar PK, Prakash O, Ahuja PS (1999) Direct rooting and hardening of tea microshoots in the field. Plant Cell Tiss Orga Cult 58:111–118

Shibata M, Kuranuki Y (1993) Improvement of medium components for *in vitro* cuttings of tea plant. 1. Effects of concentration of some components of MS medium and comparison between MS medium and woody plant medium. J Tea Sci 77:39–45

Singh A, Sharma J, Rexer KH, Varma A (2000) Plant productivity determinants beyond minerals, water and light: *Piriformospora indica*- A revolutionary plant growth promoting fungus. Curr Sci 79:1548–1554

Tabachnick L, Kester DE (1977) Shoot culture for almond and almond peach hybrid clones *in vitro* Hort Sci 12:545–547

Tahardi JS (1994) Micropropagation of tea through shoot proliferation from excised axillary buds. Menara Perkebunan 62:20–24

Tahardi JS, Shu W (1992) Commercialization of clonal micropropagation of superior tea genotypes using tissue culture technology USAID/CDR Network meeting on tea crop biotech, Costa Rica

Tian-Ling (1982) Regeneration of plantlets in cultures of immature cotyledons and young embryos of *Camellia oleifera* Abel. Acta Biol Exp Sin 15:393–403

Torres KC, Carlisi JA (1986) Shoot and root organogenesis of *Camellia sasanqua*. Plant Cell Rep 5:381–384

Tosca A, Pondofi R, Vasconi S (1996) Organogenesis in *Camellia x williamsii:* cytokinin requirement and susceptibility to antibiotics. Plant Cell Rep 15:541–544

Tukey HB (1934) Artificial culture methods for isolated embryos of deciduous fruits. Amer Soc Hort Sci Proc 32:303–322

Vieitez AM, Barciela J, Ballester A (1989a) Propagation of *Camellia japonica* cv. Alba Plena by tissue culture. J Hort Sci 64: 177–182

Vieitez AM, San-Jose MC, Ballester A (1989b) Progress towards clonal propagation of *Camellia japonica* cv. Alba Plena by tissue culture techniques. J Hort Sci 64:605–610

Vieitez AM, San-Jose MC, Vieitez J, Ballester A (1991) Somatic embryogenesis from Roots of *Camellia japonica* plantlets cultured *in vitro*. J Am Soc Hort Sci 116:753–757

Vieitez AM, Vieitez ML, Ballester A, Vieitez E (1992) Micropropagation of *Camellia* spp. In: Bajaj YPS (ed) Biotechnology in agriculture and forestry, vol 19. High tech and micropropagation III. Springer-Verlag, Berlin Heidelberg, pp 361–387

Yan MQ, Ping C, Wei M, Wang YH (1984) Tissue culture and transplanting of *Camellia oleifera*. Sci Silvae Sin 20:341–350

4 Somatic Embryogenesis and Alternative *In Vitro* Techniques

4.1 Introduction

Genetic engineering through *in vitro* manipulation of the cell seems to be an additional tool for overcoming some of the problems of tea breeding. The prerequisite step for such a technology is an efficient *in vitro* plant regeneration system. Somatic embryogenesis is considered to be the most efficient regeneration system of tea (Jain and Newton 1990). However, the efficacy of such a system for plant production depends on the efficiency of multiplication and conversion rate of somatic embryos. The advantage of somatic embryogenesis is the development of adventitious embryos from explant tissue without an intervening callus phase which helps in maintaining genetic fidelity (Bano et al. 1991). Therefore, it has a tremendous potential in clonal propagation (Mondal et al. 2001a) and most importantly through genetic transformation of tea (Mondal et al. 1999; Mondal et al. 2001b). It has been successfully used for artificial seed production (Mondal et al. 2000; Mondal 2002) and some interspecific hybrid crosses of *Camellia* (Nadamitsu et al. 1986), where immature somatic embryos were rescued and cultured before abortion. It can also be used for disease free plant production and androgenic or haploid plant production of tea (Chen and Liao 1982). In tea, somatic embryogenesis is the only alternative pathway to conventional micropropagation due to little or no success on neoformation of adventitious buds or protoplast/suspension cultures. The various factors, which govern the somatic embryogenesis of tea and related species have been summarized concisely in Table 4.1 and described below.

4.2 Induction

Induction of somatic embryogenesis depends upon several factors such as explant's type, physiological stage of the explants and media formulation including plant growth regulators (PGRs). A judicious selection of all the factors may lead to a successful embryogenesis of tea.

4.2.1 Explants

Although induction of androgenesis in tea began with pollen as explants (Chen and Liao 1983), later mature cotyledons or zygotic embryos became more popular as choice of explants (Table 4.1). Use of immature cotyledons (Abraham and Raman 1986; Nakamura 1988a; Bano et al. 1991), de-cotylenated embryos (Nakamura 1985; Paratasilpin 1990; Mondal et al. 2000), de-embryonated cotyledons (Rajkumar and Ayyappan 1992; Ponsamuel et al. 1996), nodal cuttings (Akula and Akula 1999), juvenile leaves (Sarathchandra et al. 1988) and leaf stalk (Hua et al. 1999) for the production of somatic embryos was also documented. Although somatic embryogenesis has been reported from various explants of ornamental *Camellia* variants, most of the workers have, however, used mature cotyledon slices or zygotic embryos (Table 4.1). In *Camellia ja-*

T. K. Mondal, *Breeding and Biotechnology of Tea and its Wild Species*,
DOI 10.1007/978-81-322-1704-6_4, © Springer India 2014

Table 4.1 Somatic embryogenesis in *Camellia* species

Species/cultivar	Explant	Medium				Reference
		Induction	Maturation	Germination	Multiplication	
C. sinensis	Immature cotyledon	MS+BA (0.5)+IBA (0.5)	–	MS+BA (0.5)+NAA (5)	–	Abraham and Raman (1986)
C. sinensis TRI-2025	Nodal segment	MS+BA (0.5)+GA_3 (3)	–	–	MS0+1/2 macro salts	Akula and Dood (1998)
C. sinensis	Mature seed	1/2 MS macro+Full micro MS+AHS (100)+Gln (100)	1/2 MS macro+Full micro MS+AHS (100)+Gln (100)	1/2 MS macro+Full micro MS+AHS (100)+Gln (100)	–	Akula et al. (2000)
C. sinensis (L.) O. Kuntze	Cotyledon	Modified MS+BA (10)+YE (2)	–	Modified MS+BA (10) with 2 % sucrose	–	Arulpragasam et al. (1988)
C. sinensis (L.) O. Kuntze	Callus derived from shoot apex	MS+BA (1)+Mannitol (56 g/l)	MS+BA (1)+Hygromycin (10)	–	–	Aoshima (2005)
C. sinensis	Cotyledon	MS+BA (1.13)+NAA (22.7)	MS+BA (1.13)+NAA (22.7)	MS+BA (2.26)+NAA (22.7)+GA_3 (0.12)	–	Bag et al. (1997)
C. sinensis	Mature cotyledon, leaf	MS+BA (2)+2,4-D (5)	–	MS+BA (2)+NAA (3)	–	Balasubramanian et al. (2000a)
Thea sinensis (L.), B-61	Cotyledon	MS+Kn (0.05)+2,4-D (0.5)	1/2 MS+Kn (0.05)+AC (0.2 %)+glucose (1.5 %)	1/2 MS+Kn (0.05)+AC (0.2 %)+glucose (1.5 %)	–	Bano et al. (1991)
C. japonica	Cotyledon	MS+BA (1)	–	–	–	Barciela and Vieitez (1993)
C. japonica	Cotyledon from immature seed	MS with modified vit+BA (1–2)+IBA (0–2)	–	–	MS with modified vit+BA (1)	Barciela and Vieitez (1993)
C. japonica	Mature cotyledon	Knop+BA (0.5–5)	–	–	–	Bennett and Scheibert (1982)
C. sinensis	Immature cotyledon	Modified MS+BA (3)+NAA (2)	–	MS+BA (5)	–	Haridas et al. (2000)
C. sinensis, T-78	Cotyledon	MS+BA (10)+IBA (0.5)+A (80)	MS+BA (10)+IBA (0.5)+A (80)	B_5+BA (3)+IAA (2)	–	Jha et al. (1992)
C. sinensis, Yabukita	Cotyledon	MS+BA (4)+IBA (2)	–	MS+BA (10)+IBA (0.5)	–	Kato (1982)
C. japonica	Mature cotyledon	MS+BA (0–5)+IBA (0–2)	–	MS+GA_3 (1)	–	Kato (1986a)

Table 4.1 (continued)

Species/cultivar	Explant	Medium				Reference
		Induction	Maturation	Germination	Multiplication	
C. sasanqua	Mature cotyledon	MS+BA (0–10)+IBA (0–2)	–	–	–	Kato (1986b)
C. japonica	Primary somatic embryo	MS+GA_3 (1)+colchicine (0.1 %)	–	MS+GA_3(1)	–	Kato (1989b)
C. sinensis, Yabukita, Sayamamidori, Benikaori, Akane	Immature leaves of *in vitro* grown shoots	Liquid MS+2,4-D (0.5)	MS+BA (10)+IBA (0.5) or MS+BA (4)+IBA (2)	MS+BA (10)+IBA (0.5) or MS+BA (4)+IBA (2)+GA_3 (1)	–	Kato (1996)
C. nitidissima	Mature cotyledons	WPM+2,4-D (1)	WPM+BAP (0.2)+NAA (0.2)	WPM+BAP (0.2)+NAA (0.2)	WPM supplemented with IBA (5)+NAA (0.05)	Lu et al. (2013)
C. sinensis	Mature de-embryonated cotyledon	MS+BA (2)+IBA (0.2)	Modified MS+BA (2)+IBA (0.2)+Gln (1)+K_2SO_4	Modified MS+BA (2)+IBA (0.2)+Gln (1)		Mondal et al. (2001a)
C. vietnamensis × *C. chrysantha*	Mature cotyledon	MS+BA (3)+NAA (1)	–	MS+GA_3 (1)+CM (10 %)	–	Nadamitsu et al. (1986)
C. sinensis	Cotyledonary segment	NN	NN+N-Z-amino type-A (0.1 %)	NN+N-Z amino type-A (2.6 %)	–	Nakamura (1985)
C. sinensis with 13 cultivars, C. japonica 3 cultivars, C. sasanqua, C. brevistela, C. nokoensis, C. japonica (cv. Kosyougatu) × *C. granthamiana*	Half-sliced cotyledon	MS+BA (1–5)	–	–	–	Nakamura (1988a)
C. sinensis cv. Shan Chat Tien	Somatic embryo	MS+BAP (8)	–	MS+BAP (3)	BAP (3) and IBA (0.3)	Nguyen (2012)
C. sinensis and C. assamica	De-embryonated cotyledon segments	B_5+NAA (0.1)+BA (2)	B_5+NAA (0.1)+BA (2)	B_5+NAA (0.1)+BA (2)	–	Paratasilpin (1990)
C. sinensis (L.) O. Kuntze	Mature cotyledon	MS+BAP (0.1–4.5)	MS+BAP (0.1–4.5)	MS+BAP (0.1–4.5)	MS+BAP (0.1–4.5)	Park et al. (1997)
C. japonica cv. Elegans	In vitro leaf segment	1/2 MS+DTT (2.5)+ferric citrate (2.5) in lieu of ferrous sulphate	1/2 MS+DTT (2.5) ferric citrate (2.5) in lieu of ferrous sulphate	Full strength, induction media+D-glucose (25 g/l)+BA (1)+IBA (0.1)+IAA (0.1)	–	Pedros and Pais (1993)

Table 4.1 (continued)

Species/cultivar	Explant	Medium				Reference
		Induction	Maturation	Germination	Multiplication	
C. reticulata	Mature and immature cotyledonary segment	MS+IBA (0.5–1)	MS+BA (1)+IBA (0.5)	–	MS+GA_3 (3–5)+IAA (1–2)	Plata and Vieitez (1990)
C. reticulata cv. Mouchang	Immature zygotic embryo	–	–	–	MS+Modified vit+BA (0.5)+IBA (0.1)	Plata et al. (1991)
C. sinensis, UPASI-10	De-embryonated immature cotyledonary segment	MS+PBOA (0.12)+BA (0.1)	MS+Brassin (0.5)	MS+Brassin (0.5)	–	Ponsamuel et al. (1996)
C. sinensis, UPASI-10	Cotyledon with/without embryogenic axis	Modified MS+BA (0.25)	MS+BA (3–5)+CW (10%)+GA_3 (0.25–1)	MS+BA (3–5)+CW (10%)+GA_3 (0.25–1)	–	Rajkumar and Ayyappan (1992)
C. sinensis, TRI-2024	Nodal segment and leaf	VW+CW (15%) MS+IBA (0.1)+BA (1) for nodal segment; MS+2,4-D (1)+Kn (1) for leaf	–	–	Callus, embryo-like structure	Sarathchandra et al. (1988)
C. japonica 'Alba Plena' *C. reticulata* 'Mouchang'	*In vitro* leaf segment	MS+BA (8)+IBA (0.5)	MS+BA (8)+IBA (0.5)	MS+GA_3 (2)+IAA (1)	Secondary somatic embryogenesis in MS+BA (1)+IBA (0.1)	San-Jose and Vieitez (1993)
C. sinensis, Kangra Jat	Cotyledon slice	1/2 MS+BA (2)+IAA (0.2)	–	1/2 MS+BA (2)+IAA (0.2)+GA_3 (0.2)	–	Sood et al. (1993)
C. sinensis, UPASI-9, UPASI-10 and ATK-1 and 'Cambod' variety UPASI-17	Excised cotyledons	MS+Glutamine (200)	MS+ABA (5)+PEG 3%	MS+GA_3 (5) and BAP (2)	MS+GA_3 (3)	Suganthi et al. (2012)
C. sinensis, cv. Yabukita	Deembryonated cotyledonary	1/2 MS media+BAP (2)	1/2 MS Kn (0.1)+ABA (0.1)	1/2 MS liquid	Pre-sterilized soil-sand (3:1) mix in the green-house with a low mortality rate	Tahardi et al. (2000)
C. sinensis, cv. Yabukita	Primary globular embryo	WPM liquid+BAP (1)+IAA (5)	WPM liquid+BAP (0.1)+ABA (0.5)	WPM liquid	–	Tahardi et al. (2003)
C. japonica	Immature and mature zygotic embryo	MS with modified vit+BA (1–2)+IBA (0–2)	MS with modified vit+BA (1–2)+IBA (0–2)	MS with modified vit+BA (1)+(0.1) IBA+GA_3 (5)+IAA (2)	Secondary embryogenesis in MS+GA_3(1–2)	Vieitez and Barciela (1990)

Table 4.1 (continued)

Species/cultivar	Explant	Medium				Reference
		Induction	Maturation	Germination	Multiplication	
C. japonica cv. Alba Plena	*In vitro* root segment	MS with thiamin (1)+nicotinic acid (0.1)+pyridoxine-HCl (0.1)	MS with thiamin (1)+nicotinic acid (0.1)+pyridoxine-HCl (0.1)	MS+GA_3 (5)+IAA (1)	–	Vieitez et al. (1991)
C. sinensis	Cotyledon slice	Modified MS	Modified MS	Modified MS +Kn (1.8) or BA (1)	–	Wachira and Ogado (1995)
C. sinensis, Chyi-Men and Pyng-Shoei	Mature cotyledon	MS+Kn (10)+IAA (1)	–	–	–	Wu et al. (1981)
C. sinensis	Mature cotyledon	–	–	–	–	Yan and Ping (1983)
C. oleifera	Cotyledon	–	–	–	–	Yan et al. (1984)
C. japonica × *C. chrysantha*	Immature zygotic embryos	MS+Kn (0.1–0.5)+NAA (0.5–1.0)+YE (1)	–	AM+2ip (0.2–0.5)+GA_3 (5)+PVP (5 g/l)	–	Yamaguchi et al. (1987)
C. chrysantha	Cotyledon	MS+BA (1)+NAA (0.2–0.5)	–	–	–	Zhuang and Liang (1985b)
C. reticulata	Mature whole cotyledon	MS+BA (1)+NAA (0.2)	–	B_5 or liquid MS+BA (0.1–0.2)+IAA (0.1–0.5)	–	Zuang and Liang (1985a)
C. sasanqua	Whole cotyledon	MS+BA (1)+NAA (0.2–0.5)	–	MS+BA (2)+IAA (0.5)+ABA (0.2)+Gln (500) or MS+GA_3 (1)	–	Zhuang et al. (1988)

Figures in parenthesis denote concentrations in mg/l (unless otherwise stated)
A adenine, *ABA* abscissic acid, *AC* activated charcoal, *AHS* adenine hemisulphate, *AM* Anderson (1984) basal medium, *CW* coconut water, *DTT* dithiothreitol, *Gln* glutamine, *MS0* MS (Murashige and Skoog 1962) basal medium without added growth regulators, *NN* Nitsch and Nitsch (1969) medium, *PBOA* phenylboronic acid, *PVP* polyvinyl pyrrolidone, vit vitamins, *VM* Vacin and Went (1949) medium, *YE* yeast extract

ponica and *Camellia reticulata*, immature cotyledons and zygotic embryos, roots, stems and leaves had been used for the induction of somatic embryogenesis (Plata and Vieitez 1990; Vieitez and Barciela 1990; Plata 1993; Pedroso and Pais 1993; Zhuang and Liang 1985a). Dark period of at least 14 weeks appeared to be necessary for somatic embryogenesis from *in vitro* leaf for *C. reticulata* (San-Jose and Vieitez 1993). Vieitez et al. (1991) reported somatic embryogenesis from the *in vitro* roots of *C. japonica* clones cv. 'Alba Plena'.

4.2.2 Physiological Stages and Genotypic Variations

Seeds of different tea varieties mature at different time. Therefore, the physiological stage and genotypic variation are interrelated, hence both play important roles for the successful induction of somatic embryos from cotyledons (Vieitez 1994). However, systematic study on the effect of physiological maturity for different cultivars of tea is scanty. Nakamura (1988a), found that the optimum timing for cotyledon culture in *C. sinensis* was late September to mid-October. This was when seeds were physiologically matured in Shizuoka, Japan. In that report, a high percentage of embryogenesis (45–50 %) was observed during this period in contrast to 15 % from immature seeds collected during August. Paratasilpin (1990) reported that while China type produced 28 % variation of induction of somatic embryogenesis, Assam type could produce 2 % only. Mondal et al. (2000) found that seed maturity of four different Indian tea cultivars was attained at different periods. Primary somatic embryos were induced from de-embryonated cotyledons of tea on MS supplemented with naphthalene acetic acid (NAA, 2.5 mg/l) and BAP (0.2 mg/l). Among them, the best time for induction of somatic embryogenesis for cultivar TG 270/2/B and UPASI-9 was during the monsoon (July–August) with embryogenic responses of 11.1 and 50.0 %, respectively. For the cultivar Tuckdah-78, the best time was during autumn (September–October) with the response of 44.1 %, whereas the highest number of somatic embryo inductions in cultivar Kangra Jat was registered at 66.7 % during winter (November–December). Nakamura (1988a) screened 13 Japanese tea cultivars and observed a variation between 0 and 50 % somatic embryogenesis. Among the cultivars, the best response was reported in 'Yabukita'. She also reported that the formation of 28 % somatic embryos in *C. sinensis*, against 2 % in *C. assamica*. Genotype also plays a major role in the induction of embryogenesis. In the same study, she also screened four *Camellia* species including Japanese tea cultivars among which the embryogenic response varied between 0 and 50 %. Of all the cultivars screened, the best response was reported in 'Yabukita' and 'Kurasawa' tea cultivars. Among the other *Camellia* species, high differentiation rates of somatic embryos were obtained with *C. japonica* (48–58 %), *C. sasanqua* (9–81 %) and *C. brevistela* (93 %). Kato (1996) also used various cultivars of tea and reconfirmed the highest frequency of embryogenesis in cultivar 'Yabukita'. All these works collectively indicated that somatic embryogenesis was controlled by genetic factors in *Camellia*.

Successful induction of somatic embryos from cotyledon explants depends upon the physiological maturity of the cotyledons. In *C. japonica,* Vieitez and Barciela (1990) collected seeds in July, September and October to determine the right stage for induction of somatic embryogenesis. They achieved 94 % embryogenesis in seeds collected in September in contrast to 20 % of those collected in October under the climatic condition prevailing at Spain. The seeds collected in September were fully grown but were still immature suggesting a transitory dormancy stage.

A protocol for the induction of direct somatic embryogenesis was developed for *C. japonica* L. to achieve a controlled induction of somatic embryogenesis on the specific leaf region. Leaves isolated from 12-week-old shoots were immersed in indole-3-butyric acid (IBA, 1.0 g/l) for 20 min, followed by incubation for 11 days in darkness on diluted MS modified basal medium (Pedroso and Pais 1993). The cells from a cultured leaf responded differently to the same culture conditions forming embryos, roots, non-morphogenic and calluses. Shoot and leaf ages also affected *in vitro* leaf response. So, not all leaves from the same shoot were induced for embryogenesis after the induction treatment. Those that did not, were referred as non-embryogenic induced leaves. Direct embryo formation only occurred in defined regions of the leaf blade. Direct root formation only occurred in a well-defined region of midrib, whereas calluses were preferentially formed on the leaf bases. The results suggested the existence of differences in morphogenic competence according to leaf regions. A leaf region-specific response was also observed when leaves from 1-year-old shoots were sectioned transversely in slices (2–4 mm) or sections. Explant region-specific embryogenic competence also occurred in stem segments and cotyledons of this species (Pedroso and Pais 1995a). Plantlet regeneration was successfully achieved from all the culture systems developed (Pedroso and Pais 1993; Pedroso and Pais 1995a, 1995b). The *C. japonica* leaf culture system enabled a comparison between embryogenic and non-embryogenic leaves isolated from the same shoot and the study of embryogenic and non-embryogenic leaf regions within the same leaf. For these reasons, this culture system was used as an experimental model system with the subsequent studies searching for markers of direct somatic embryogenesis in dicotyledonous species.

4.2.3 Basal Media and Growth Regulators

The type, concentration, and time of application of different growth regulators in culture media have been extensively worked out. In general, a high cytokinin-to-low auxin or low cytokinin alone was found to be necessary for induction of somatic embryos in *Camellia*, but cytokinins had been often reduced or omitted in subsequent subculturing. Though MS was the most commonly used medium for the induction of tea somatic embryos (Table 4.1), WPM (Woody Plant Medium, Lloyd and McCown 1980) and Nitsch and Nitsch (1969) media had also been used by researchers. Type, concentration, and time of application of different PGRs have also been extensively worked out. In general, a high cytokinin-to-low auxin ratio or low cytokinin alone was found to be necessary for the induction of somatic embryos in tea, even reduction or omission of cytokinin in subsequent sub-culturing is also known. 6-Benzylaminopurine (BAP, 0–10 mg/l) had been widely used for tea and related species such as *C. japonica* (Bennett and Scheibert 1982; Beretta et al. 1987; Barciela and Vieitez 1993), *C. sasanqua* (Kato 1986b; Zhuang et al. 1988), *C. chrysantha* (Zhuang and Liang 1985b) and *C. reticulata* (Zhuang and Liang 1985a), but in certain reports, Kn (0.05–10 mg/l) had also been used in the induction medium (Wu et al. 1981; Bano et al. 1991; Wachira and Ogado 1995). Among the different auxins, IBA (0–2 mg/l) was used mostly for somatic embryo induction in tea. However, different concentrations of NAA (Paratasilpin 1990; Bag et al. 1997; Balasubramanian et al. 2000a), 2,4-D (Das and Barman 1988; Bano et al. 1991) and indole-3-acetic acid (IAA, Wu et al. 1981; Sood et al. 1993) were also used to elicit a better response for induction of somatic embryos in *C. sinensis*. Ponsamuel et al. (1996) used some novel auxins, namely, tetraphenylboron (TPB), phenylboronic acid (PBOA), etc., for induction of somatic embryos from immature cotyledons of *C. sinensis* cultivar UPASI-10. They found that TPB (0.35 g/l) and PBOA (0.12 g/l) were capable of inducing somatic embryos. (Mondal et al. 2000) also found that half strength of MS salts containing NAA (2.5 mg/l) and BAP (0.2 mg/l) were the best for induction of somatic embryo. Among them, a maximum of 66.7 % somatic embryos were formed by cultivar Kangra Jat. Addition of abscisic acid (ABA, 7.5 mg/l) to the induction medium significantly enhanced the rapid formation of somatic embryos without callusing from mature cotyledon of tea cultivar 'TRI-2025' within 2 weeks of culture initiation (Akula et al. 2000).

In general, direct somatic embryos in *Camellia* can be obtained on a wide range of culture conditions: full to one-third strength modified-MS, sucrose (15–30 g/l), D-glucose or combinations of both, auxin (0–10 mg/l) and cytokinin (0–10 mg/l) in liquid, semisolid, or solid medium. Light was an important parameter for somatic embryo formation, especially from stem and leaf explants (direct and indirect). Somatic embryos did not either form in the dark or their number was significantly lower than that for cultures under photoperiod. Successful conversion of cotyledon-derived embryos into plants ranged from 35 to 79%, depending on the culture medium used (Pedroso-Ubach 1994). Identical results were obtained for leaves, cultured on modified Murashige and Skoog (MS) medium supplemented with sucrose (20 g/l) or D-glucose (25 g/l), 2,4-D (1.0 mg/l), and kinetin (0.1 mg/l) (Pedroso-Ubach 1991). Leaves cultured on MS liquid medium with BAP (1.0 mg/l) and 2,4-D (0.5 mg/l) produced clusters of 3–23 somatic embryos/leaf. Less frequently (2–3%), globular embryos detached from the remaining differentiated leaf tissues and developed singularly in the liquid culture. Only the embryos arising in clusters developed into plantlets. Among the PGRs, cytokinins such as BAP (0–10 mg/l) had been widely used for *Camellia*, though Vieitez et al. (1991) claimed that zeatin (1 mg/l) in combination with BAP and IBA was essential for the induction of somatic embryos in *C. japonica* cv. Alba Plena. In auxins, IBA (0–2 mg/l) was widely used in the induction medium for *Camellia*, though NAA had also been used in different concentrations in different species such as 0–2 mg/l in *C. reticulata* (Zhuang and Liang 1985a), 0.2–0.5 mg/l in *C. sasanqua* (Yamaguchi et al. 1987), 0.5–1.0 mg/l in *C. japonica*, *C. chrysantha* and 1 mg/l in *C. vietnamensis* and *C. chrysantha* (Nadamitsu et al. 1986). The other species on which embryogenesis was carried out are *C. brevistela* (Nakamura 1988a), *C. chrysantha* (Zhuang and Liang 1985b), *C. oleifera* (Yan et al. 1984), and *C. sasanqua* (Nakamura 1988b; Zhuang et al. 1988), and with new hybrids whose development was hindered by poor fertility or embryo viability (Nadamitsu et al. 1986; Yamaguchi et al. 1987). In all these reports, somatic embryogenesis was achieved using MS medium containing a cytokinin (usually BAP) with or without auxin (usually NAA). However, none of these studies involved comprehensive experiments to determine optimum conditions for embryogenesis.

4.2.4 Growth Adjuvants

Growth adjuvants play an important role, nevertheless it seems that in tea their requirement is low, which may be due to the fact that cotyledons of tea have a high degree of inherent embryogenic capacity (Vieitez 1994), and hence does not demand any growth adjuvant. However, this had not restrained researchers from using an adjuvant like yeast extract (Arulpragasam et al. 1988), coconut milk (Sarathchandra et al. 1988; Rajkumar and Ayyappan 1992), adenine sulfate (Jha et al. 1992), betain (Akula et al. 2000), etc. The positive result of yeast extract had also been found in some *Camellia* hybrids (Yamaguchi et al. 1987).

4.3 Secondary Embryogenesis

Exploitation of embryogenesis through a biotechnological approach demands a system of repetitive/recurrent or secondary somatic embryogenesis, which is especially important for genetic transformation as foreign genes can be introduced into primary embryos and made to multiply subsequently as had been reported by Mondal et al. (2001a). Two different growth patterns for secondary embryogenesis have been reported in *Camellia*: (1) somatic embryo-to-somatic embryo, commonly known as repetitive embryogenesis, more frequent in tea as well as other *Camellia* species; (2) callus-to-somatic embryo, in which the multiplication of somatic embryos depends upon subculturing of callus (Vieitez 1994). However, the various works of synchronized secondary somatic embryo formation and their long-term maintenance of embryogenic capacity are discussed below.

Abraham and Raman (1986) used a single medium containing different concentrations of BAP and NAA for the induction, germination and maintenance of somatic embryos. Kato (1986a) used Gibberellic acid (GA_3, 1 mg/l) for induction of secondary embryos from isolated primary embryos. However, Jha et al. (1992) obtained secondary embryos on cotyledons of primary embryo within 6–8 weeks on MS or B_5 medium supplemented with BAP (0.1–3 mg/l),

IBA (0.1 mg/l), IAA (0.1–2 mg/l), and GA_3 (5.0 mg/l). Histological observations indicated that an embryogenic callus was induced from a special zone around vascular bundle of primary embryo (Balasubramanian et al. 2000a). Apart from growth regulators, nutritional effect of nitrate salts of potassium and ammonium, together with different concentrations of sulphate salts of aluminum, potassium, magnesium and ammonium on secondary somatic embryogenesis was investigated (Mondal et al. 2001a). The primary embryos, which were induced on de-embryonated cotyledon, were used for secondary embryogenesis. Each responsive explant (primary embryo) evinced about 10–15 secondary embryos. Among the different concentrations of nitrates, it had been found that MS with reduced potassium nitrate (800 mg/l) as well as ammonium nitrate (825 mg/l) along with potassium sulphate (260 mg/l) produced maximum number of secondary embryos within 2 weeks (i.e. 20–25 secondary embryos per primary embryo in 91.6% responsive explants). Each of these yellow, heart-shaped secondary embryos weighed about 4–5 mg and developed synchronously to the shiny, green, torpedo-shaped structure within 3 weeks of culture. In a further experiment, different factorial combinations of L-glutamine, BAP and IBA were investigated and maximum number of synchronous secondary embryos developed on MS medium supplemented with BAP (2 mg/l), IBA (0.2 mg/l), and L-glutamine (1 g/l). Following this medium formulation, synchronous globular, heart and torpedo-shaped somatic embryos were developed, which later germinated to an extent of 52%. Furthermore, high embryogenic capacity of this culture could be maintained on the above medium for more than 4 years, highest so far among the published reports.

In *C. japonica*, secondary embryogenesis had been reported primarily from embryos derived from a wide range of initial explants, i.e. from cotyledons or from excised embryos (Kato 1986a; Vieitez and Barciela 1990), roots (Vieitez et al. 1991) or from *in vitro* leaves (San-Jose and Vieitez 1993). Primary embryo, upon transfer to MS medium with or without growth regulators, gave rise to secondary embryogenesis within 3–4 weeks. In general, growth regulators used for *C. japonica* had higher concentration of BAP along with lower concentration of IBA.

In *C. reticulata*, high frequency (65%) secondary embryogenesis was found on cotyledonary and hypocotyl region of isolated primary embryos (Plata and Vieitez 1990). This response was evinced on MS medium containing BAP (0.5 mg/l) and IAA (0.5 mg/l). Plata et al. (1991) studied the anatomical sequence of events, which led to the differentiation of secondary embryogenesis in *C. reticulata* cv. Mouchang. They found that embryogenesis occurred mainly on the hypocotyl region of primary embryos. Histological monitoring revealed that secondary embryos apparently had a multicellular origin from embryogenic areas originating in both epidermal and subepidermal layers of hypocotyl region. This morphogenic competence was related to the presence of relatively undifferentiated cells in superficial layers of the hypocotyl of the primary embryo.

4.4 Bioreactor Technology for Secondary Embryogenesis

Applications of bioreactor technology in micropropagation facilitate large-scale continuous production of propagules. Although there are reports on various aspects of embryogenesis, bioreactor technology was not properly explored initially. Later, Akula et al. (2000) were pioneers in this field with the only approach until now for developing a bioreactor system for repetitive embryogenesis in tea. Uniform globular embryos were induced on nodal explants of cultivar 'TRI-2025' for developing a modified temporary immersion system (TIS). The relative efficiency of different conventional methods namely roller drum, liquid medium shaker for multiplying embryos were compared with the temporary immersion method. The highest rate of multiplication for secondary embryos (24-fold) was achieved using the TIS. By controlling the immersion cycles, they achieved more consistent, synchronized multiplication and embryo development with a high level of plant recovery. A one-step computer-programmed immersion protocol based on a single, simple medium with no growth regulators was developed, enabling multiplication, maturation, germination and plant recovery within 17 weeks. Plantlets recovered through this method were hardy, with 2–5 cm long shoots with a minimum of 2–4 lush green leaves with well-formed taproot. Callus formation, hyperhydricity and other developmental abnormalities were not observed at any stage of the process. Harvesting of secondary embryos was continued up to 2 years from the original set of experiment by regular sub-culturing after every 6 weeks. Plantlets produced using this method was successfully acclimatized to glasshouse conditions. This protocol avoided culture transfers and thus minimized the risk of contamination and reduced labour costs. Therefore, this technique was the first significant step for the commercial implication of bioreactor technology for large-scale production of tea somatic embryos.

4.5 Maturation and Germination

The success of embryogenesis also depends on high germination rate and subsequent plant regeneration, which occurs through series of events, such as synchronized embryo maturation followed by cotyledon expansion, hypocotyl-root axis elongation and emergence of the shoot (Webster et al. 1990). However, it had been noticed that as compared to other woody plants, induction of somatic embryos on tea cotyledon was easier due to inherent capacity but they had low germination rates, especially on hormone-free culture media (Vieitez 1994). Perhaps, this is due to the lack of physiological understanding of the somatic embryogenic pathway. Two common problems that often encounter with tea somatic embryos are precocious and abnormal germination due to lack of storage reserve, improper balance of endogenous hormone profile coupled with vulnerability of somatic embryo to desiccation. Therefore, various factors such as supply of external nutrient and PGRs, which will improve the conversion rate of tea somatic embryos, are discussed below.

4.5.1 Sugars

Source and concentration of sugars play an important role for maturation and germination of somatic embryos. Sugars served dual purpose of acting as a carbon source as well as an inducer of desiccation tolerance during maturation and germination of somatic embryos (Lecouteux et al. 1993) by acting as an osmotic agent (Tremblay and Tremblay 1995). One way to resist desiccation by preventing osmotic changes could be achieved by addition of an osmotic stabilizer to the culture medium. This was generally accomplished by inclusion of an osmoprotectant such as sucrose. The sucrose concentration in the culture medium and duration of culture treatment affected the germination rate of tea somatic embryos (Mondal et al. 2002). They registered significantly higher germination percentage (35.2 %) at 3 % sucrose after 5 weeks of treatment, whereas no response was observed at any of the higher sucrose concentrations (6 or 9 %) with various periods of treatment (3, 4 and 5 weeks). This may due to the fact that food reserves in the form of readily available carbohydrates during maturation are important for subsequent successful germination of zygotic embryos within seeds. Thus, it is not surprising that the embryos cultured on 3 % sucrose for periods of 5 weeks germinated normally to the extent of 35.2 % but declined thereafter.

4.5.2 Desiccation

Desiccation treatment promotes germination capacity for somatic embryos (Roberts et al. 1990). In tea, somatic embryos were still found to be desiccation sensitive and were adversely affected when desiccated. Probably the recalcitrant behaviour of tea seed was reflected in the somatic embryos also. In tea, drying and death coupled with poor germination (1.2–1.6 %) of somatic embryos were observed within a week when the globular embryos were desiccated for 3 weeks under high relative humidity created by different chemicals i.e. 60 % relative humidity by $MgNO_3$, $6H_2O$ or 90 % relative humidity by $ZnNO_3$, $7H_2O$. Therefore, it was clearly evident that to achieve the successful maturation one should use the osmotic protectant in the culture media to prevent the tea somatic embryo from desiccation (Mondal et al. 2002).

4.5.3 Plant Growth Regulators and Additives

Among the different PGRs, ABA had been widely used for the maturation of somatic embryos of woody plants particularly because it inhibited precocious germination by conferring desiccation tolerance (Roberts et al. 1990) and promoting accumulation of storage lipids (Avgioglu and Knox 1989) and proteins (Roberts et al. 1990). The report of Mondal et al. (2002), however, indicated that none of the tea somatic embryos germinated with ABA (0.5–62.5 mg/l) treatment applied for 3–5 weeks. Similarly, GA_3 (0.5–5 mg/l), which promoted germination in several woody species and hybrids of *Camellia*, was not significantly effective for tea. Although, Kato (1986a) found that BAP (10 mg/l) along with IBA (0.5 mg/l) were useful for induction of the concomitant development of shoots and roots in tea embryos, a much lower concentration of BAP (1 mg/l) and IBA (0.1 mg/l) could still induce germination of *C. japonica* somatic embryos (Vieitez and Barciela 1990; Vieitez et al. 1991; Pedroso and Pais 1993), and BAP (1 mg/l) along with IBA (0.5 mg/l) could induce germination of *C. reticulata* somatic embryos (Plata and Vieitez 1990; Plata et al. 1991). However, Jha et al. (1992) obtained 20 % germination of tea somatic embryos on B_5 medium (Gamborg et al. 1968) containing BAP (3 mg/l) and IAA (2 mg/l). Thus it seems that a high cytokinin stringently BAP to low auxin such as IBA or IAA is the best for somatic embryo germination of tea as well as related species. Influence of other growth regulators such as Brassin, a synthetic analog of a naturally occurring brassionoide (Maugh 1981) was tested by Ponsamuel et al. (1996). They reported about 50 % germination of *C. sinensis* somatic embryos on 0.8 % agar solidified-MS medium containing brassin (0.48 g/l). The positive influence of brassin on germination rate may be due to the inherent sensitivity of the tissue to this compound as brassinoide had been reported to be present in tea leaves (Ikekawa 1991). The effect of different concentration of maltose and trans-cinamic acid (t-CA) alone or in combination was evaluated to improve the germination efficiency of tea somatic embryo (Mondal et al. 2002). While none of the untreated somatic embryos germinated, significantly a low percentage (3–6 %) of germination was observed at all concentrations of either maltose or t-CA alone. On the other hand, normal and significantly high germination (70.6 %) was observed when the embryos were cultured on a medium supplemented with both maltose (4 %) and t-CA (3 mg/l) for 4 weeks followed by their transfer to MS medium containing GA_3 (1.5 mg/l). The success achieved with maltose and its superiority over sucrose is probably due to difference in its breakdown products. Since maltose is broken down more slowly than sucrose, it can provide a readily metabolizable carbon source (glucose) to the embryos, which probably lack any reserved carbohydrate. This further reflects on the inefficiency of the developing embryos in accumulating reserves during the embryo maturation phase. Cinamic acid is an important compound on a wide variety of metabolic pathways such as flavonoid biosynthesis, the phenolic synthesis and most important in the synthesis of malonyl-Co-A, the key precursor of fatty acid synthesis pathway. However, the inability of either maltose or t-CA alone for germination indicated that the requirement of both sugars and fatty acid which was essential during embryo maturation and subsequent germination of tea somatic embryo.

4.6 *In Vivo* Embryogenesis

Hitherto, emphasis has been given to manipulate the nutrient composition, growth regulators in culture medium, physical conditions of incubation, and other stress treatments to induce somatic embryos. However, induction of *in vivo* embryogenesis of tea could be achieved without using the conventional tissue culture media. In an attempt to germinate tea seeds under sterile conditions inside steel boxes containing moist sand for the use of protoplast isolation, (Mondal et al. 2001c) observed the induction of embryogenesis on cotyledon surface of mature tea seeds under *in vivo* conditions at 28 °C. The induction was however, influenced by genotype, time of seed collection and temperature of incubation. Selected regime of temperature (24, 28 and 32 °C) for incubating the seeds was examined so as to determine the effect on embryogenesis *in vivo* among the three cultivars of tea namely Kangra Jat, UPASI-9 and T-78. At 32 °C, no response of embryogenesis was evident in any of the three genotypes irrespective of the time of seed collection, except a very low germination frequency (1.6 %), which occurred in Kangra Jat seeds. At 24 °C also, small percentage (4.2 %) of Kangra Jat seeds germinated but there was no development of somatic embryos. However, when the seeds were incubated at 28 °C, both germination and adventive embryos were observed. The percentage of germination was lower in comparison to that of embryo formation. Incidentally, a considerable amount of seeds were remained unresponsive irrespective of the treatments. Further, the formation of somatic embryos per seed also depends upon genotype. While the maximum number of somatic embryos (15–16 per seed) were observed after 90 days of incubation at 28 °C in UPASI-9, that of Kangra Jat and T-78 showed 10–12 and 6–8 such embryos per seed respectively (data not shown). The maximum number of seeds showing formation of embryogenic response was observed in UPASI-9 compared to two other cultivars. The response in UPASI-9 seeds collected in September was 23.3 % while that of T-78 collected in the month of October was 16.5 % and Kangra Jat collected in the month of November was 11.7 %. After another 42 days, the embryos underwent to a typical embryogenic pathway as evident by cup-shaped structures. One important observation was that the seeds, which follow normal germination, did not show embryogenesis. Histological evidence confirmed somatic embryogenesis. However, a considerable amount of seeds remained unresponsive irrespective of the treatments. Therefore, we concluded that induction of embryogenic response was not only depended upon the external supply of nutrients and growth regulators, but also on temperature. At a particular temperature, mature tea seeds are capable of expressing their somatic embryogenic potential even under *in vivo* conditions. Bhatia et al. (1986) observed *in vivo* response of callusing and adventitious shoot formation from de-embryonated *Arachis hypogea* cotyledons inoculated on moist sand and cotton wool in enamel trays. The greater influence of temperature on embryogenesis, forced us to propose a hypothesis that, in *Camellia* the induction of *in vivo* embryogenesis must be regulated by temperature-dependent gene. Though reason for this observation is not clear presently, yet seeds of *Camellia* appear to have a considerable inherent capacity for embryogenesis. Thus, at the right physiological stage, with appropriate levels of internal hormone and appropriate moisture profile and temperature of the substrate under sterile conditions, the tea seeds are able to produce embryos without any exogenous nutrient.

4.7 Hardening and Field Transfer

Information on hardening of tea somatic embryo derived plants is scant in the literature. One reason perhaps, establishment of somatic-embryo derived plantlets is easy compared to micropropagated plants due to the presence of tap root system in the former. Generally, sand or soil with some additives such as peat or vermiculite in presence of high humid condition are sufficient for initial establishment. Although Wu et al. (1981) were the first to transfer the plantlets to soil, the composition of soil mixture and other conditions used by them, was not mentioned clearly. Kato (1989a) grew healthy tea somatic seedlings under natural conditions by transferring them into a mixture of vermiculite and soil (1:1) and they were covered with the plastic. On the other hand, Jha et al. (1992) hardened tea plantlets for 8 weeks in quarter-strength MS salts before transplanting to pots containing a mixture of peat and soil (1:1) and achieved an acclimatization rate of 70 %. Wachira and Ogada (1995) reported that multiple shoots differentiated from the germinated embryos were successfully rooted in autoclaved mixture of sand:peat (3:1) in small pots. Shoots were watered, kept under polytunnel to prevent from excessive moisture loss as well as to maintain a high relative humidity. After 8 weeks, the small plants had developed root system and were transferred into 10 cm long layflat polythene sleeves containing a mixture of soil and sand in a ratio of 3:1. Subsequently, plants were shifted to a shaded nursery bed. In order to improve rooting, Ponsamuel et al. (1996) treated the plantlets with indole-3-acetonitrile (0.34 g/l), brassin (0.48 g/l) and phloroglucinol (1.2 g/l) in liquid MS medium for 15 days and after profuse root proliferation, the plantlets were acclimatized in pots containing vermiculite. Eventually the plants were transplanted in the greenhouse. Akula and Akula (1999) transferred the small plantlets with a strong tap root and 4–6 leaves into small pots filled with pre-sterilized potting mixture (sand:peat:vermiculite::1:2:1) and kept into a greenhouse with misting facility, at 80–95 % humidity under low light. The new leaves were observed within 5–6 weeks when they were taken into bigger pots. Following this procedure, they achieved 90–95 % survival rate and more than 200 plantlets were transferred to the field at Indonesia. Tea plantlets with young leaves and stout roots, with a height of 4–5 cm, were taken from culture room to Hikko trays containing pre-sterilized sand and cowdung (1:1). These Hikko trays were then kept in poly tunnel with intermittent watering for 60 days inside indigenously developed poly house (90 % survival rate). Later, they were transferred to polythene sleeves filled with black virgin soil and kept for further 1 year in the same poly-house (Mondal et al. 2002). Following these techniques, 3,000 somatic seedlings of tea were produced at Research and Development Department of Tata Tea Ltd, Kerala, India, which had been transferred to the field (Mondal et al. 2004).

4.8 Somaclonal and Gametoclonal Variation

Despite the fact that stable somaclones may be desirable for crop improvement especially in perennials, no effort has been made to generate the useful somaclones in tea. The reasons perhaps tea being cross-pollinated, variations already exist among natural populations and worldwide most of the breeders were engaged for identifying the superior variants. Secondly, in tea, the number of tissue culture derived plants, which had been transferred to the field, might not represent a sizable amount to study the somaclonal or gametoclonal variants. Rajkumar et al. (2001) undertook a detailed study about somaclonal variation of tea. The plants were derived from somatic embryos, which were induced on cotyledonary tissues of clone UPASI-10. They identified five somaclonal variants in the field on the basis of morphological, physiological and biochemical parameters. Interestingly, based on the 'membrane stability index' study, they concluded that the somaclone variants would be tolerant to low temperature stress and thus suitable to grow at high altitude. There were few preliminary attempts earlier the work of Rajkumar et al. (2001) to identify the somaclonal variants among the micropropagated tea. A wide range of chromosomal variations was observed in callus derived from *in vitro* leaf and cotyledon explants of tea (Das 1992). The study revealed that more than 60 % cells were diploids in 8-week-old calli with remaining cells being of triploids, tetraploids or aneuploids with more than 60 chromosomes. However, Jha et al. (1992) reported that embryo-derived plants had chromosome number of $2n=30$, suggesting an apparent genetic uniformity. Variation in plants regenerated from gametophytic tissue had been reported in some cases due to uncovering 'residual heterozygosity' (Evans and Sharp 1986). Variation in the chromosome number of gametes or gametophytic tissue plays an important role in gametoclonal variation. This is evident from a range of aneuploids and mixoploids recovered from anther cultures of wheat, maize and sexual hybrid of wheat and triticale (Hu 1983). In tea, Chen and Liao (1983) examined the ploidy level of anther-derived plantlets. They found that 9 out of 12 such plantlets were aneuploids and most of them had chromosome number of $2n=18$–22. Further, morphological variations were not mentioned.

A study has been undertaken to characterize 15 field grown somaclonal variants derived from cotyledonary tissues of UPASI-10 using morphological, physiological and biochemical characters. Although variants were derived from UPASI-10, a very few variants possessed unique 'Chinary' characters while others exhibited 'Assam' characters. However, no variant showed identical morphological characters aligning with the parent. Somaclonal variants showed distinct variation in terms of photosynthetic carbon assimilation, stomatal conductance and diffusion resistance. Proline accumulation and water use efficiency showed marginal variations among the variants. Two variants namely SE 8 and SE 10 recorded higher values of 'membrane stability index' denoting their tolerant nature against stress. Variants SE 2 and SE 13 were segregated distinctly representing their black tea characters. (Thomas et al. 2006).

4.9 Origin and Morphology of Somatic Embryos

The somatic embryo originates from either single cell or group of cells, which depends upon the plant species. The histological and anatomical aspects of somatic embryo origin in *Camellia* had been studied (Plata and Vieitez 1990; Vieitez et al. 1991). Barciela and Vieitz (1993) made a detailed study on the origin and anatomical development of *C. japonica* somatic embryos differentiated on cotyledon sections. They studied through computer-aided image analysis for cytological quantification, and the measurements of cell starch and protein contents as the stained cell areas by staining with periodic acid-Schiff (PAS) as well as by mercuric bromophenol blue, respectively. It had been observed that small protuberances or nodules began to appear on the abaxial epidermis of the cotyledons from 7-day-old *in vitro* tissue. The nodules continued to develop and by 30th day, they were 4–6 mm in diameter and became moderately prominent. After 2 months of culture, embryos were 6–8 mm long and became ready to isolate either for germination or secondary embryogenesis. It was found that only the abaxial surface of the cotyledon explants was morphologically competent and had multicellular origin. To determine whether the embryogenic nodules could be maintained indefinitely in culture, they were isolated from the initial cotyledons, removed from *in vitro* produced somatic embryos, and then cultured for 6 months with monthly transfer to a fresh medium. The parenchymatic tissue of the nodule failed to proliferate or grow and turned progressively necrotic. The above histological analysis suggested that the nodules associated with the occurrence of somatic embryogenesis can be considered as a small localized callus tissue, which was necessary for the redetermination of embryogenic cells. Histological observations of embryogenesis in *C. reticulata* (Plata and Vieitez 1990) suggested that somatic embryos developed directly from cotyledon without any apparent callus phase. However, the differentiation of the embryos was nevertheless related to the developing swollen parts, swollen whitish areas, or compact bulging tissue of cotyledon explant (Plata and Vieitez 1990). Such swellings might be equivalent to the nodules observed in *C. japonica* (Vieitez and Barciela 1990; Barciela and Vieitz 1993).

The morphology of the *Camellia* somatic embryos was influenced by the concentration of cytokinin in the medium. In case of *C. japonica* (Vieitez and Barciela 1990; Vieitez et al. 1991), most embryos could be classified into the following two clearly distinct types, (1) seed-like embryos, which were yellowish-white with large cotyledons alike to mature zygotic *Camellia* embryos, and (2) bud-like embryos, which were green with cotyledons resembling true leaves. They generally developed in media with relatively high BAP concentration. Ammirato (1985) stated that reasonably high levels of cytokinins partially or totally inhibited the development of somatic embryo cotyledons and the shoot apex grows out to form the first mature leaves so that the somatic embryo looked more like a shoot. The observed bud-like embryos may be an example of such cytokinin-induced premature shoot emergence. Anomalies such as polycotyledon, hypertrophy, or fascination were also observed to various extents among both seed-like and bud-like embryos, but both kinds were genuinely bipolar having both shoot as well as root meristems.

4.10 Biochemical Changes of Somatic Embryogenesis

Low germination and poor conversion into normal plantlets were the major limitations to somatic embryogenesis in tea (Mondal et al. 2002). The main cause for poor germination of embryos was their failure to complete normal stages of embryogeny that was generally common to zygotic embryos (Sharma et al. 2004). Embryos in angiosperm develop through several changes namely histo-differentiation and endosperm formation, followed by embryo maturation and finally enter into the desiccation stage. The maturation on the other hand starts with cessations of cell division, increasing of cell size due to reserve accumulation. This is followed by gradual depletion of metabolism and finally terminates by drying as the embryos enter into the desiccation stage. This then becomes metabolically inactive and tolerant to desiccation. However, the ability to tolerate desiccation is gradually acquired during the maturation phase itself (Vieitez 1994). It had been found several biochemical changes took place during the development of somatic embryos which are discussed below.

Changes of ABA, starch, total soluble sugars (TSS), proteins, and phenols were studied in the somatic embryos at different stages of development (globular, heart, torpedo and germinating embryos) in order to investigate the accumulation of storage reserves to prevent the precocious germination. It had been found that after ABA treatment with 5.0 mg/l for 14 days, the starch and protein contents of the embryo were increased by several folds with a simultaneous increase in TSS. ABA treatment at the heart stage improved the germination. ABA treatment prior to or after heart stage did not improve somatic embryo germination (Sharma et al. 2004). In the normal seed formation of tea, ABA content was found to be maximum at maturation than in any other stages (Nagar and Sood; 2006; Preeti et al. 2004). Therefore, it was concluded that external application of ABA at a specific dose was capable of circumventing the developmental blocks of improper reserve accumulation to a certain extent, provided that ABA was supplemented only at the early maturation stage. Further, Bhattacharya et al. (2002) showed that ABA accumulation was maximum at heart stage of tea somatic embryo. This analysis suggested that ABA application earlier or later than heart stage was ineffective. Free ABA was converted into the inactive or bound form during late maturation in zygotic embryos of tea (Bhattacharya et al. 2002).

Mondal et al. (2002) observed adverse effects of ABA application at the both globular and torpedo stages of somatic embryo development. Thus, while no, or negligible, germination was observed in the globular and torpedo stages of somatic embryos, more than 50 % of the heart-stage somatic embryos developed normally, germinated, and finally converted into healthy plantlets after ABA treatment. Therefore, they concluded that while tea somatic embryos failed to accumulate storage reserves during early stage of embryo development, ABA at later stages might have helped in triggering the process of reserve accumulation, thereby leading to high germination. This finding was also in accordance with studies of other researchers examining different plant species (Tian and Brown 2000).

The role of endogenous IAA, soluble proteins, and RNA in the development of tea seeds was also investigated. Unlike other orthodox seeds, the level of free IAA in tea embryos also remained high even at full maturity. The total RNA content remained high in the stages with high moisture content but declined with progressive decline in moisture content (Bhattacharya et al. 2004). Various polyamines (PA) namely putrescine (Put), spermidine (Spd), and spermine (Spm), had been played important role in embryogenic processes. Studies on PA pattern and biosynthesis had shown that somatic embryo formation, development, and conversion were affected by free PA levels in *C. japonica* (Pedroso and Pais 1993). Further, it had been suggested that the levels of endogenous PAs present in the embryogenic and non-embryogenic regions were not responsible to determine the response of embryogenesis in *C.*

japonica (Pedroso and Pais 1993). It had been found that PA pattern were identical for both embryogenic and non-embryogenic leaf regions until the formation of proembryo (day 20) of the *in vitro* leaf, differing from each other in soluble and insoluble conjugated Put and soluble conjugated Spd contents upon formation of globular embryos (day 45). The absence of significant differences in the ratios of free Put/free PA between both leaves was due to the fact that the non-embryogenic leaf regions were also morphogenically competent (rhizogenic). Endogenous PA-proteins binding was highest when mitotic activity was greatest, suggesting a positive relation between PA binding proteins and mitotic activity. In fact, conjugated PA in the embryogenic leaf regions kept on increasing during the entire period of culture, while in the non-embryogenic leaf regions, conjugated PA did not change significantly after 20 days of culture (Pedroso and Pais 1994). The slightly higher free Put/free PA ration recorded for embryogenic leaf regions in *Camellia* at 20th and 45th day, with compared with the non-embryogenic ones, might be associated with the embryonic development occurring in these leaf regions. Apart from this, starch also played critical role in embryo development in *Camellia* (Pedroso and Pais 1994). Primary embryo cortical cells underlying globular secondary embryos contained abundant starch, but these deposits declined as the secondary embryo developed. The starch deposits of the secondary embryos itself was also declined as it developed, and were found only around the apical meristems in the cotyledonary stage.

A strongly starch-positive zone was always detected during the formation of the nodules (7–9 days) and the PAS-staining proportion of these areas was about four times that of the equivalent region in the initial cotyledon. Starch deposits declined in the nodule parenchyma adjacent to preembrygenic cells, reaching a minimum upon the development of somatic embryos, when the parenchyma had few starch grains and thinner cell walls than the initial cotyledon tissue. On the other hand, in nodules producing no embryos, very few starch grains were observed (Pedroso and Pais 1994).

4.11 Histological and Ultrastructural Changes During Embryogenesis

4.11.1 Direct Somatic Embryogenesis

Histological study of direct somatic embryogenesis from leaf tissue enabled the establishment of a time sequence of physiological and ultrastructural changes occurring during the induction of embryogenesis in the *C. japonica* (Pedroso-Ubach 1994). It had been noticed that Ca^{++} concentration played major role during embryogenesis as its concentration varied during embryogeneis (Pedroso and Pais 1994). Cell wall became thick in embryogenic cell followed by accumulation of starch and calcium oxalate crystals in the surrounding parenchyma cells, up to 15 days after induction (until initial divisions for proembryo formation). However, both starch and calcium oxalate crystals disappeared after the onset of somatic embryogenesis. The changes in Ca^{++} level recorded during induction of somatic embryogenesis in *C. japonica* which could be related to changes in the levels of linked and precipitated Ca^{++}. An increase in free calcium and calmodulin after induction of direct embryogenesis was observed which became maximum at globular stages but decreased thereafter (Roberts et al. 1992). Histochemical tests, including enzymatic digestion with PGS-lipase (a cutinase), showed the presence of cutin and callous on the embryogenic parenchyma cells of cell walls after 3 to 15 days (Pedroso-Ubach 1994).

4.11.2 Secondary Embryogenesis

Barring the increased sizes, primary embryos did not undergo any changes till 12th day, when the newly formed secondary embryos became visible. Secondary embryos occurred mainly in cluster on the hypocotyl of the mother (primary) embryos, without any intermediate callus formation. After that, secondary embryogenesis passed through typical globular, heart and torpedo stages before reaching the cotyledonary stage. After 6–7 weeks of culture, 70 % of the primary embryos produced secondary embryos at an average rate of 10 per productive embryo. Secondary embryos over 7 mm long was used for a further embryogenic cycle or transferred to germination medium to develop into plantlets. Smaller secondary embryos had much less embryogenic capacity, and tend to suffer necrosis in both embryogenic and germination media. Histological monitoring revealed that secondary embryos apparently had a multicellular origin from embryogenic areas originating in both epidermal and subepidermal layers of the hypocotyl region. Microcomputer image analysis was applied for quantifying cytological events associated with somatic embryogenesis. This method showed an increasing gradient in the nucleus-to-cell area ratio from differentiated cells passing through pre-embryogenic cells to embryogenic cells. The formation of embryogenic areas was preceded by accumulation of starch in the surrounding cortical cells. The cells underlying globular secondary embryos still contained abundant starch, but it declined as the secondary embryos developed (Plata et al. 1991).

4.12 Electron Probe X-ray Microanalysis: A Tool for Early Diagnosis of Embryogenesis

A protocol to study the early embryogenic competence of *C. japonica* leaf was developed by electron probe X-ray microanalysis. *In vitro* derived leaves isolated from 12-week-old shoots were immersed in IBA (1 g/l) for 20 min followed by incubation of 11 days in darkness on diluted MS basal medium (Pedroso and Pais 1993). These leaves upon culturing produced seven different regions with varying response such as embryogenesis, roots, callus, and non-morphogenic regions. The results suggested the existence of difference in morphogenic competence in different leaf regions. Plantlets regeneration was successfully achieved from all the culture systems developed (Pedroso and Pais 1993; Pedroso and Pais 1995a; Pedroso and Pais 1995b). Later Pedroso and Pais (1994) demonstrated that significant fluctuations of Ca, C, O, K, Na, P, Fe, S and Mg levels were detected in the cells from embryogenic and non-emberyogenic region of the same leaf which were analyzed with X-ray spectra. Thus, a positive correlation was established between the elemental levels in the tissue region and *in vitro* responses such as embryogenesis, rhizogenesis, calluses were obtained. Therefore, in *C. japonica*, EPMA (electron probe X-ray microanalysis) can be used for early detection of embryogenic competence or its absence.

4.13 Alternative *In Vitro* Techniques

Apart from micropropagation and somatic embryogenesis, several others *in vitro* approaches were standardized which have been summarized in Table 4.2 and discussed here.

4.13.1 Storage of *In Vitro* Culture

The potential of using *in vitro* systems for germplasm collection and conservation as well as for multiplication had been broadly discussed in several reviews (Kartha 1985; Engelmann 1997). A summary of different *in vitro* works on this aspect is tabulated in Table 4.3. The application of *in vitro* techniques to germplasm storage is of particular interest for the conservation of plants such as *Camellia* species that are normally propagated vegetatively and/or have recalcitrant seeds. The storage of *Camellia* seeds in gene banks was problematic because *Camellia* seeds were classified as recalcitrant (Pence 1995). They were sensitive to low temperatures as well as desiccation and were unable to retain their viabilities through long-term storage (Kato 1989a). Even when they were maintained under moist conditions at 3–5 °C, their viabilities were relatively short-lived (Salinero and Silva-Pando 1986). The most common method for preserving the genetic resources of species with recalcitrant seeds or those vegetatively propagated, is as plants under field gene banks which is popularly known as *ex situ* conservation. Limited works have been carried out on cold storage and cryopreservation of *Camellia* with both material obtained *ex vitro* (seeds) and material cultured *in vitro* (somatic embryos, embryonic axes, and shoot apices).

Table 4.2 A brief description of some cell culture techniques used in tea

Name of the technique	Objectives	Remarks	Reference
Somaclonal and gametoclonal variation	Development of mutant resistant to different stresses	No commercial success	Rajkumar et al. (2001)
Artificial seed	Storage of propagule	Maximum 60 days storage was possible without loss of germination	Janeiro et al. (1995); Mondal et al. (2000)
Protoplast culture	Hybrid plant production	Regeneration was not possible	Balasubramanian et al. (2000b)
Anther culture	Haploid plant production	Regeneration was not achieved	Saha and Bhattacharya (1992); Raina and Iyer (1992)
Suspension culture	Secondary metabolite production	Commercially exploited	Orihara and Furuya (1990); Matsuura et al. (1991)
Cryopreservation	Long-term storage of propagule	Not exploited further either academic or commercial purposes	Kuranuki and Yoshida (1991)

Table 4.3 Summary of studies on the cold storage and cryopreservation of embryogenic and embryonic axis cultures of *Camellia*

Species	Explant	Storage method	Response/remarks	Reference
C. japonica	Somatic embryo clusters	Short- to medium-term storage at 2–4 °C	Reduced embryogenic competence after 6 months improved germination capacity after 2 months	Janeiro et al. (1995)
	Encapsulated somatic embryos	Short- to medium-term storage at 2–4 °C	Reduced embryogenic competence after 2 months. 30–40 % plant recovery after 2 months	Janeiro et al. (1997)
	Somatic embryos and encapsulated somatic embryos	Cryopreservation	No survival of frozen material	Janeiro et al. (1996)
	Embryonic axes from mature seeds	Cryopreservation after 2–3 h desiccation	100 % survival, 40 % plant recovery, and 18 % somatic embryogenesis rate	Janeiro (1996)
C. reticulata	Somatic embryo clusters	Cold storage at 2–4 °C	76 % germination after 2 months	Chaudhury et al. (1991); Chandel et al. (1995)

4.13.2 Low Temperature and Short-Term Storage

The potential use of low temperature storage is numerous for *in vitro* conservation of germplasm, their exchange, hand-pollinated hybrids with reduced seed fertility and genetically modified plants with sterile or unstable genotypes. However, efficacy of storage depends upon efficiency with which synthetic seeds can be stored at low temperature without reducing multiplication rate or germination efficiency. In tea, the first successful plant regeneration from syn-seed was established by Mondal et al. (2000). They investigated the effect of low temperature/cold storage (4 °C) on germination efficiency of encapsulated somatic embryos up to 75 days derived from cotyledons of cultivar, Kangra Jat. Pale yellow to white coloured somatic embryos (0.2–0.5 mm diameter) developed directly on the surface of cotyledons were taken for syn-seed production. Among the different alginate concentration, 3 and 4 % alginate were found to be equally good for formation of clear, transparent, round-shaped beads whereas increase or decrease in the alginate concentration resulted in abnormal beads. A maximum germination efficiency of 59 % was achieved with naked somatic embryos kept in germination media in culture room at 25±2 °C. However, germination rate of nonencapsulated somatic embryos reduced more rapidly from 59 to 9.7 % with increasing periods of storage time. Upon transfer to MS medium supplemented with GA_3 (1.5 mg/l), the beads cracked and bud sprouted within 8 weeks. Similar observation was made earlier in *C. japonica* somatic embryos (Janeiro et al. 1995). It was postulated that this perhaps due to naked embryo just excised from embryo clusters which was unable to recover 'excision shock' while detaching from mother tissue coupled with less capability to resist drying than the embryo clusters. On the other hand, germination efficiency of unstored encapsulated tea somatic embryos was 34.3 %, which was reduced to 7.7 % up to 60 days of storage. No germination was registered in either nonencapsulated or encapsulated somatic embryo stored for 75 days. The lower germination rate of encapsulated somatic embryos may be related to both oxygen deficiency and rapid drying in the gel bead (Redenbaugh et al. 1991). The reduction of germination efficiency recovered in alfalfa (Redenbough et al. 1987), *Asparagus cooperi*, (Ghosh and Sen 1994) and *Eucalyptus citrodora* (Muralidharan and Mashcarenhas 1995) from cold storage was documented earlier. However, considerable amount of secondary somatic embryogenic response was observed irrespective of the duration of cold treatment. Therefore, it was concluded that tea somatic embryos were susceptible to cold. The germinated somatic embryos were grown into plants, successfully hardened, and transferred to the field.

It is well-documented that cold-treated somatic embryos of certain woody perennials were known to promote both normal maturation and germination capacity (Tulecke 1987). While 2 months of cold-treated tea somatic embryos at 4 °C improved germination, the plantlet conversion ability of secondary embryos increased significantly (up to 100 %) as compared to control embryogenic lines, which accounted for 32–65 % germination rate. The incidence of secondary somatic embryogenesis during germination also decreased after the cold treatment. While cold storage reduced the competence for secondary somatic embryogenesis in *Camellia*, it increased their capacity for conversion to plantlets (Mondal et al. 2000). In order to evaluate the effect of cold storage on bud sprouting response, encapsulated nodal explants were stored under both at 25±2 °C (tissue culture room condition) as well as at 4 °C for varying sets of 15 days interval up to 60 days. Storage of artificial seeds at 4 °C did not show any adverse effects on shoot proliferation efficiency, rather the time taken for initial sprouting was reduced by 15 days as compared to the ones that were stored at 25 °C. The bud sprouting efficiency (90.2–92 %) of both nonencapsulated and encapsulated nodal explants was maintained up to 45 days but declined to 45.9 and 50 %, respectively, when stored up to 60 days (Mondal et al. 2000). Therefore, it was concluded that while somatic embryos were unsuitable for low temperature storage, nodal explants were not only suitable, but also sprouted buds with cold treatment earlier than untreated one in tea.

In an another study, it had been found that syn-seeds of tea stored up to 30 days under room temperature germinated but their viability declined significantly thereafter. Half strength MS medium was found to be better over full strength MS medium on germination while during growth and development of the germination, the trend was reversed. Among the PGRs, BAP influenced germination rate. Physical state of the nutrient medium showed a significant variation between liquid and solid state. Germination rate of syn-seeds was significantly higher in acidic pH while near neutral pH enhanced the growth rate of germinated synthetic seeds (Mariya John and Raj-Kumar 2006).

The effects of short- to medium-term cold storage on the maintenance of embryogenic capacity and germination of somatic embryos of *Camellia* were investigated (Janeiro et al. 1995). Four embryogenic lines were used: three belonging to *C. japonica* (1, 2 and SY-89) and one to *C. reticulata* cv. Mouchang. Lines 1 and 2 of *C. japonica*, which were used to study survival and the preservation of embryogenic capacity of somatic embryos induced directly on the roots of *in vitro* grown plantlets (Vieitez et al. 1991). The effect of cold storage on the germination of *Camellia* somatic embryos into whole plants was studied in greater detail (Janeiro et al. 1995). Cold treatment for 8 weeks significantly improved the secondary embryogenesis, but improvement was varied on genotypes. Depending on the genotypes, the shoot and root length of the germinated plantlets were also significantly increased by 2 months cold treatment. However, the incidence of secondary embryogenesis during germination also decreased after cold treatment.

Janeiro et al. (1995) found that 2 months storage somatic embryos clusters at 4 °C significantly enhanced germination and plant production whereas, germination of individually cold-stored naked embryos was reduced (Janeiro et al. 1997). This apparent discrepancy might be due to individual embryos responded relatively poorly because having been excised from embryogenic clusters just before cold storage, they had not yet recovered from the stress of excision. Alternatively, individual embryos may be more susceptible to damage by drying than embryo clusters.

The potential uses for artificial seeds are numerous including storage, handling, and delivery of elite germplasm. The possibility of using cold storage to preserve synthetic *Camellia* seeds was investigated (Janeiro et al. 1996). In that study, the effects of cold storage of *C. japonica* somatic embryos on the maintenance of embryogenic competence and germination of encapsulated embryos were determined. Somatic embryos were encased in sodium alginate (3 %) beads made in MS basal medium with 3 % sucrose. The beads were then stored for 1–2 months in darkness at 2–4 °C. After 1 month, the encapsulated embryos exhibited a significant reduction in both survival rate and competence for secondary embryogenesis, however, further, 60 days cold storage had little reduction effect. The survival and secondary embryogenesis rates were 68 and 69 % respectively, when placed in the maintenance medium following 60 days storage at 4 °C which were still acceptable. However, the productivity (number of secondary embryos per responsive encapsulated embryo) was dramatically reduced from 62.6 % for unstored encapsulated embryos (control) to 5.4 % secondary embryos indicating the negative influence of cold on secondary embryo formation. The reduced competence for secondary embryogenesis of cold-stored encapsulated embryos of *Camellia* appears to reflect increased maturity, since their capacity for germination is better preserved than their embryogenic competence. In this respect, short- or medium-term cold storage of syn-seeds destined for germination appears to be feasible as long as a 30–50 % fall in plant recovery rate could be tolerable. In contrast, cold storage alone could not be used to maintain embryogenic competence, since the productivity of cold-stored encapsulated embryos was seriously reduced.

Ballester et al. (1997) reported almost 100 % survival frequencies in seven of the eight clonal shoot cultures of *C. japonica* tested, when stored at 2–4 °C for up to 12 months. Shoot tips of *C. japonica* encapsulated in alginate beads and stored at 2–4 °C survived for a shorter period of time than uncapsulated ones. Encapsulated material had survival rates of 75, 50 and 10 % after 30, 60 and 75 days, storage at 4 °C respectively.

4.13.3 Cryopreservation

It has been suggested that tissues may be frozen and stored in liquid nitrogen at − 196 °C for long-term storage of germplasm. This would be of great value in the conservation of woody plants for which vegetative propagation is prime important such as tea. The preservation of embryonic axes in liquid nitrogen was attempted in *C. sinensis* (Chaudhury et al. 1990; Chandel et al. 1995). While the intact seeds could not be cryopreserved due to their large size as well as high moisture content, the excised embryonic axes could be preserved successfully by desiccating them up to 13 % moisture (Chaudhury et al. 1991). The cryopreserved embryonic axes could be revived and germinated by placing them simply on a moisten filter paper for 20–25 days followed by culturing on nutrient medium leading to healthy seedlings of 5–6 cm height. Kuranuki and Yashida (1991) reported that the excised embryonic axes were not only able to tolerate the frost but also desiccation.

The feasibility of using cryopreservation techniques to maintain the embryiogenic competence was investigated (Janeiro 1996; Janeiro et al. 1996). In these studies, somatic embryos of *C. japonica* (2–5 mm) were subjected to several protective pretreatments to prevent the formation of ice crystals inside the cells. Following pretreatment, half of the somatic embryos in each experiment (controls) were placed directly in maintenance medium (MS medium supplemented with 1 g/l BAP and 0.1 g/l IBA) and the other half was placed in sterile 2 ml polypropylene cryovials and immersed in liquid nitrogen for 24 h. After 10 weeks, they found that no frozen somatic embryos survived regardless of the desiccation period; however, survival rate of unfrozen embryos was 100 % after 15 min of desiccation treatment, which was further reduced to 53 % after 2 h of desiccation indicating an acceptable tolerance of these somatic embryos to dehydration (Janeiro 1996). Janeiro et al. (1996) also investigated the feasibility of cryopreservation of *C. japonica* embryonic axes. The explants were isolated from mature seeds. After sterilization of seeds, the embryonic axes were excised from the cotyledons with 1–2 mm of petiole to protect the plumule, and were either used as such or dehydrated for 1.5 or 3 h in sterile laminar air flow. Half the material was placed directly in MS maintenance medium (controls), and the other half was placed in cryovials and frozen in liquid nitrogen for 24 h before transfer to the same MS medium. They found that the capacity of *Camellia* embryogenic axes to produce somatic embryos, especially on the hypocotyl region, was maintained and even enhanced after the stress produced by cryoexposure.

4.14 Organogenesis

Adventitious shoot regeneration via callus phase from *in vitro* leaf explants of tea has been standardized. Callus as well as rhizogenesis were obtained on MS medium supplemented with varied concentrations of 2,4-D (2.5, 5.0, 7.5 and 10.0 mg/l). Adventitious shoot buds developed indirectly on leaf explants after prolonged culture for 16 weeks on MS medium supplemented with 2,4-D (10 mg/l). GC analysis of the medium as well as the tissues at different stages of development concluded that critical level of 2,4-D in the tissue was responsible for morphogenesis. Shoot buds developed on rhizogenic calli, only when 2,4-D declined to undetectable or negligible concentrations in the tissue probably due to detoxification and metabolism. Alternatively, shoot buds could also be evoked when rhizogenic calli were transferred to medium supplemented with low concentration of 2,4-D (1.5 mg/l). The adventitious nature of the shoots was confirmed through histological studies (Sandal et al. 2005).

4.15 Caulogenesis

In order to produce Callus, *in vitro* derived leaves of Assam tea were cultured on modified MS medium containing 2,4-D (0.1–1.0 mg/l) and BAP (0.1–1.0 mg/l) or Kn (0.1–1.0 mg/l). Cultures were incubated for 6 week at 30 ± 2 °C and 16 h photoperiod at light intensity of 2000 Lux. It was found that the cultured leaves produced compact callus. The amount of callus produced varied from 45 to 65 %, depending on the concentrations of 2,4-D, BAP and Kn in the medium. The highest percentage (65 %) of explants producing callus was obtained on modified MS medium supplemented with 2,4-D (1 mg/l) in combination with Kn (0.1 mg/l) within 6 weeks of treatment (Tripetch et al. 2009).

The effect of ABA and excess BAP on the formation of callus was investigated. Callus was formed and grew well when explants were cultured on MS basal medium supplemented with thiamine-HCl (1.25 mg/l); pyridoxine-HCl (0.625 mg/l); nicotinic acid (0.625 mg/l); IAA (30 mg/l); NAA (30 mg/l); Kn (0.1 mg/l); myoinositol, (100 mg/l); as well as 3 % (w/v) sucrose. After 2 months, the calli were transferred to a modified B_5 medium in order to regenerate plant. As a result, rhizogenesis was occurred in the transferred calli to B_5 medium after one month. Subsequently, the calli were transferred to the aforesaid B_5 medium supplemented with ABA (2 mg/l) and excess BAP (400 mg/l) to form shoot. The calli turned green and showed differentiation of globular and heart shaped embryos when transferred to the modified B_5 medium, without formation of shoot. These findings showed that the applied concentration of ABA may cause inhibition of conversion of globular and heart-shaped embryos to shoot. The increased level of BAP, however, was not able to ameliorate the effect of ABA (Ghanati and Ishka 2009).

4.16 Protoplast Culture

Trans-generic gene transfer through protoplast culture and production of somatic hybrids or cybrids is of prime importance to tea. However, the success of achieving a good yield of protoplasts depends upon the source material, which is mostly a 'cell suspension culture'. Although, Bagratishvili et al. (1979) was first to establish tea suspension cultures, Nakamura (1983) successfully isolated the protoplasts with a yield of 3.1×10^5/g of fresh weight from the suspension culture of tea leaves and flower petals. This was further improved by Kubai et al. (1991) who attained a yield of 87% viable protoplasts with a yield 3×10^7/g of fresh weight from young leaves of cultivar 'Yabukita'. The problems of endogenous bacterial contamination and polyphenol oxidation during protoplast preparation from *in vitro* leaves were solved with the addition of PVP in the isolating media (Purakayastha and Das 1994). Additionally, Purakayastha and Das (1994) reported that macerozyme including cellulase along with other factors such as enzyme concentration, osmotica and their concentrations were important for increasing yield. However, further development of micro-calli could not be achieved. Effect of different synthetic auxins were tried to induce the callus from tea stem tissue and found that Picloram (241 mg/l) alone induced most callus formation (Frisch and Camper 1987). Balasubramanian et al. (2000b) used different explants such as leaves, cotyledons, embryogenic and morphogenic calli for culture and isolation of protoplasts, and reported *in vitro* leaves to be the best explants for a good yield of viable protoplasts provided fructose was used as an osmoticum. Despite the fact that success has always eluded protoplast culture and somatic hybridization, these techniques have tremendous potential in the improvement of tea. Many of the wild relatives of tea have agronomically important biotic and abiotic stress tolerant trait, which can be incorporated into the cultivated variety of tea through somatic hybridization or cybridization. Somatic hybridization also has potential for the production of caffeine-free tea through the fusion of protoplasts of caffeine-free but aroma rich *C. luetacense* or *C. irrawadiensis* with that of cultivated tea. Thus protoplast culture has tremendous potential for varietal improvement of tea (Sriyadi 1987).

4.17 Anther Culture

Microspore culture presents a number of potential advantages mainly in relation to *in vitro* selection strategies and to genetic studies for developing doubled-haploid mapping population, etc. As tea is highly heterozygous and heterogeneous, its production of homozygous diploid plants is of great importance in tea improvement. The pioneer attempts of anther culture of tea were made by Katsuo (1969) and by Okano and Fuchinone (1970) who could produce roots from anther derived callus. However, the first successful plants from tea anthers of cultivar Fuyun No-7 were produced by Chen and Liao (1982). Though they worked on nine different tea cultivars, plants were produced only from 'Fuyun-7' on N6 medium supplemented with Kn (2 mg/l), 2,4-D (0.5 mg/l), L-glutamine (800 mg/l) and serine (100 mg/l), followed by subculturing on N6 medium supplemented with zeatin (2 mg/l), adenine (20 mg/l) and lacto-albumin hydrolysate (10 mg/l). On this medium the calli continued to proliferate into shiny masses or shoots. These shoots were subsequently rooted on medium containing IAA (0.1 mg/l). While three out of four plants were haploids, the rest were aneuploids with a chromosome number $2n=18$. Later, Saha and Bhattacharya (1992) used NAA (0.1 mg/l), 2,4-D (0.1 mg/l), Kn (0.1 mg/l), sucrose (7%) and glutamine (400 mg/l) which produced globular structures of protoplast but failed to differentiate further. However, the differentiation of true pollen embryos and regeneration of haploid plants were described by Raina and Iyer (1992) and Shimokado et al. (1986). A systematic study was made at TES, Assam, India on haploid production of tea plants using anthers of TV-1 and TV-13 cultivars. Uninucleated stage of the pollens was found to be the best for induction of haploid calli. They established a correlation between developmental stage of pollens and colour of the anthers for selecting appropriate stage of pollen for optimum response. Though regeneration from the calli was not reported but cytologically confirmed to be haploid ($n=15$) (Laskar et al. 1993). Pedroso and Pais (1994) tested 17 different media combination based on MS and N6 with various concentrations of carbon source, growth regulators, and amino acids such as serine as well as glutamine for *C. japonica*. The embryogenic calli were achieved when microspores were cultured on 2,4-D (1 mg/l) and Kn (0.1 mg/l) and subsequently on MS supplemented with BAP (0.5 mg/l). However, further growth was ceased at maturation stage. In an attempt to regenerate haploid plants from anthers of five different Sri Lankan tea clones, Seran et al. (1999) reported that highest response (98 %) in terms of micro-calli formation was achieved on half strength MS supplemented with 2,4-D and BAP in presence of light with the cultivar TRI-2043. Other cultivars which produced callus were TRI-2023, TRI-2024, TRI-2025 and TRI-777 cultivars in a descending order. Determination of ploidy levels in the callus cells showed that the frequency of haploid cells was greater (68 %) than that of diploid cells (6 %). However, plantlets could not be regenerated. Thus it seems that although several attempts have been made to regenerate haploid plant in tea but success remain up to mere development of micro-calli.

In *C. japonica*, embryogenesis was induced from microspore. Among the various media composition and PGR formulation, MS along with 2,4-D (10 mg/l) and Kn (0.1 mg/l) were reported to be the best. The development of microspore derived proembryos was obtained in MS medium supplemented with BAP (0.5 mg/l) and reached the highest level when the microspores were cultured in this medium. However, the development of microspore-derived embryos ceased at maturation stage (Pedroso and Pais 1994) and no further work on this area had been reported further. Recently, Hazarika and Chaturvedi (2013) induced haploid calli with significant antioxidant activity from unpollinated ovary of tea. A high cytokinin/auxin ratio, provided by BAP and 2,4-D, and high-temperature treatment (33°C) for 10 days in the dark promoted maximum callus induction. Callus was maintained on MS medium containing BAP (5 mg/l) and IBA (2 mg/l) in the light at 25 °C. Well-developed tracheids were formed within 4 weeks in callus subcultured on MS medium containing TDZ (0.4 mg/l) and 2,3,5-triiodobenzoic acid (2.5 mg/l). Flow cytometric analysis revealed that most of the cells were haploid.

4.18 Secondary Metabolites Production

Tea is valued for pleasant taste and aroma, which is due to the presence of alkaloids, caffeine and other methyl xanthenes such as theobromine and theophylline. These alkaloids have also been used as therapeutic and drugs which have been discussed in earlier section. The exploration of production of secondary metabolites in tea is very important so much so that the work started way back in 1821, when caffeine (1,3,7-trimethylxanthine) was first prepared in pure form from tea leaves (Spedding and Wilson 1964). Much later, Ogutuga and Northcote (1970) also produced caffeine from callus tissue of tea. Catechins too were produced in the cell cultures of tea as high as 30 % (Hao et al. 1994). The formation of polyphenols in intact tea plants as well as in callus tissue was first reported by Forrest (1969) who found that the synthesis of simplest catechins and leuco-anthocyanin was highly dependent on the original explants and inversely co-related with the growth rate of cultured cells. Koretskaya and Zaprometov (1975) also reported that polyphenol content depends on addition of two kinds of precursor namely shikimic and quinic acids. However, higher level of gallic acid inhibited the formation of phenolic compounds.

Plant growth regulators such as NAA found to be better than 2,4-D to increase total soluble phenolics, flavans as well as the phenolic polymer lignins in callus culture (Zaprometov and Zagoskina 1979). Application of kinetin to the auxin-containing medium enhanced the synthesis of total soluble phenolics and flavans (Zaprometov and Zagoskina 1987). Bagratishilli et al. (1979) reported that NAA stimulated both cell growth and polyphenol production. ABA inhibited cell growth as well as the formation of all forms of soluble phenolic compounds (Zaprometov and Zagoskina 1987). The best carbon source for catechins and proanthocyanin production was reported to be 5 % glucose. Orihara and Furuya (1990) reported that tea cells cultured on B_2K medium (MS basal with 2 mg/l IBA and 0.1 mg/l Kn) produced L-glutamine as a main free amino acid and that addition of ethylamine to the medium increased theanine production. Matsuura et al. (1991) reported that a theanine accumulation level was remarkably high in the stem-derived callus when ethylamine-HCl was added to solidify MS along with BAP (4 mg/l). The effect of main inorganic constituents in medium on theanine accumulation was also investigated. Biotransformations of primary amines by *C. sinensis* cell culture were performed and their γ-glutamyl derivatives (products C_{1-8}) were obtained in the yield. Their molecular weight showed that primary amines were combined with glutamic acid and the characteristic fragment peaks suggested that primary amines were connected with γ-carboxylic acid in glutamic acid not to carboxilic acid (Furuya et al. 1990). Takemoto and Tanaka (2001) developed a method for producing (*S*)-(+)-α-phenyl-2-pyridylmethanol in 83 % chemical yield with 86 % optical yield by the repetitive use of immobilized tea cell culture. The tea cell culture showed similar capability for asymmetric bioreduction to that of *Catharanthus roseus* cell culture. *In vitro* cell culture of tea with high peroxidase activity for producing various chemically active compounds was also developed in tea (Takemoto et al. 2002a, 2002b).

Callus and root suspensions from tea had been established to produce and accumulate caffeine and theobromine as secondary metabolites. Leaf fragments from a mature greenhouse tea plant, grown on MS medium supplemented with IAA (1 mg/l) produced roots, while leaf explants of the same plant cultured on MS medium supplemented with 2,4-D (1 mg/l) and BAP (0.1 mg/l), gave rise to the formation of friable callus. Both callus and roots when transferred to MS liquid medium supplemented with 2,4-D (1 mg/l) and BAP (0.1 mg/l), produced caffeine and theobromine, which were detected by TLC, UV and GC (Shervingto et al. 1998). The effect of UV-B radiation on the accumulation and tissue localization of phenolic compounds in two strains of callus cultures of tea plant were investigated. The strains differed in their morphological and physiological characteristics and biosynthetic capacity. UV-B radiation hampered culture growth, decreased the size of callus-forming cells and promoted the accumulation of soluble and, to a lesser extent, polymeric forms of phenolic compounds, such as lignin. This accumulation was accompanied by an increase in the phenolic compound deposition in cell walls and intercellular space and by deposition of a lignin like material on the surface of callus cultures. The strain characterized by an increased formation of phenolic compounds was more resistant to UV-B radiation as compared to that with lower phenolic productivity (Zagoskina et al. 2003).

The effects of cadmium on the growth of tea callus cultures derived from leaves, stems, and roots and on the formation, in these cultures, of phenolic compounds, including flavans and lignin, which are characteristic of the tea plant, were investigated. In the calli derived from leaves and stems, cadmium treatment decreased the biomass increment, while in the calli derived from roots, growth characteristics remained at the control level. Under the effect of cadmium, the content of phenolic compounds, including flavans, in the leaf calli decreased, while in the stem and root calli, it either increased or was close to a control one. The lignin content in the root and stem calli increased, but it did not change in the leaf calli. All this data demonstrated that the cadmium-induced changes in phenolic metabolism of the tea plant callus culture depended both on the cadmium concentration in the medium and on the origin of calli (Zagoskina et al. 2007).

To determine whether caffeine biosynthesis is controlled by the availability of purine precursors and/or methyl-donors, Deng et al (2008) examined the effect of some purine compounds on purine alkaloid accumulation, using tea callus cultures. No stimulation of caffeine biosynthesis was observed when the calli were cultured with adenosine, guanosine or hypoxanthine for 3 weeks. However, paraxanthine (0.1 mg/l) doubled the caffeine level relative to controls. Adenosine stimulated the growth of callus and reduced the caffeine concentration 3 months after treatment. These results indicated that methylation of xanthosine by 7-methylxanthosine synthase was the most plausible rate-limiting step of caffeine biosynthesis.

Sikkimic acid can act as inducer to produce secondary metabolites under *in vitro* condition. It had been found that shikimic acid (3.5 g/l) increased the synthesis of polyphenols, catechins, caffeine, and other secondary components in the suspension culture of tea (Muthaiya et al. 2013).

4.19 Embryo Rescue

Natural hybridization among the different species of *Camellia* is not very common due to cross incompatibility. Therefore embryo rescue is very important for *Camellia* for developing new hybrids. Breeders from all over the world have desired to develop yellow flowered *Camellia*. The discovery of *C. chrysantha* generated great excitement among the *Camellia* growers and breeders as a potential source for a new range of *Camellia* floral colours. Although numerous interspecific hybridizations had been attempted, crossing of *C. chrysantha* with some other species was very difficult. In this regard, several cultivars of *C. japonica* that contributed to about 70% of the current horticultural needs and *C. chrysantha* with its potential for new colour were thought to be especially important. However, probably due to the high phylogenetic distance between the two species, the interspecific hybridization was extremely difficult (Yoshikawa and Yoshikawa 1990). Hwang et al. (1992), therefore, did a systematic investigation to understand the nature of reproductive barrier between *C. japonica* and *C. chrysantha* with intra and interspecific crosses using two different lines of each species. They found that zygote formation and early embryo development were similar in intra and interspecific crosses. Full size but empty ovules in mature capsules resulted from embryo abortion. Liang et al. (1986) reported that interspecific hybrid embryos of *C. pitardii* var. yunnanensis × *C. chrysantha* developed normally, reached torpedo stage, and differentiated normally. However, a complete successful protocol of embryo rescue will be immensely helpful to develop the long-awaited yellow coloured *Camellia* using *C. chrysantha* as a source.

4.20 Conclusion

Somatic embryogenesis is an inherent character of tea cotyledon, so much so that, it is being produced *in vivo*, without any culture media. There are excellent reports on induction, maturation, germination, long-term multiplication including bioreactor approach of tea and other *Camellia* species. The advantage of somatic embryogenesis over conventional micropropagation is the presence of tap roots in somatic embryo derived seedlings, which help them better to combat drought. Significant amount of works also been done related to development of primary embryogenesis, secondary embryogenesis, their origin, morphology as well as histochemical and biochemical mechanisms so much so that embryogenic region within a leaf have been identified. Cold storage and their recovery have also been studied in detail for both tea as well as *Camellia*. It is particularly important as tea seeds are recalcitrant in nature. However, efforts should be directed towards the commercialization of somatic embryogenesis, bioreactor technology, and field performance of somatic embryo-derived plants. Additionally regeneration from protoplast, anther-derived tissue for haploid plant production though important in tea breeding, yet little success has been achieved.

References

Abraham GC, Raman K (1986) Somatic embryogenesis in tissue culture of immature cotyledons of tea (*Camellia sinensis*). In: Somers DA, Gengenbach BG, Biesboor DD, Hackett WP and Green CE (eds) 6th international congress on plant tissue and cell culture. Univ. Minnesota, Minneapolis, p 294

Akula A, Akula C (1999) Somatic embryogenesis in tea (*Camellia sinensis* (L.) O Kuntze). In: Jain SM, Gupta PK, Newton RJ (eds) Somatic embryogenesis in woody plants, vol 5. Springer, Berlin, pp 239–259

Akula A, Akula C, Bateson M (2000) Betaine a novel candidate for rapid induction of somatic embryogenesis in tea (*Camelia sinensis* (L.) O.Kuntze). Plant Growth Regu 30:241–246

Akula A, Dodd WA (1998) Direct somatic embryogenesis in a selected tea clone, TRI-2025 (*Camellia sinensis* (L). O. Kuntze) from nodal segment. Plant Cell Rep 17:804–809

Ammirato PV (1985) Patterns of development in culture. In: Henke RR, Hughes KW, Constantin MP, Hollaender A (eds) Tissue culture in forestry and agriculture. Plenum, New York, pp 9–29

Anderson W. C. (1984). A revised tissue culture medium for shoot multiplication of rhododendron. Journal of the American Society for Horticultural Science, 109: 343-347.

Aoshima Y (2005) Efficient embryogenesis in the callus of tea (*Camellia sinensis*) enhanced by the osmotic stress or antibiotics treatment. Plant Biotechnol 22:277–280

Arulpragasam PV, Latiff R, Seneviratne P (1988) Studies on the tissue culture of tea (*Camellia sinensis* (L.) O. Kuntze). 3. Regeneration of plants from cotyledon callus. Sri Lank J Tea Sci 57: 20–23

Avgioglu A, Knox RB (1989) Storage lipid accumulation by zygotic and somatic embryo in culture. Ann Bot 63:409–412

Bag N, Palni LMS, Nandi SK (1997) Mass propagation of tea using tissue culture methods. Physiol Mol Biol Plant 3:99–103

Bagratishvili DG, Zaprometov MN, Butenko RG (1979) Obtaining a cell suspension culture from the tea plant. Fiziol Rast 26:449–451

Ballester A, Janeiro LV, Vieitez AM (1997) Cold storage of shoot cultures and alginate encapsulation of shoot tips of *Camellia japonica* L. and *Camellia reticulata* Lindley. Sci Hort 71:67–78

Bano Z, Rajaratnam S, Mohanty BD (1991) Somatic embryogenesis in cotyledon culture of tea (*Thea sinensis* L.). J Hort Sci 66:465–470

Barciela J, Vieitez AM (1993) Anatomical sequence and morphometric analysis during somatic embryogenesis oncultured cotyledon explants of *Camellia japonica* L. Ann Bot 71:395–404

Balasubramanian S, Marimuthu S, Rajkumar R, Haridas V (2000a) Somatic embryogenesis and multiple shoot induction in *Camellia sinensis* (L.) O. Kuntze. J Plant Crop 28:44–49

Balasubramanian S, Marimuthu S, Rajkumar R, Balasaravanan T (2000b) Isolation, culture and fusion of protoplast in tea. In: Muraleedharan N, Rajkumar R (eds) Recent advance in plants crops research. Allied Publishers, Mumbai, pp 3–9

Bennett WY, Scheibert P (1982) *In vitro* generation of callus and plantlets from cotyledons of *Camellia japonica*. Camellia J 37:12–15

Beretta D, Vanoli M, Eccher T (1987) The influence of glucose, vitamins and IBA on rooting of *Camellia* shoots *in vitro*. Abstracts from symposium on vegetative propagation of woody species, Pisa, Italy, p 105

Bhatia CR, Murty GSS, Mouli C, Kale DM (1986) Nuclear techniques and *in vitro* culture for plant improvement. In: Proceedings of international symposium on nuclear techniques and *in vitro* culture for plant improvement. IAEE, FAO and UN, Vienna, 19–23 August, pp 419–427

Bhattacharya, A., Nagar, P.K. and Ahuja, P.S. (2002) Seed development in Camellia sinensis (L.) O. Kuntze. Seed Science Research 12, 39–46

Bhattacharya A, Nagar PK, Ahuja PS (2004) Changes in endogenous indole-3-acetic acid and some biochemical parameters during seed development in *Camellia sinensis* (L.) O. Kuntze. Acta Physiol Plant 26:399–404

Chandel KPS, Chaudhury R, Radhamani J, Malik SK (1995) Desiccation and freezing sensitivity in recalcitrant seeds of tea, cocoa and jackfruit Annal Bot 76:443–450

Chaudhaury R, Lakhanpal S, Chandel KPS (1990) Germination and desiccation tolerance of tea (*Camellia sinensis* (L.) O. Kuntze) seeds and feasibility of cryopreservation. Sri Lank J Tea Sci 59:89–94

Chaudhury R, Radhamani J, Chandel KPS (1991) Preliminary observation in the crypreservation of desiccated embryonic axes of tea (*Camellia sinensis*) L. O. Kuntze) seeds for genetic conservation. Cryo-Lett 12:31–36

Chen Z, Liao H (1982) Obtaining plantlet through anther culture of tea plants. Zhongguo Chaye 4:6–7

Chen Z, Liao H (1983) A success in bringing out tea plants from the anthers. China Tea 5:6–7

Das SC, Barman TS (1988) Current state and future potential of tissue culture in tea. Proceedings of 30th Tocklai Conference, TRA, Jorhat, India, pp 90–94

Das SC (1992) Non-conventional techniques of regenerating polyploids in tea. Proceedings of 31st Tocklai Conference, TRA, Jorhat, India, pp 26–30

Deng WW, Li YY, Ogita S, Ashihara H (2008) Fine control of caffeine biosynthesis in tissue cultures of *Camellia sinensis*. Phytochem Lett 1:195–198

Engelmann F (1997) *In vitro* conservation research activities at the International Plant Genetic Resouces Institute (IPGRI). Plant Tissue Cult Biotech 3:46–52

Evans DA, Sharp WR (1986) Somaclonal and gametoclonal variation. In: Evans DA, Sharp WR, Ammirato PV (eds) Hand book of plant cell culture, vol 4. Technique and applications. Macmillan, New York, pp 97–132

Forrest GI (1969) Studies on the polyphenol metabolism of tissue culture derived from the tea plant (*C. sinensis* L.). Biochem J 113:765–772

Frisch CH, Camper ND (1987) Effect of synthetic auxins on callus induction from tea stem tissue. Plant Cell Tissue Orga Cult 8:207–213

Furuya T, Orihara T, Tsuda Y (1990) Caffeine and theanine from cultured cells of *Camellia sinensis*. Phytochem 29:2539–2547

Gamborg O, Miller R, Ojima K (1968) Nutrient requirements of suspension cultures of soyabean root cells. Exp Cell Res 50:157–158

Ghanati F, Ishka MR (2009) Investigation of the interaction between abscisic acid (ABA) and excess benzyladenine (BA) on the formation of shoot in tissue culture of tea (*Camellia sinensis* L.). Int J Plant Prod 3:7–14

Ghosh B, Sen S (1994) Plant regeneration from alginate encapsulated somatic embryos of *Asparagus cooperi* Baker. Plant Cell Rep 13: 381–385

Hazarika RR, Chaturvedi R (2013) Establishment of dedifferentiated callus of haploid origin from unfertilized ovaries of tea (*Camellia sinensis* (L.) O. Kuntze) as a potential source of total phenolics and antioxidant activity. *In Vitro* Cell Dev Biol-Plant 49:60–69

Hao C, Wang Y, Yang S (1994) Effects of macroelements on the growth of tea callus and the accumulation of catechins. J Tea Sci (China) 14:31–36

Haridas V, Balasaravanan T, Rajkumar R, Marimuthu S (2000) Factor influencing somatic embryogenesis in *Camellia sinensis* (L.) O. Kuntze. In: Muraleedharan N, RajKumar R (eds) Recent advances in plantation crops research. Allied Publishers, Mumbai, pp 31–35

Hu H (1983) Genetic stability and variability of pollen derived plants. In: Sen SK, Giles KL (eds) Plant cell culture in crop improvement. Plenum Press, New York, pp 145–157

Hua LD, Dai ZD, Hui X (1999) Studies on somatic embryo and adventitious bud differentiation rate among different tissues of *Camellia sinensis* L. Acta Agron Sin 25:291–295

Hwang YJ, Okubo H, Fujieda K (1992) Pollen tube growth, fertilizer and embryo development of *Camellia japonica* L. × *C. chrysantha* (Hu) Tuyama. J Jap Soc Hort Sci 60:955–961

Ikekawa N (1991) In: Physiology and biochemistry of sterols. Patterson GW, Nes WD (eds). American Oil Chemists Society, Champaign, pp 347 -360

Jain SM, Newton RJ (1990) Prospects of biotechnology for tea improvement. Proc Indian Natl Sci Acad 6:441–448

Janeiro LV (1996) Almacenamiento en frio de especies lenosas propagadas *in vitro*. Aplicación de las tecnologías de semilla artifical y criopreservación en el género *Camellia*, Doctoral Thesis, University of Santiago de Compostela, Santiago de Compostela

Janeiro LV, Ballester A, Vieitez AM (1995) Effect of cold storage on somatic embryogenesis systems of *Camellia*. J Hort Sci 70:665–667

Janeiro LV, Ballestter A, Vieitez AM (1996) Cryopreservation of somatic embryos and embryonic axes of *Camellia japonica* L. Plant Cell Rep 15:699–703

Janeiro LV, Ballester A, Vieitez AM (1997) *In vitro* response of encapsulated somatic embryo of *Camellia*. Plant Cell Tissu Orga Cult 51:119–125

Jha TB, Jha S, Sen SK (1992) Somatic embryogenesis from immature cotyledon of an elite Darjeeling tea clone. Plant Sci 84:209–213

Kato M (1982) Results of organ culture on *Camellia japonica* and *C. sinensis*. Jap J Breed 32:267–277

Kartha KK (1985) Meristem culture and germplasm preservation. In: Kartha KK (ed) Cryopreservation of plant cells and organs. CRC Press, Boca Raton, pp 115–134

Kato M (1986a) Micropropagation through cotyledon culture in *Camellia japonica* L. and *Camellia sinensis* L. Jap J Breed 36:31–38

Kato M (1986b) Micropropagation through cotyledon culture in *Camellia sasanqua*. Jap J Breed 36:82–83

Kato M (1989a) *Camellia sinensis* L. (Tea): *In vitro* regeneration. In: Bajaj YSP (ed) Biotechnology in agriculture and forestry, vol 7. Medicinal and aromatic plants II. Springer, Berlin, pp 83–98

Kato M (1989b) Polyploids of *Camellia* through culture of somatic embryos. Hort Sci 24:1023–1025

Kato K (1996) Somatic embryogenesis from immature leaves of *in vitro* grown tea shoots. Plant Cell Rep 15:920–923

Katsuo K (1969) Anther culture in tea plant (a preliminary report). Study Tea 4:31

Koretskaya TF, Zaprometov MN (1975) Phenolic compounds in the tissue culture of *Camellia sinensis* and effect of light on their formation. Fiziol Rast 22:941–946

Kuboi T, Suda M, Konishi S (1991) Preparation of protoplasts from the leaves. In: Proceedings of International Symposium on Tea Science, Shizuoka, Japan, pp 427–431

Kuranuki Y, Yoshida S (1991) Cryopreservation of tea seeds and excised embryonic axes in liquid nitrogen. In: Proceedings of International Symposium on Tea Science, Shizuoka, Japan pp 419–420

Laskar MA, Das SC, Deka PC (1993) Tissue culture of tea: anther culture for haploid plant production. In: Proceedings of the international symposium tea science and human health. Tea Research Association, Calcutta, India, 11–14 January

Lecouteux CG, Lai FM, Mckersie BD (1993) Maturation of alfalfa (*Medicago sativa* L.) somatic embryos by abscisic acid, sucrose and chilling stress. Plant Sci 94:207–213

Liang H, Zhang Z, Zhang X (1986) Investigation of the sexual process in interspecific crosses between *Camellia pitardii* var. *yunnanesis* and *C chrysantha*. Acta Bot Yunnan 8:147–152

Lloyd G, McCown B (1980) Commercially feasible micropropagation of mountain laurel, *Kalmia lalifolia* by use of shoot tip culture. Comb Proc Int Plan Prop Soc 30:421–427

Lu J. Chen R., Zhang M. Teixeira da Silva JA, Ma G (2013) Plant regeneration via somatic embryogenesis and shoot organogenesis from immature cotyledons of Camellia nitidissima Chi. J Plant Physiol. 170: 1202–1211

Mariya John KM, RajKumar R (2006) Factors influencing synthetic seed germination in tea (*Camellia sinensis* (L.) O. Kuntze). J Plan Crop 34:40–42

Matsuura T, Kakuda T, Kinoshita T, Takeuchi N, Sasaki K (1991) Production of theanine by callus culture of tea. In: Proceedings of the International Symposium on Tea Science, Shizuka, Japan, pp 432–436

Maugh TH (1981) The natural occurring brassionoide in the plant species. Science 212:33–34

Mondal TK, Bhattacharya A, Sood A, Ahuja PS (1999) An efficient protocol for somatic embryogenesis and its use in developing transgenic tea (*Camellia sinensis* (L) O. Kuntze) for field transfer. In: Altman A, Ziv M, Izhar S (eds) Plant biotechnology and *in vitro* biology in 21st century. Kluwer Academic, Dordrecht, pp 101–104

Mondal TK, Bhattacharya A, Sood A, Ahuja PS (2000) Factor effecting induction and storage of encapsulated tea (*Camellia sinensis* L (O) Kuntze) somatic embryos. Tea 21:92–100

Mondal TK, Bhattacharya A, Ahuja PS (2001a) Induction of synchronous secondary embryogenesis of tea (*Camellia sinensis*). J Plant Physiol 158:945–951

Mondal TK, Bhattacharya A, Ahuja PS, Chand PK (2001b) Factor effecting *Agrobacterium tumefaciens* mediated transformation of tea (*Camellia sinensis* (L). O. Kuntze). Plant Cell Rep 20:712–720

Mondal TK, Bhattacharya A, Sharma M, Ahuja PS (2001c) Induction of *in vivo* somatic embryogenesis in tea (*Camellia sinensis*) cotyledons. Curr Sci 81:101–104

Mondal TK (2002) Camellia biotechnology: a bibliographic search. Int J Tea Sci 1:28–37

Mondal TK, Bhattacharya A, Sood A, Ahuja PS (2002) Factors affecting germination and conversion frequency of somatic embryos of tea. J Plant Physiol 59:1317–1321

Mondal TK, Bhattacharya A, Laxmikumaran M, Ahuja PS (2004) Recent advance in tea biotechnology. Plant Cell Tissue Orga Cult 75:795–856

Muralidharan EM, Mascarenhas AF (1995) Somatic embryogenesis in *Eucalyptus*. In: Jain SM, Gupta PK, Newton R (eds) Somatic embryogenesis in woody plants, vol 2. Angiosperms. Kluwer Academic, Dordrecht, pp 23–40

Murashige T, Skoog F (1962) A revised medium for rapid growth and bioassay with tobacco tissue cultures. Physiol Plant 15:473–497

Muthaiya MJ, Nagella P, Thiruvengadam M, Mandal AKA (2013) Enhancement of the productivity of tea (*Camellia sinensis*) secondary metabolites in cell suspension cultures using pathway inducers. J Crop Sci Biotechnol 16:143–149

Nadamitsu S, Andoh Y, Kondo K, Segawa M (1986) Interspecific hybrids between *Camellia vietnamensis* and *C. chrysantha* bycotyledon culture. Jap J Breed 36:309–313

Nagar PK and Sood S (2006) Changes in endogenous auxins during winter dormancy in tea (Camellia sinensis L.) O. Kuntze 28: 165-169

Nakamura Y (1983) Isolation of protoplasts from tea plant. Tea Res J 58:36–37

Nakamura Y (1985) Effect of origin of explants on differentiation of root and its varietal difference in tissue culture of tea plant. Tea Res J 62:1–8

Nakamura Y (1988a) Efficient differentiation of adventitious embryos from cotyledon culture of *Camellia sinensis* and other *Camellia* species. Tea Res J 67:1–12

Nakamura Y (1988b) Effects of the kinds of auxins on callus induction and root differentiation from stem segment culture of *Camellia sinensis* (L.) O. Kuntze. Tea Res J 68:1–7

Nguyen VT (2012) Regeneration plantlets from somatic embryos of tea plant (*Camellia sinensis* L.) J Agri Tech 8:1821–1827

Nitsch JP, Nitsch C (1969) Haploid plants from pollen grains. Science 163:85–87

Okano N, Fuchinone Y (1970) Production of haploid plants by anther culture of tea *in vitro*. Jap J Breed 20:63–64

Ogutuga DBA, Northcote DH (1970) Caffeine formation in tea callus tissue. J Expt Bot 21:258–273

Orihara Y, Furuya T (1990) Production of theanine and other γ-glutamine derivatives by *Camellia sinensis* cultured cells. Plant Cell Rep 9:1215–1224

Park YG, In-Suk AHN, Bozhkov P (1997) Effect of exogenous plant growth regulator on morphogenetic response *in vitro* by embryo and leaf cultures of *Camellia sinensis (L.)* O. Kuntze. Korean J Plant Tissu Culture 24:129–135

Paratasilpin T (1990) Comparative studies on somatic embryogenesis in *Camellia sinensis* var. *sinensis* and *C. sinensis* var. *assamica* (Mast.). Pierre J Sci Soc Thailand 16:23–41

Pedroso-Ubach MC (1991) Contribuicao para a preservacao e o melhoramento de *Camellia japonica* L. Master's Thesis (English abstract). Faculdade de Ciencias da Universidade de Lisboa, Lisboa, Portugal, pp 23–50

Pedroso MC, Pais MS (1993) Direct embryo formation in leaves of *C. japonica* L. Plant Cell Rep 12:639–643

Pedroso-Ubach MC (1994) Somatic embryogeneisis in *Camellia japonica* L. a search for markers. Ph.D. thesis, Faculdade de Ciencias da Universidade de Lisboa, Lisbon, Portugal

Pedroso MC, Pais MS (1994) Induction of microspore embryogenesis in *Camellia japonica* cv. Elegans. Plant Cell Tiss Orga Cult 37:129–136

Pedroso MC, Pais MS (1995a) Factor controlling somatic embryogenesis: cell wall changes as an *in vivo* marker of embryogenic competence. Plant Cell Tissue Orga Cult 43:147–154

Pedroso MC, Pais MS (1995b) Plant regeneration from embryogenic suspension cultures of *Camellia japonica*. *In Vitro* Cell Dev Biol-Plant 31:31–35

Pence VC (1995) Cryopreservation of recalcitrant seeds. In: YSP Bajaj (ed) Biotechnology in agriculture and forestry. Vol 32: crypreservation of plant germplasm I. Springer, Berlin, pp 29–50

Plata E, Vieitez AM (1990) *In vitro* regeneration of *Cammelia reticulata* by somatic embryogenesis. J Hort Sci 65:707–714

Plata E, Ballester A, Vieitez AM (1991) An anatomical study of secondary embryogenesis in *Camellia reticulata*. *In Vitro* Cell Dev Biol Plant 27:183–189

Plata E (1993) Morphogenesis *in vitro* de *Camellia reticulata*: Proceson de embryogenesis somatica regeneration de plants. Doctoral Thesis, University of Santiago de Compostela, Santiago de Compostela

Ponsamuel J, Samson NP, Sathyprakash V, Abraham GC (1996) Somatic embryogenesis and plant regeneration from the immature cotyledonary tissues of cultivated tea (*Camellia sinesis* (L.) O. Kuntze). Plant Cell Rep 16:210–214

Preeti S, Pandey S, Bhattacharya A, Nagar PK, Ahuja PS (2004) ABA associated biochemical changes during somatic embryo development in *Camellia sinensis* (L.) O Kuntze. Plant Physiol 161:1269–1276

Purakayastha A, Das SC (1994) Isolation of tea protoplast and their culture. In: Proceedings of 32nd Tocklai Conference

Raina SK, Iyer RD (1992) Multicell pollen proembryoid and callus formation in tea. J Plant Crop 9:100–104

RajKumar R, Ayyappan P (1992) Somatic embryogenesis from cotyledonary explants of *Camellia sinensis* (L.) O. Kuntze. Plant Chron May:227–229

Rajkumar R, Balasusaravanam S, Jayakumar D, Haridas V, Marimuthu S (2001) Physiological and biochemical feathers of field grown somaclonal variants of Tea. UPASI Tea Res Found Bull 54:73–81

Redenbaugh K, Slade D, Viss P, Fujii JA (1987) Encapsulation of somatic embryos in synthetic seed coats. Hort Sci 22:803–809

Redenbaugh K, Fujii JA, Slade D (1991) Synthetic seed technology. In: Vasil KI (ed) Scale-up and automation in plant propagation. Cell culture and somatic cell genetics of plants, vol 8. Academic Press, New York, pp 35–74

Roberts DR, Sutton BCS, Flinn BS (1990) Synchronous and high frequency germination of interior spruce somatic embryo following partial drying at high relative humidity. Can J Bot 68:1086–1093

Roberts PM, Weaver CD & Oh S-H (1992) Intracellular receptor proteins for calcium signals in plants. In: Gresshoff PM (Ed) Plant Biotechnology and Development: Current Topics in Plant Molecular Biology (pp 129–113). CRC Press Inc., Florida

Saha SK, Bhattacharya NM (1992) Stimulating effect of elevated temperatures on callus production of meristemmoids from pollen culture of tea (*Camellia sinensis (L.)* O. Kuntze). Indian J Expt Biol 30:83–86

Salinero MC, Silva-Pando FJ (1986) La multiplication delas camellias. In: Diputacion provincial de Pontevedra (ed) La Camellia. C Peon. Pontevedra, Spain, pp 175–184

Sandal I, Kumar A, Bhattacharya A, Sharma M, Shanker A, Ahuja PS (2005) Gradual depletion of 2,4-D in the culture medium for indirect shoot regeneration from leaf explants of *Camellia sinensis* (L.) O. Kuntze. Plant Growth Reg 47:121–127

San-Jose MC, Vieitez AM (1993) Regeneration of *Camellia* plantlets from leaf explant cultures by embryogenesis and caulogenesis. Sci Hort 54:303–315

Seran TH, Hirimburegama K, Hirimburegama WK, Shanmugarajha V (1999) Callus formation in anther culture of tea clones, *Camellia sinensis* (L) O Kuntze. J Nat Sci Found Sri Lanka 27:165–175

Shervington A, Shervington LA, Afifi F, El-omari MA (1998) Caffeine and theobromine formation by tissue cultures of *Camellia sinensis* 47:1535–1536

Sarathchandra TM, Upali PD, Wijeweardena RGA (1988) Studies on the tissue culture of tea (*Camellia sinensis* (L.) O. Kuntze) 4. Somatic embryogenesis in stem and leaf callus cultures. Sri Lank J Tea Sci 52: 50–54

Sharma P, Pandey S, Bhattacharya A, Nagar PK, Ahuja PS (2004) ABA associated biochemical changes during somatic embryo development in *Camellia sinensis* (L.) O. Kuntze. J Plant Physiol 161:1269–1276

Shimokado TT, Murata, Miyaji Y (1986) Formation of embryoid by anther culture of tea. Jap J Breed 36:282–283

Sood A, Palni LMS, Sharma M, Rao DV, Chand G, Jain NK (1993) Micropropagation of tea using cotyledon culture and encapsulated somatic embryos. J Plant Crop 21:295–300

Spedding DJ, Wilson AT (1964) Caffeine metabolism. Nature 204:73

Sriyadi B (1987) Tissue culture a method to overcome hand-pollinated difficulties in tea plant. Warta Balai Peneli The Dan Kina 13:105–111

Suganthi M, Arvinth S, Raj Kumar R (2012) Impact of osmotica and abscisic acid on direct somatic embryogenesis in tea. Int J Plant Res 2:22–27

Tahardi JS, Raisawati T, Riyadi I, Dodd WA (2000) Direct somatic embryogenesis and plant regeneration in tea by temporary liquid immersion. Menara Perkebunan 68:1–9

Tahardi JS, Riyadi I, Dodd WA (2003) Enhancement of somatic embryo development and plant recovery in *Camellia sinensis* by temporary immersion system. J Bioteck Perta 8:1–7

Takemoto M, Tanaka K (2001) Synthesis of optically active α-phenylpyridylmethanols by *Camellia sinensis* cell culture. J Mol Cat B Enzym 15:173–176

Takemoto M, Aoshima Y, Stoynov N, Kutney JP (2002a) Establishment of *Camellia sinensis* cell culture with high peroxidase activity and oxidative coupling reaction of dibenzylbutanolides. Tetr Lett 43:6915–6917

Takemoto M, Suzuki Y, Tanaka K (2002b) Highly selective oxidative coupling of 2-naphthol derivatives catalyzed by *Camellia sinensis* cell culture. Tetra Lett 43:8499–8501

Thomas J, Kumar RR, Mandal AKA (2006) Metabolite profiling and characterization of somaclonal variants in tea (*Camellia* spp.) for identifying productive and quality accession. Phytochem 67:1136–1142

Tian LN, Brown DCW (2000) Improvement of soybean somatic embryo development and maturation by abscisic acid treatment. Can J Plant Sci 80:271–276

Tripetch P, Kerdchoechuen O, Laohakunjit N, Pitchayangkura R (2009) Induction of callus in Assam tea leaves (*Camellia sinensis var. Aassamica*). Agricultural Sci J 40:209–212

Tremblay L, Tremblay FM (1995) Maturation of black spruce somatic embryos: sucrose hydrolysis and resulting osmotic pressure of the medium. Plant Cell Tissue Orga Cult 42:39–46

Tulecke W (1987) Somatic embryogenesis in woody perennials. In: Bonga JM, Durzan DJ (eds) Cell and tissue culture in forestry, vol 2. Martinus Nijhoff, Dordrecht, pp 61–69

Vacin EF, Went FW (1949) Some pH changes in nutrient solutions. Bot Gaz 110:605–613

Vieitez AM (1994) Somatic embryogenesis in *Camellia* spp. In: Jain S, Gupta P, Newton R (eds) Somatic embryogenesis in woody plants. Kluwer Academic, Dordrecht, pp 235–276

Vieitez AM, Barciela J (1990) Somatic embryogenesis and plant regeneration from embryonic tissues of *Camellia japonica* L. Plant Cell Tiss Orga Cult 21:267–274

Vieitez AM, San-Jose C, Vieitez J, Ballester A (1991) Somatic embryogenesis from roots of *Camellia japonica* plantlets cultured *in vitro*. J Am Soc Hortic Sci 116:753–757

Wachira F, Ogado J (1995) *In vitro* regeneration of *Camellia sinensis* (L.) O. Kuntze by somatic embryo. Plant Cell Rep 14:463–466

Webster FB, Roberts DR, Mclnnis SM, Sutton BCS (1990) Propagation of interior spruce by somatic embryogenesis. Can J For Res 20:1757–1762

Wu CT, Huang TK, Chen GR, Chen SY (1981) A review on the tissue culture of tea plants and on the utilization of callus derived plantlets In: Rao AN (ed) Tissue culture of economically important plants. Proc Costed Symp Singapore, pp 104–106

Yamaguchi S, Kunitake T, Hisatomi S (1987) Interspecific hybrid between *Camellia japonica* cv. choclidori and *C. chrysantha* produced by embryo culture. Jap J Breed 37:203–206

Yan MQ, Ping C (1983) Studies on development of embryoids from the culture cotyledons of *Thea sinensis* L. Sci Silv Sin 19:25–29

Yan MQ, Ping C, Wei M, Wang YH (1984) Tissue culture and transplanting of *Camellia oleifera*. Sci Silvae Sin 20:341–350

Yoshikawa K, Yoshikawa N (1990) Inter-specific hybridization of *Camellia*. Bull Seibu Maizuru Bot Inst 5:56–75

Zaprometov MN, Zagoskina MV (1979) One more evidence for chloroplast involvement in the biosynthesis of phenolic compounds. Plant Physiol (Russian) 34:165–172

Zaprometov MN, Zagoskina MV (1987) Regulation of phenolic compounds formation in cultured cells of tea plant (*Camellia sinensis*). In: Proc Inter Tea Quality Human Health Symp China, pp 62–65

Zagoskina NV, Dubravina GA, Alyavina AK, Goncharuk EA (2003) Effect of ultraviolet (UV-B) radiation on the formation and localization of phenolic compounds in tea plant callus cultures. Russ J Plant Physiol 50:270–275

Zagoskina NV, Goncharuk EA, Alyavina AK (2007) Effect of cadmium on the phenolic compounds formation in the callus cultures derived from various organs of the tea plant. Russ J Plant Physiol 54:237–243

Zhuang C, Liang H (1985a) *In vitro* embryoid formation of *Camellia reticulata* L. Acta Biol Exp Sin 18:275–281

Zhuang C, Liang H (1985b) Somatic embryogenesis and plantlet formation in cotyledon culture of *Camellia chrysantha*. Acta Bot Yunn 7:446–450

Zhuang C, Duan J, Zhou J (1988) Somatic embryogenesis and plantlets regeneration of *Camellia sasanqua*. Acta Bot Yunn 10:241–244

5 Genetic Transformation

5.1 Introduction

Due to the non-availability of new land for expansion of cultivation along with the objective of reducing the cost of cultivation, tea breeders need to develop improved tea cultivars as per the requirements of the industry. On the other hand, natural variability of important traits is not available in tea. In this context, transgenic technology offers several advantages for perennial plants like tea. They are: (1) possibility of introducing genes of any origin, (2) greater speed for developing transgenic material, (3) free from epistasis interaction for transgene expression, and (4) being targeted in nature, it is free from linkage drag (Bhattacharya et al. 2004; Mondal 2007, 2008). Therefore, gene transfer through biotechnological means appear to be not only a time effective but also an advantageous and better alternative. A summary of research work on transgenic tea is listed in Table 5.1 and is discussed in this chapter.

5.2 *Agrobacterium tumefaciens*

The first protocol for production of transgenic tea, cv. Kangra jat, was developed (Fig. 5.1) via *Agrobacterium*-mediated genetic transformation (Mondal et al. 1999, 2001a). Two disarmed *Agrobacterium tumefaciens* strains, EHA 105 and LBA 4404, both carrying the binary plasmid p35SGUS-INT with the *neomycin phosphotransferase II* (*nptII*) gene and *gus* (β-glucuronidase) intron were evaluated as vector systems. Several parameters were evaluated to maximize the transformation efficiency. While pre-culture, wounding, and acetosyringone (AS) treatment in different concentrations were inhibitory, the bacterial growth phase (OD 600 = 0.6), cell density (10^9/ml), co-cultivation period (5 days), and pH of the co-cultivation medium (5.6) had positive effects on transformation efficiency. AS is a low-molecular-weight phenolic compound naturally released by wounded plant cells and acts as an inducer of the virulence genes. Following co-cultivation, globular somatic embryos were placed on a multiplication medium and stressed with kanamycin (50 mg/l). Further selection occurred in the maturation and germination medium at an elevated kanamycin level (75 mg/l). An average of 40 % transient expression was registered based on the GUS histological assay. Then, the kanamycin-resistant, GUS-positive embryos were germinated and the resulting micro-shoots were multiplied *in vitro* on Murashige and Skoog (MS) medium fortified with thidiazuron (TDZ, 1.1 mg/l) and α-naphthaleneacetic acid (NAA, 2 mg/l). Later, they were micro-grafted onto seed-grown rootstocks (Mondal et al. 2005) of cv. Kangra jat and eventually hardened in a walk-in poly-house. Integration of the transgenes into the tea nuclear genome was confirmed by polymerase chain reaction (PCR) analysis using *nptII*- and *gus*-specific primers (Fig. 5.2) and by southern hybridization using an *nptII*-specific probe. However, use of phenolic inducer, AS, did not enhance the efficiency of transformation (Mondal et al. 2005). In several other experi-

T. K. Mondal, *Breeding and Biotechnology of Tea and its Wild Species*,
DOI 10.1007/978-81-322-1704-6_5, © Springer India 2014

Table 5.1 Summary of transgenic tea research

Techniques	Remarks	Transgene	References
A. tumefaciens	Antibiotic selection for *Camellia* species was reported	–	Tosca et al. (1996)
A. rhizogenes	First attempt to induce hairy root formation	*rolB*	Zehra et al. (1996)
A. rhizogenes	Transformation protocol was exploited to promote roots to facilitate hardening of the micropropagated tea plants	*rolB*	Konwar et al. (1998)
A. tumefaciens	Preliminary study for gene transfer to tea plants	*gus intron*	Matsumoto and Fukui (1998)
A. tumefaciens	Standardization of somatic embryogenesis and transient expression of *gus* gene	*npt-II*	Mondal et al. (1999)
Particle bombardment	First attempt for standardization of the biolistic-mediated transformation protocol	*npt-II*	Akula and Akula (1999)
A. tumefaciens	Transgenic calli were produced utilizing the phenolic inducer AS at an effective range (20–100 mg/l)	*npt-II*	Matsumoto and Fukui (1999)
A. tumefaciens	Detailed study on Bt gene transformation was reported	*Bt*	Luo and Liang (2000)
A. tumefaciens	Standardization of transformation	*npt-II*	Aoshima et al. (2001)
A. tumefaciens	Development of selection system for putative transformants	*npt-II*	Mondal et al. (2001a)
A. tumefaciens	Production of transgenic plants from transformed somatic embryos	*npt-II* and *gus intron*	Mondal et al. (2001b)
A. tumefaciens	Tea leaves with glabrous surface having lower phenol and wax content were identified to be more suitable for infection	*npt-II*	Kumar et al. (2004)
A. tumefaciens and particle bombardment	Green florescence protein gene was transferred with organelle target signals	*gfp*	Kato et al. (2004)
*A. tumefaciens*and particle bombardment	Attempt was made for standardization of the protocol	*npt-II*	Wu et al. (2003)
A. tumefaciens	Attempts to overcome the bactericidal effect of tea leaf polyphenol during transformation	*npt-II*	Sandal et al. (2007)
A. tumefaciens	Silencing of glutathione synthetase gene in callus	*gs*	Mohanpuria et al. (2008)
A. rhizogenes	High production of catechin through hairy root culture	*rolb*	John et al. (2009)
A. tumefaciens	Caffeine-free tea plant production	*cs*	Mohanpuria et al. (2011)
Particle bombardment	Optimization of the protocol with the introduction of osmotin gene	*osmotin*	Saini et al. (2012)

rolB rooting-locus gene B, *gus-intron* beta-glucuronidase, *npt-II* neomycin phosphotransferase II, *gfp* green florescence protein, *gs* glutathione synthetase, *cs* caffeine synthase

ments on woody plants, it had been found that AS could not increase transformation efficiency (Confalonieri et al. 1997). The inability of AS to improve the transformation efficiency could be due to the inherent prevalence of high amounts of phenolics in tea. Contrary to this, Matsumoto and Fukui (1999) found that AS has a positive role in tea transformation as this plant is considered to be one of the recalcitrant plants for *Agrobacterium* infection. Naturally occurring crown galls are hardly observed on tea plant and this bacterium is not cited as an economically important pathogen in Japan. Therefore, they concluded that the low efficiency of the *Agrobacterium* infection was overcome by AS. Resistant calli emerged only on the explants treated with a higher concentration of AS. Thus, they concluded that application of AS (20 mg/l) to a co-culture medium was effec-

Surface disinfection of mature sinker seeds

↓

Germination on ½ MS + sucrose (30 g/l) +Agar (0.7%)

↓ 2 weeks

Induction of somatic embryos from cotyledon slices on ½ MS+ sucrose (20 g/l) + NAA (2.5 g/l) + BAP (0.2 g/l)

↓ 6-8 weeks

Multiplication of primary embryos on (EMM) MS+ BAP (2 g/l) + IBA (0.2 g/l) + L-glutamine (1 g/l)

↓ 6 weeks

Infection of globular embryos with *Agrobacterium* suspension (O.D. 600=0.6 X 10^9 cells/ml) 20 min

↓

Co-cultivation in EMM medium (pH=5.6) in darkness

↓ 5 d

Washing 2 times with sterile water and 2 times with liquid EMM containing sporidex (400 mg/l) and blot drying

↓

Transfer of embryos to EMM + sporidex (250 mg/l) + carbenicillin (250mg/l)

↓ 2 weeks

Selection on above medium + kanamycin (50 mg/l)

↓ 8-12 weeks

Transfer to MS + maltose (40 g/l) + trans-cinamic acid (3 g/l) + kanamycin (75 mg/l)

↓ 4 weeks

Transfer to MS + 1.5 mg/l GA_3 + kanamycin (75 mg/l)

↓ 7 weeks

Multiplication of transgenic shoots on MS+ TDZ (0.001 mg/l) + NAA (2 mg/l)

↓ 8 weeks

Micrografting on seedling rootstocks and growth in green house

Fig. 5.1 Protocol for *Agrobacterium*-mediated genetic transformation of tea. (Mondal et al. 2001b)

tive for tea transformation. In tea, AS probably played important role in T-DNA transfer, as reported in a wide range of plant species (Godwin et al. 1992). Another important factor that affects transformation efficiency is the selection of transgenic tissue on an antibiotic medium. In order to select transformed explants, selective agents are added into the medium to produce selective pressure. In general for tea, among the different antibiotics, hygromycin was more effectively used at a low concentration (20 mg/l) for somatic embryos followed by kanamycin with a range from 50 to 200 mg/l as a selection dose. In a wild species of tea, Tosca et al. (1996) found that 75 mg/l kanamycin was lethal to the tissues. Similarly, Mondal et al. (2001b) reported that 50 mg/l kanamycin followed by an elevated dose of 75 mg/l were the levels to be used for the effective selection of transgenic somatic embryos of tea. On the other hand, Matsumoto and Fukui (1998, 1999) found that 200 mg/l kanamycin was effective for tea when leaves were used as explants. Another prerequisite for genetic transformation is the molecular characterization of putative transformants. While Matsumoto and Fukui (1998, 1999) reported stable transformations in callus on the basis of PCR analysis and southern hybridization, Mondal et al. (2001b) used various tools such as gus staining, kanamycin selection, RT-PCR, and southern hybridization to confirm the stable transformation. This indicated the presence of the marker kanamycin (*nptII*) linked to the *gus* gene as a single "transfer-DNA (T-DNA) strand" in the genomic DNA of the transformed plants. The failure of the rest of the lines in yielding PCR-amplification products may be attribut-

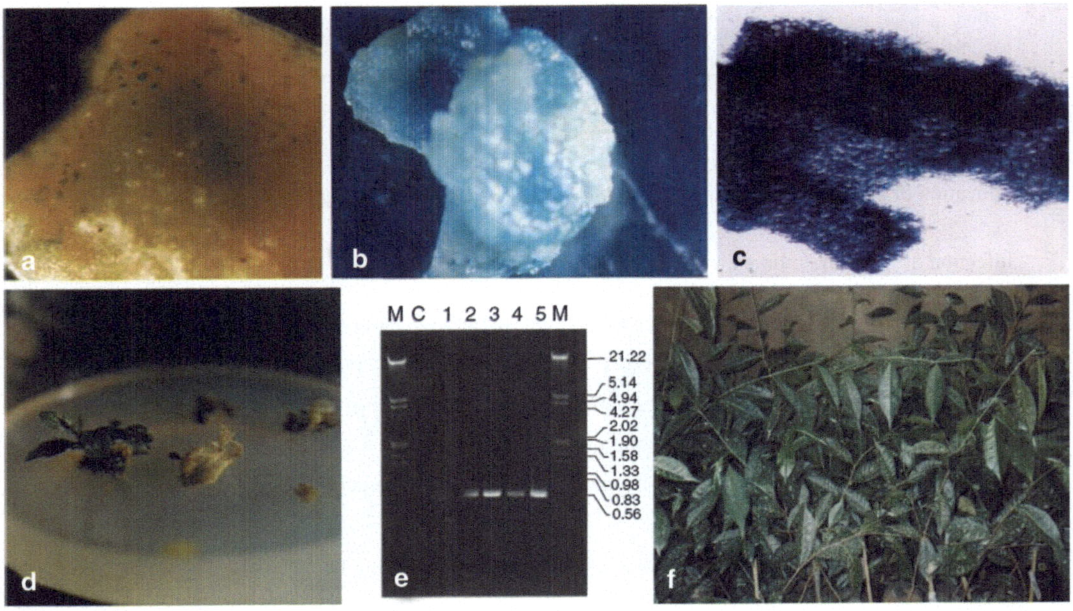

Fig. 5.2 *Agrobacterium*-mediated genetic transformation of tea somatic embryos: **a** *blue* spot indicating *gus* activity after 48 h of infection, **b** a *gus*-positive secondary somatic embryo of 4 months, **c** a hand section of leaf tissue of transformed shoot of 12 months, **d** kanamycin selection, **e** PCR amplification of a 693 bp fragment of *nptII* gene (lane *M*: DNA double digested with EcoR1 and HindIII; lane *C*: DNA from untransformed tea plant (control); lanes *1–5*: DNA from independently transformed plants), **f** the greenhouse-grown transgenic tea plants. (Adopted from Mondal 2008)

ed to the presence of "false positives" during the antibiotics selection. Further, when the leaves of these five independent transgenic plants were subjected to southern blot analysis, PstI-digested genomic DNA from each of the four putative transgenic lines generated an internal transgene fragment of 1.6 kb that hybridized to the *nptII* probe. Additional shorter fragments produced in some transgenic lines further indicated a deletion of a part of the T-DNA containing *nptII*. The deletion perhaps occurred during transformation or regeneration. Different banding patterns observed in the southern hybridization could be due to multiple insertions, rearrangement, and/or deletions of the integrated transgenes in the regenerated plants as is common to *A. tumefaciens*-mediated transformation (Mercuri et al. 2000). Interestingly, the efficiency of transformation was increased when the leaf explants of tea were pre-cultured on medium containing polyvidone at 16 g/l for 2–3 days before being infiltrated by *A. tumefaciens* (Wu et al. 2003). Although 80–90 % survival of transgenic plants was achieved under greenhouse conditions, the stability of the transgene remains to be elucidated as tea plants take years to flower and set seeds. The host range specificity between the bacterium and five different tea cultivars as well as an unrelated plant, *Artemisia parviflora*, having extreme surface characteristics was evaluated (Kumar et al. 2004). The degree of *Agrobacterium* infection in the five cultivars of tea was affected by leaf wetness, micro-morphology, and surface chemistry. Wet leaf surfaces of TV-1, UPASI-9, and Kangra jat showed a higher rate (75 %) of infection. This indicated that the leaves with a glabrous surface having lower Ø (large surface area covered by water droplet) and higher phenol and wax content were suitable for *Agrobacterium* infection. Caffeine fraction of tea promoted *Agrobacterium* infection even in leaves containing less wax (UPASI-10), whereas caffeine-free wax inhibited both *Agrobacterium* growth and infection. Thus, this study suggested the importance of leaf surface features in influencing the *Agrobacterium* infection in tea leaf explants. Considering the fact that lepidopteran and coleopteran insects account for 31.5 and 18.8 % crop loss in tea, respectively, in China, Luo and Liang (2000) transformed tea plants with *Bacillus thuringiensis* (*Bt*) toxin gene. The vector pGA471 containing *Bt* gene was digested with HindIII and BglII and inserted into the vector pCAMBIA2301. The engineered plasmid with *Bt* gene (*cryIAc*), *gus* intron gene, and *nptII* was transformed into *Escherichia coli* and introduced into *Agrobacterium* strains LBA 4404 and EHA 105 through a triparental cross. They detected the transient expression of *gus* gene in calli and leaves of putative transgenic tea plants. However, no transgenic plants were recovered. Three genes including *rolb, Bt,* and *chitinase* had been transferred to tea cv. TTL-1 at the Research and Development Department, Tata Tea Ltd., Kerala, India for developing a tea plant with better yield and resistance to pest as well as to blister blight disease. Preliminary study indicated that there was no difference in quality of made tea in the transgenic plant but transgenic plant with *rolb* gene produced more root and shoot biomass in 2-year-old young transgenic plants (data not shown) (Mondal et al. 2004). Later, Lopez et al. (2004) also produced transgenic tea plants. Cotyledon-derived embryogenic callus cultures were co-cultivated with *A. tumefaciens* harbouring a binary vector carrying the hygromycin phosphotransferase (*hptII*), glucuronidase (*gus*), and green fluorescent protein (*gfp*) genes. Following co-cultivation, embryogenic calli were cultured in a medium containing carbenicillin (500 mg/l) for a week and then transferred on an antibiotic selection medium containing hygromycin (75 mg/l) for 8–10 weeks. Hygromycin-resistant somatic embryos were selected. The highest production efficiency of hygromycin-resistant calli occurred with co-cultivation for 6–7 days in the presence of AS (80 mg/l). Hygromycin-resistant somatic embryos developed into complete plantlets in a regeneration medium containing half-strength MS salts with 6-benzylaminopurine (BAP, 1 mg/l) and gibberellic acid (GA_3, 9 mg/l). Transformants were subjected to *gfp* expression analysis, GUS histochemical assay, PCR analysis, and southern hybridization to confirm transgene integration. The GFP is a stable, cell-autonomous fluorescent protein derived from jellyfish (*Aequorea victoria*). It has been extensively used as a non-destructive reporter gene for plant transformation (Chiu et al. 1996). After 30 days of co-cultivation, *gfp* expression was evident in calli. Non-transformed tissue from original explants did not express green fluorescence, and explant growth was suppressed. Leaves of transformed plantlets also expressed *gfp*, and the green fluorescence was easily detectable with blue light illumination of calli. Production of transgenic tea tissue was also reported by Jayaraman and Nithya (2005). Embryogenic tissues of tea were co-cultivated with *A. tumefaciens* strain LBA4404. The plasmid pBi121, with *nptII* gene providing kanamycin resistance as a selectable marker and the *β*-glucuronidase (*gus*) reporter gene, was used as binary vector. The highest transformation frequency was obtained with 5-day-old tissues grown in a liquid medium and co-cultivated with *Agrobacterium* for 2 days in the same medium but containing AS (10 mg/l). There was improvement in the recovery of kanamycin-resistant tissues when tissues were first grown for 10 days on a medium containing timentin (350 mg/l) to prevent bacterial overgrowth, before application of the selection pressure. Resistant tissues obtained after 6 weeks on the kanamycin selection medium showed a stable *gus* expression. PCR analysis demonstrated the presence of the transgenes, while southern hybridization confirmed their integration into the genome. Following this protocol, they could produce transgenic plants within 4 months after co-culture. As reported for other species (Bergmann and Stomp 1992), the physiological status of the tissue was considered to be an important factor for production of transgenic plants.

5.3 *Agrobacterium rhizogenes*

Several groups also transformed the tea plant with *Agrobacterium rhizogenes*. Zehra et al. (1996) infected *in vitro* raised tea leaves with *A. rhizogenes* A_4 strain. A cell density of 10^8/ml for 2 min infection followed by blotting on sterile filter papers and co-cultivation in dark for 2 days induced hairy roots after 35 days. The isolated hairy roots grew healthy in the liquid medium. Mannopine from these roots were analyzed through paper electrophoresis, which confirmed stable integration of the gene. Later, Konwar et al. (1998) also transformed 4–6-month-old *in vitro* grown tea shoots by infecting at the basal portion followed by co-cultivation in a liquid MS medium supplemented with indole-3-butyric acid (IBA, 5 mg/l) and rifampicin (100 mg/l). Root initiation from the basal portion of 66 % explants after 32–45 days of *in vitro* culture enabled convenient hardening of the shoots in pots or nursery beds.

John et al. (2009) transformed the *in vitro* raised leaves of UPASI-9 cultivars with *A. rhizogenes* strain MTCC 532. It had been found that use of AS (60 mg/l) had maximum positive effects which increased the transformation efficiency up to 70 %. MS supplemented with maltose (30 mg/l) along with indole-3-acetic acid (IAA, 5 mg/l) was found suitable for hairy root culture and accumulation of phenolic compounds. They also reported the higher accumulation of catechin in transgenic cell lines. However, no transgenic plants were produced.

Callus and hairy roots were obtained *in vitro* from the roots of five tea cultivars. The hairy roots were characterized by rapid growth, absence of geotropism, growth of dense white fibrils, and strong capacity for diversification. Two types of branching of hairy roots were observed: direct branching and branching around the stromata that initially appeared on the hairy roots. These results suggested the significant effect of the composition of the culture medium on the differentiation and multiplication of hairy roots. The level of callus induction significantly varied among the cultivars (Peng et al. 2004).

5.4 Biolistic-Mediated Transformation

Although no transgenic tea plants have produced via particle gun, a preliminary study on transient expression was reported by Akula and Akula (1999). Somatic embryos were bombarded with plasmid DNA (p2k7)-coated gold particles (1.5–3 μm diameter). The transformation vector p2k7, which was used in their study, was originally derived from binary vector pBI221 with *nptII* gene and *gus* gene. Both genes were driven individually by the cauliflower mosaic virus 35S promoter. The gold particles were coated with DNA (1 μg/μl) by precipitation with $CaCl_2$ (111 mg/ml) and spermidine (14.52 mg/ml). Optimization of various factors such as the distance between the site of delivery of the micro-projectile and the target tissue, helium pressure, and the state of target tissue to obtain transient expression were evaluated on the basis of GUS assay after 30–40 h of bombardment. Following bombardment, the highest transient expression levels (up to 1,085 blue spots/shot) were obtained in the somatic embryos using a helium pressure of 550 kPa with target tissue at a distance of 9.5 cm from the site of delivery of DNA. Mannitol pretreatment did not influence the transient expression as both control and treated cultures gave the same level of expression. However, further details of regeneration of transgenic somatic embryo were not mentioned.

The *gfp* gene with organelle target signals was introduced by particle bombardment or co-cultivation with *A. tumefaciens* using embryogenic callus in tea plant. Putative transformed embryos were obtained from the embryogenic callus grown on the medium with kanamycin. The *gfp* expression was observed by a spectral imaging system to eliminate the yellow-green auto-fluorescence in tea plant and Q-PCR analysis had been done to confirm the *gfp* gene integration into the tea genome and its expression (Kato et al. 2004).

5.5 Applications

Apart from developing the transgenic tea, the technique has been also attempted to silence the gene. Glutathione is an important antioxidant compound that restores redox balance inside the living cell which is produced by glutathione synthetase (*gs*). Silencing of *gs*, that reduced glutathione content in the somatic embryos of tea had been reported by Mohanpuria et al. (2008). A 457 bp *gs* fragment had been cloned to RNAi construct which was utilized for tea somatic embryo transformation via *A. tumefaciens*. In order to alleviate the constraints related to transgenic tea development, Mohanpuria et al. (2011) developed a rapid *Agrobacterium*-mediated root transformation system for tea. Later, that was used to develop the caffeine-free transgenic tea plants. They cloned 376 bp of *caffeine synthase* (*cs*) cDNA fragment in an RNAi construct (pFGC1008-CS), infected by wounding at the root elongation zone. The suppressed expression of *cs* gene and a marked reduction in caffeine and theobromine contents in young tea shoots were obtained after root transformation through *Agrobacterium*.

5.6 Conclusion

Despite the fact that the transgenic technology has tremendous scope for tea, no transgenic plants have been released commercially so far. Among the many reasons, a few could be as follows: (1) tea plant is not easily amenable for *Agrobacterium* infection, (2) non-availability of generic protocol, which can be applied for a wide range of varieties of tea plant, (3) being woody perennial, production of transgenic tea is time consuming. However, it is evident now that among the different techniques *A. tumefaciens*-mediated transformation has been attempted with different groups but transgenic tea plant is yet to be commercialized. On the other hand, though *A. rhizogenes* transformation in tea has been demonstrated, the technique has not been exploited commercially to produce the secondary metabolites so far, which will be immensely useful for a crop like tea. Contrary to the above techniques, work on particle bombardment is at very initial stage. It is noteworthy to mention here that perhaps transgenic tea has a unique advantage as, during processing, green tea leaves are exposed to a temperature of 120 °C at the drying step during which foreign gene product such as *Bt* toxins will be destroyed, which is critical for genetically modified foods globally for some agricultural crops.

References

Akula A, Akula C (1999) Somatic embryogenesis in tea (*Camellia sinensis* (L) O Kuntze). In: Jain SM, Gupta PK, Newton RJ (eds) Somatic embryogenesis in woody plants, Vol. 5. Kluwer Academic Publishers, Dordrecht, pp 239–259

Aoshima Y, Ugaki M, Niwa Y (2001) Investigation of gene delivery condition in tea callus by *Agrobacterium*-mediated transformation using high level expressing reporter gene. Bull Shizuoka Tea Exp Stat 23:29–36

Bergmann BA, Stomp AM (1992) Effect of host plant genotype and growth rate on *Agrobacterium tumefaciens* mediated gall formation in *Pinus radiata*. Phytopathology 82:1457–1462

Bhattacharya A, Mondal TK, Sandal I, Prakash O, Kumar S, Ahuja PS (2004) Genetic transformation of Tea. In: Curtis IS (ed) Transgenic crops of the world. Kluwer Academic Publishers, Dordrecht, pp 245–255

Chiu W, Niwa Y, Zeng W, Hirano T, Kobayashi H, Sheen J (1996) Engineered GFP as a vital reporter in plants. Curr Biol 6:325–330

Confalonieri M, Balestrazzi A, Cella R (1997) Genetic transformation of *Populus deltoides* and *Populus* x *euramericana* clones using *Agrobacterium tumefaciens*. Plant Cell Tissue Organ Cult 48:53–61

Godwin ID, Ford-Lloyd BV, Newbury HJ (1992) *In vitro* approaches to extending the host-range of *Agrobacterium* for plant transformation. Aust J Bot 40:751–763

Jayaraman PR, Nithya MS (2005) *Agrobacterium tumefaciens*-mediated transformation of embryogenic tissues of tea (*Camellia sinensis* (L.) O. Kuntze). Plant Mol Biol Rep 23:299–302

John KMM, Joshi DS, Mondal AKA, Kumar SR, Rajkumar R (2009) *Agrobacterium rhizogenes* mediated hairy root production in tea leaves (*Camellia sinensis* (L) O. Kuntze). Indian J Biotech 8:430–434

Kato M, Uematu K, Niwa Y (2004) Transformation of green fluorescent protein in tea plant. In: Proceedings of 2004 international conference O-Cha (tea) culture and science, pp 219–221

Konwar BK, Das SC, Bordoloi BJ, Dutta RK (1998) Hairy root development in tea through *Agrobacterium rhizogenes*-mediated genetic transformation. Two Bud 45: 19–20

Kumar N, Pandey S, Bhattacharya A, Ahuja PS (2004) Do leaf surface characteristics affect *Agrobacterium* infection in tea (*Camellia sinensis* (L.) O. Kuntze). J Biosci 29:309–317

Lopez JS, Raj Kumar R, Pius RK, Muraleedharan N (2004) *Agrobacterium tumefaciens*-mediated genetic transformation in tea (*Camellia sinensis* (L.) O. Kuntze). Plant Mol Biol Rep 22:201–204

Luo YY, Liang YR (2000) Studies on the construction of *Bt* gene expression vector and its transformation in tea plant. J Tea Sci 20:141–147

Matsumoto S, Fukui M (1998) *Agrobacterium tumefaciens* mediated gene transfer in tea plant (*Camellia sinensis*) cells. Jap Agrl Res Quart 32:287–291

Matsumoto S, Fukui M (1999) Effect of acetosyringone application on *Agrobacterium*-mediated transfer in tea plant (*Camellia sinensis*). Bull Nat Res Inst Veg Orna Tea. Shizuoka 14:9–15

Mercuri A, Beneditti LD, Burchi G, Schiva T (2000). *Agrobacterium*-mediated genetic transformation of African violet. Plant Cell Tissue Organ Cult 60:39–46

Mohanpuria P, Rana NK, Yadav SK (2008) Transient RNAi based gene silencing of glutathione synthetase reduces glutathione content in *Camellia sinensis* (L.) O. Kuntze somatic embryos. Biol Plant 52:381–384

Mohanpuria P, Kumar V, Ahuja PS, Yadav SK (2011) *Agrobacterium*-mediated silencing of caffeine synthesis through root transformation in *Camellia sinensis* L. Mol Biotech 48:235–243

Mondal TK (2007). Tea. In: Devey MR, Pua P (eds) Biotechnology in agricultural and forestry: transgenic crops V, Vol. 60. Springer, Germany, pp 519–536

Mondal TK (2008) Tea. In: Kole C, Hall TC (eds) A compendium of transgenic crop plants: plantation crops, ornamentals and turf grasses. Blackwell, London, pp 804–838

Mondal TK, Bhattachraya A, Sood A, Ahuja PS (1999) An efficient protocol for somatic embryogenesis and its use in developing transgenic tea (*Camellia sinensis* (L) O. Kuntze) for field transfer. In: Altman A, Ziv M, Izhar S (Eds) Plant biotechnology and *In Vitro* biology in 21st century. Kluwer Academic Publishers, Dordrecht, pp 101–104

Mondal TK, Bhattacharya A, Ahuja PS (2001a) Development of a selection system for *Agrobacterium*-mediated genetic transformation of tea (*Camellia sinensis*). J Plant Crops 29:45–48

Mondal TK, Bhattacharya A, Ahuja PS, Chand PK (2001b) Factor effecting *Agrobacterium tumefaciens* mediated transformation of tea (*Camellia sinensis* (L). O.Kuntze). Plant Cell Rep 20:712–720

Mondal TK, Bhattacharya A, Laxmikumaran M, Ahuja PS (2004) Recent advance in tea Biotechnology. Plant Cell Tissue Organ Cult 75:795–856

Mondal TK, Parathiraj S, Mohan Kumar P (2005) Micrografting-a technique to shorten the hardening time of micropropagated shoots of tea (*Camellia sinensis* (L.) O. Kuntze). Sri Lank J Tea Sci 70:5–9

Peng ZY, Liu DH, Xiao HJ, Zhang LX, Peng ZY, Liu DH, Xiao HJ, Zhang LX (2004) On the induction frequency of callus and hairy root in roots of tea plant. J Hunan Agril Univ 30:138–141

Sandal I, Saini U, Lacroix B, Bhattacharya A, Ahuja PS, Citovsky V (2007) *Agrobacterium*- mediated genetic transformation of tea leaf explants: effects of counteracting bactericidity of leaf polyphenols without loss of bacterial virulence. Plant Cell Rep 26:169–176

Saini U, Kaur D, Bhattacharya A, Kumar S, Singh RD, Ahuja PS (2012) Optimising parameters for biolistic gun-mediated genetic transformation of tea (*Camellia sinensis* (L.) O. Kuntze). J Hort Sci Biotech 87:605–612

Tosca A, Pondofi R, Vasconi S (1996) Organogenesis in *Camellia x williamsii*: cytokinin requirement and susceptibility to antibiotics. Plant Cell Rep 15:541–544

Wu S, Liang YR, Lu JL, Kim HS, Wu Y, Wu S, Liang YR, Lu JL, Kim HS, Wu Y (2003) Optimization of *Agrobacterium*-mediated and particle bombardment-mediated transformation systems in tea plant (*Camellia sinensis*). J Tea Sci 2:5–9

Zehra M, Banerjee S, Mathur AK, Kukreja AK (1996) Induction of hairy roots in tea (*Camellia sinensis* (L.) using *Agrobacterium rhizogenes*. Curr Sci 70:84–86

6 Molecular Markers

6.1 Introduction

In addition to the shortfalls described earlier in Chap. 2, progress of tea breeding had also been slowed down due to the lack of reliable selection criteria (Kulasegaram 1980). Various morpho-biochemical markers had been reviewed in past (Wachira 1990; Singh 1999; Ghosh-Hazra 2001), and it had been seen that they had marginally improved the efficacy of selection for desired agronomic traits in tea. This was mainly due to the fact that most of the morphological markers defined so far were influenced greatly by the environmental factors and hence showed a continuous variation with a high degree of plasticity. Therefore, these markers could not be separated into discrete groups for identification (Wickramaratna 1981). Recently, development of the molecular biology had resulted in alternative DNA-based markers for crop improvement of tea. These markers can assist the process of traditional breeding with several efficacies. The greatest advantages of molecular markers are: (1) free from the environmental influences and (2) detection of polymorphisms at an early stage of growth. The different markers, which have been employed for varietal improvement of tea and its wild relatives, are reviewed in this chapter.

6.2 Morphological Markers

Morphological traits controlled by a single locus can be used as genetic markers if their expression is reproducible over a range of environments (Staub et al. 1982). Tea had been classified into three different types based on morphological characters. Barua (1963) provided morphological and anatomical descriptions of these three types of tea, which were later elaborated by Bezbaruah (1971). Since then, several morphological as well as biochemical markers had been identified and used in the tea-breeding programme (Table 6.1). While leaf architect, pollen morphology (Chen et al. 1997b), growth habits, and floral biology were important criteria used by tea taxonomists (Banerjee 1992), the bush vigour, pruning weight, periods of recovery from prune, plant height, root mass, root-shoot ratio, plucking point density, dry matter production, and partitioning were considered as yield indicators in tea (Banerjee 1992). In tea, morphological markers had been used to study genetic diversity (Wickramaratna 1981; Toyao and Takeda 1999), variation (Gunasekara et al. 2001; Piyasundara et al. 2006; Su et al. 2007), phylogeny, and classification (Chen et al. 2005a; Vo 2006; Piyasundara et al. 2008; Pi et al. 2009). For example, in Sri Lanka, 20 genotypes of tea were characterized using 13 morphological descriptors and were grouped into

T. K. Mondal, *Breeding and Biotechnology of Tea and its Wild Species*,
DOI 10.1007/978-81-322-1704-6_6, © Springer India 2014

Table 6.1 Pre-genomics markers used in tea genetic improvement

Criterion	Reference
Quantity and shape of the scleriods	Barua (1958)
Bush vigour	Barua and Dutta (1971)
Leaf geometry	Banerjee (1987)
Volatile flavour compounds	Borse et al. (2002)
Leaf pose, colour, serration of the margin and angle	Eden (1976)
Chlorophyll content/photosynthesis rate	Ghosh-Hazra (2001)
Quantitative changes in chlorophyll *a* and chlorophyll *b* and carotenoids	Hazarika and Mahanta (1984)
Epicuticular waxes	Kabir et al. (1991)
Dry matter production and partitioning	Magambo and Cannell (1981)
Green leaf catechin and ratio of dihydroxylated to trihydroxylated catechins	Magoma et al. (2000)
Root lengths	Nagarajah and Ratnasurya (1981)
Catechins, Caffeine, and volatile flavour compounds	Owuor and Obanda (1998); Singh et al. (2013)
Leaf, floral biology, and growth morphology	Sealy (1958)
Chloroform test	Sanderson (1964)
Pruning litter weights	Satyanarayan and Sharma (1982)
Anthocyanin pigmentations in young leaves	Satyanarayan and Sharma (1986); Kerio et al. (2012)
Evenness of the flash, plucking density, and recovery time of pruning	Singh (1999)
Terpene index	Takeo (1981)
Leaf pubescences	Wight and Barua (1954)
Phloem index	Wight (1954)
Isozymes	Hairong et al. (1987); Xu et al. (1987); Ikeda et al. (1991); Chengyin et al. (1992); Anderson (1994); Singh and Ravindranath (1994); Yang and Sun (1994); Borthakur et al. (1995); Chen (1996); Sen et al. (2000); Fernández-Cáceres et al. (2001); Pedro et al. (2001); Jeyaramraja et al. (2002); Magoma et al. (2003); Neog et al. (2004); Chen et al. (2005a); Otaghvari et al. (2010); Rajanna and Ramakrishnan (2010)
Metal contents	McKenzie et al. 2010

4 classes based on principal component analysis (PCA). Of these, leaf width, leaf shape, leaf pigments, and petiole pigmentation were found to be the maximum contributors for morphological variation (Piyasundara et al. 2006).

Morphological diversity of three main taxa was studied, and the importance of different descriptors in categorizing accessions into distinct groups was also examined. Twelve accessions were characterized using 15 morphological descriptors of International Plant Genetic Resource Institute, Rome guidelines (IPGRI 1997). The results PCA on morphological characters showed that the first two principal components accounted for 44.77 % of the total variance. In the evaluated quantitative characters, all three taxa had a coefficient of variation (CV) greater than 24.85 %, and within the taxon the CV was greater than 9.59 %. The qualitative characters showed a wide range of variations and yielded significant differences ($p<0.05$). Phenotypic data had high contributing component loadings from characters such as leaf area, weight of harvested shoots, stem colour, leaf pubescence, and young shoot colour. Cluster analysis delineated the accessions into three groups. The implications of the results hold promise for assessing genetic diversity in germplasm collections, which was a prerequisite for their utilization, effective management, and crop improvement (Rajanna et al. 2011).

Variations in leaf anatomy were compared among the diploid, triploid, and tetraploid clones of tea. There were significant variations in the different attributes within and between the different groups. The number of stomata and epidermal cells per unit leaf area decreased with increasing ploidy, reflecting larger cells in the tetraploids than in the diploids and triploids. Specific leaf mass was also higher in the tetraploids (Ng Etich and Wachira 2003). Similarly, pollen morphology (Chen et al. 1992; Shu and Chen 1996) and pistal type (Yamaguchi 2001) were used to find evolutionary relationship among the different genotypes of tea and species of *Camellia*.

6.3 Artificial Neural Network (ANN): A Digital Marker

Use of artificial neural networks (ANNs) in identification of plants is gaining popularity. Applications of using ANN-based morphological traits for plant identification are well documented now (Clark and Warwick 1998). Genetic erosion assumes an alarming significance especially in those species in which genetic improvement has originated an extremely high number of cultivars, with a consequent loss or oversight of the ancient ancestors. Among these, *C. japonica* L. (Theaceae) represents an excellent example, totalling currently more than 30,000 cultivars. Thus, morphological characterization of 22 ancient *C. japonica* L. cultivars of historical garden Villa Orsi at Compito (Lucca, Italy) was done by using quantitative morphological (i.e. phyllometric) and fractal spectra traits originated from leaves image analysis, and their discrimination was done by building a specific ANN for data modelling. Results showed that the construction of a back-propagation neural network (BPNN) based on phyllometric and fractal analysis could be effectively and successfully used to discriminate *C. japonica* genotypes using simple dedicated instruments, such as a personal computer and an easily available optical scanner (Mugnai et al. 2008).

Low-cost, rapid, and accurate identification of tea varieties is always challenging. Fourier transform near-infrared (FT-NIR) spectroscopy coupled with the pattern recognition was used as a rapid and non-invasive analytical tool to identify individual tea varieties. Linear discriminant analysis (LDA) and ANN were compared to construct the identification models based on PCA. The experimental results showed that the performance of ANN model was better than LDA models. The overall results demonstrated that FT-NIR spectroscopy technology with ANN pattern recognition method was successfully applied as a rapid method to identify tea varieties (Chen et al. 2008).

Leaf characters had been successfully utilized to classify *Camellia* (Theaceae) species; however, leaf characters combined with supervised pattern recognition techniques had been explored to classify tea. Lu et al. (2012) studied the leaf morphological and venation characters of 93 *Camellia* species to assess the effectiveness of several supervised pattern recognition techniques for classifications such as learning vector quantization neural network (LVQ-ANN), dynamic architecture for ANN (DAN2), and support vector machines (SVM) and compared their accuracy. Among them, SVM offered the best classification accuracy. A hierarchical dendrogram based on leaf architecture data had confirmed the morphological classification of the studied species. The overall results suggested that leaf architecture-based data analysis using supervised pattern recognition techniques, especially DAN2 and SVM discrimination methods, was excellent for the identification of *Camellia* species (Lu et al. 2012). Following a similar approach, morphological diversity of 17 tea accessions belonging to Chinese, Assamese, and Shan tea (*C. sinensis* var. *pubilimba*) groups was studied. The ANN data were able to perform a correct identification for almost all the accessions using simple dedicated instruments (Pandolfi et al. 2009).

6.4 Biochemical Markers

To overcome the problems of morphological markers, attentions were given for searching alternative biochemical markers. Presence and quantity of calcium oxalate crystals in paranchymatous cells of leaf petiole of tea, known as the phloem index, had been suggested to be a suitable criterion to classify the hybrids (Wight 1958). The variations in the quantity and morphology of sclereids in the leaf lamina had also been utilised in differentiating tea taxa (Barua 1958; Barua and Dutta 1959). Caffeine, volatile compounds (Seurei 1996), green leaf pigmentation (Banerjee 1992), and leaf pubescence (Wight and Barua 1954) were some of the parameters which had been used as potential determinants of tea quality. Despite the disadvantages due to environmental influences, these markers were mostly adopted and were frequently used by tea breeders globally. Considerable success had been achieved in identification of tea biochemical indicators (Owuor et al. 1986; Takeo 1981). These indicators were also used widely to distinguish two species of tea, namely *C. sinensis* and *Camellia assamica* and their respective clones (Owuor 1989). Takeo (1983) suggested a chemotaxonomic method of classifying tea clones based on a ratio referred to as the Terpene index (T.I.), which expresses the ratio between linalools and linalools plus geraniols. Although not fully exploited, the polyphenol oxidase activity, individual polyphenol, amino acids, and chlorophyll content varied between the hybrids and are used as potential parameters in tea taxonomy (Sanderson 1964).

Paper chromatography was also used to investigate the taxonomy of different species of *Camellia* under the section *Thea*. It had been found that species within the *Thea* section of the genus *Camellia* were closely similar in chemical composition, the general pattern of which did not have any relation with the chemical composition of non-*Thea* camellias (Roberts et al. 1958). Although *Camellia saluenensis* was found to cross readily with *C. japonica,* identification of their F_1 hybrid known as *Camellia williamsii* at the young stage was a challenge for breeders. Interestingly, on the basis of flavour and other colouring compounds using paper chromatography, Parks and Case (1968) were successful in identifying the true hybrid seedlings of *C. williamsii,* indicating the suitability of this simple technique. Presence or absence of certain phenolic substances in the shoots had also been used in establishing relationship between different taxa (Roberts et al. 1958; Xu et al. 2004). Quantitative changes in chlorophyll *a* and chlorophyll *b* and four carotenoids (carotene, lutein, violaxanthine, and neoxanthine) were used for characterization of three types of tea (Hazarika and Mahanta 1984). Total green leaf catechin concentrations and ratio of dihydroxylated to trihydroxylated catechins were used to establish genetic relationship among the 102 Kenyan tea genotypes (Magoma et al. 2000). Accumulation and ratio of the various catechins were used to classify the three types of tea. Based upon the biochemical differentiation, it was found that there was a potential for broadening the genetic base of Assam teas more than the putative China and Cambod teas in Kenya.

A parameter of fresh tea leaf that correlates with black tea quality is highly desired. Twenty good and 20 poor quality tea clones were selected from the breeding programme at the Tea Research Foundation (Central Africa). The flavan-3-ol profile of fresh tea leaves was analysed by capillary electrophoresis while total theaflavin (TF) content was determined in the black tea manufactured from the same leaves for each clone. The above parameters were correlated with total scores and valuation from two tea tasters with regression analysis (Wright et al. 2000).

Similarly, TFs and total colour (TC) and high caffeine, total soluble solids (TSS), volatile compounds, and viscosity were used to carry out the fingerprinting of Indian tea. It had been found that teas of different origins had characteristic chemical constituent. This fingerprinting will help to trace the origin of teas (Borse et al. 2002). Catechin and its components had been widely utilized in studying the diversity of tea germplasm (Magoma et al. 2000; Gulati et al. 2009); yet, it was found that total catechin of tea leaves increased with the increase in sunlight exposure, suggesting that catechin biosynthesis was also dependent on the environment (Mariya et al. 2003) and geographical location of a plant (Wei et al. 2011).

Diversity of amino acids variation was also studied among the different tea genotypes. Differentiation of green, black, oolong, white, and pu-erh teas had been carried out according to their free amino acid contents. Using back-propagation multilayer perceptron ANN, 100% success in the classification was obtained. The most differentiating amino acids were found to be glutamic acid, asparagine, serine, alanine, leucine, and isoleucine (Alcázar et al. 2007).

Li et al. (2010) demonstrated that the polyphenols of tea leaves can be used as chemotaxonomic markers. They established the genetic relationship among 89 wild, hybrid, and cultivated tea trees from China and Japan with the following types of catechins: (1) (−)-epigallocatechin 3-O-gallate (EGCG), (2) (−)-epigallocatechin (EGC), (3) (−)-epicatechin 3-O-gallate (ECG), (4) (−)-epicatechin (EC), (5) (+)-catechin (CA), (6) strictinin (STR), and (7) gallic acid (GA). Of the 13 polyphenol patterns observed, PCA indicated that the structure types of the flavonoid B-rings, such as the pyrogallol-EGCG and EGC and catechol-ECG and EC types, greatly influenced the classification. A study based on morphological characteristics, geographical information, and historical information on tea trees showed good agreement with the current theory of tea-tree origins and concluded that the Xishuangbanna district and Puer City of China were among the original sites of the tea-tree species. The contents of gallic acid, epigallocatechin gallate, epigallocatechin, epicatechin, epicatechin gallate, catechin, caffeine, theophylline, and theobromine were determined to differentiate the 45 tea samples according to their geographical origins. Catechins, gallic acids, and tea alkaloids were found to be adequate chemical descriptors to distinguish tea samples cultivated in different geographical areas (Pedro et al. 2001).

The Indian tea germplasm comprising 'China', 'Assam', 'Cambod', and their hybrids, were evaluated using biochemical markers such as total catechins and their fractions for varietal identification and characterization of various made tea samples from northern India (Gulati et al. 2009) as well as southern India (Saravanan et al. 2005). Morphologically indistinct large-leaved 'Cambod' and 'Assam' varieties could not be differentiated using biochemical markers, since both varietal types taxonomically belong to a single species. Clones of 'China' type showed low total catechin content and catechin ratio which were distinctly grouped. The 'China-Assam' and 'China-Cambod' hybrids formed intermediate groups between 'China' and 'Cambod'/'Assam' PC groups. Tea clones which were differentially positioned in the PC group could be explained on the basis of genetic contribution by other varietal types as parents (Wright et al. 2002).

The variations of the main quality components such as polyphenols, catechins, amino acids, caffeine, and water extract of 596 Chinese tea accessions were analyzed. A wide range of biochemicals was determined. It had been reported that polyphenols were found to be from 13.6 to 47.8%, catechins ranged from 81.9 to 262.7 g/kg, amino acids content ranged from 1.1 to 6.5%, caffeine content from 1.2 to 5.9%, and water extract content from 24.4 to 57.0% (Chen and Zhou 2005).

The theanine content of the leaves of 27 species of *Camellia* plants was investigated. Although theanine was present in 21 species only, it was in much lower amount than the quantity detected in tea leaves. The major free amino acids in leaves of four species belonging to the genera *Schima* and *Eurya* were glutamic acid, aspartic acid, glutamine, asparagine, alanine, and proline, and the content of these amino acids was similar to or higher than theanine. Accumulation of free amino acids in these plants was generally lower than in *C. sinensis* var. *sinensis*. The theanine biosynthetic activity in roots was higher than that of leaves (Deng et al. 2010).

The inheritance of anthocyanin pigmentation was studied. Based on chemical analysis, in conjunction with geographical and literary information, it was suggested that the Xinan district, China was the site/centre of origin of the red-flowered (caused by anthocyanin pigmentation) *Camellia* species of which both *C. saluenensis* and *C. reticulata* had the most primitive anthocyanin content (Li et al. 2013). The variations of EGCG and EGC were the most prominent among the different tea varieties. High EGC content was found to be a characteristic of the Assam variety which was further corroborated through multivariate analysis (Sabhapondit et al. 2012). Although accuracy was higher than morphological markers, accumulations of chemicals were subjected to post-transcriptional modification which often restricted their utility as markers (Staub et al. 1982).

6.5 Metallic Markers

Tea is an infusion made from dried leaves and a good dietary source of essential trace metals for us. Therefore, it is necessary to consider the variations in elemental content of tea leaves among tea cultivars. Hence, several workers attempted to classify and to establish the metal fingerprint of tea (Pedro et al. 2001; Fernández-Cáceres et al. 2001). Thus, elemental fingerprinting technique, based on elemental contents (Al, B, Ca, Cd, Co, Cr, Cu, Fe, K, Mg, Mn, Na, P, Pb, and Zn) determined by inductively coupled plasma atomic emission spectrometry (ICP-AES) and multivariate statistical analysis, had been used to differentiate tea cultivars into various groups of specific geographical region. The ranges of element concentrations in the cultivars were in good agreement with those obtained in previous studies, and the level of most elements in the tea leaves was significantly different among the cultivars. The classifications of tea cultivars by PCA, hierarchical cluster analysis (HCA), LDA, and BPNN analysis were 100 % accurate in total. Each cultivar presented a distinctive elemental fingerprint, and the elements in tea leaves can be significant predictors in differentiating tea cultivars (Chen et al. 2009). Similar approaches were also used to fingerprint white, green, black, oolong, and pu-erh teas. Using probabilistic neural networks (PNN), it was demonstrated that the overall classification performance of metal contents was successful in tea (McKenzie et al. 2010).

6.6 Isozyme Markers

Genetic analysis of isozyme variation was used for cultivar identification in a wide range of plants (Ferguson and Grabe 1986; Hirai and Kozaki 1986). Similarly, in tea also, isozymes had been analysed by several workers for their genetic tendencies, cultivar identifications, and implications in hybrid breeding (Table 6.1). Among the isozymes, peroxidase and esterase had been the most widely studied on a wide range of tea cultivars and different *Camellia* species (Chen 1996; Borthakur et al. 1995; Chengyin et al. 1992; Yang and Sun 1994; Singh and Ravindranath 1994; Sharma and Deka 2002; Chung et al. 2003). Isozyme analyses of tetrazolium oxidase, aspartate aminotransferase, and alpha-amylase were studied among seven different tea cultivars and three different species (Sen et al. 2000). The electrophoretic analysis revealed both the qualitative and quantitative variations in the isozyme-banding pattern among different species of tea and their clones. The tetrazolium oxidase enzyme system showed the highest variability among all the enzymes studied. Cluster analysis using isozyme-banding pattern produced a dendrogram, which clearly differentiated characteristics of both clones and species studied. However, in general, the applications of isozyme study in tea were limited to few enzymes with inadequate polymorphisms (Wachira et al. 1995). Besides, various isozymes were also used to study the diversity of *C. japonica* (Wendel and Parks 1982; Wendel and Parks 1983; Ikeda et al. 1991). Interestingly, Wendel and Parks (1984) did a classical study to conclude that alcohol dehydrogenase isozymes in *C. japonica* were encoded by two genes *Adh*-1 and *Adh*-2. Both loci were expressed in seeds and their products were randomly associated with intragenic and intergenic dimmers. Electrophoresis of leaf extracts produced only the products of *Adh*-2. Formal genetic analysis indicated that the two *Adh* loci were tightly linked. Most segregation fit the expected Mendelian ratios, but in some families distorted segregation was also observed at *Adh*-1, *Adh*-2, or both. Further, starch gel electrophoresis was used to score allelic variation at 20 loci in *C. japonica* collected from 60 genotypes distributed throughout Japan. In comparison with other plant species, the genetic diversity within the population was very high, i.e. 66.2% of loci were polymorphic per population, which gave an average mean number of 2.16 alleles per locus. They also reported genotypic proportions at most of the loci in majority of all the populations and found a good fit of the Hardy–Weinberg expectations (Wendel and Parks 1985).

6.7 Cytological Markers

Cytological markers of the genus *Camellia* were elaborately studied in the early 1970s with many interesting features. Chromosome number had been established for the most available species of *Camellia* including tea (Bezbarua 1971), which was reviewed by Kondo (1977). Generally, tea chromosomes are small in size and tend to clump together due to 'stickiness'. Tea is diploid ($2n=30$; basic chromosome number, $x=15$) and karyotype ranges from 1.28 to 3.44 μ (Bezbaruah 1971). The *r* value (ratio of long arm to short arm) for all the 15 pairs of chromosomes range from 1.00 to 1.91. This consistency in diploid chromosomes number suggests a monophyletic origin of all *Camellia* species. However, few higher ploidy levels such as triploids, e.g. TV-29, HS-10 A, UPASI-3, UPASI-20 ($2n=45$), tetraploids ($2n=60$), pentaploids ($2n=75$), and aneuploids ($2n\pm1$ to 29) had also been identified (Singh 1980). Karyotypic data had also been accumulated in past for the other species of this genus (Fukushima et al. 1966; Ackerman 1971; Kondo 1975; 1978a, b; 1979). In karyotype analysis, unfortunately, grouping by chromosome size was difficult in the *Camellia* taxa, since the chromosomes grade imperceptibly from the largest to the smallest. Furthermore, even in the best preparation, homologous chromosome pairs do not appear identical in *Camellia* (Kondo 1975; 1978a, b). Relatively little intraspecific karyotypic variation had been observed in the cultivated species of *Camellia* studied (Kondo 1975). Sat-chromosomes in karyotypes within mass accessions of certain *Camellia* species were morphologically and quantitatively variable. Thus, karyotypes including characteristics of sat-chromosomes were not of taxonomic significance for *Camellia* taxa. Among the diploid species of *Camellia* studied, *C. japonica* L. sensu lato showed the greatest karyotypic variation; many of the accessions studied indicated similar karyotypic patterns to each other (Kondo 1975). For instance, *C. japonica* L. var. *Spontanea* (Makino), *C. japonica* L. var. *macrocarpa* Masamune, *C. japonica* L. subsp. *rusticana* (Honda) Kitamura, and four cultivars, namely 'Aka-Wabisuke', 'Fukurin-Wabisuke', 'Kuro-Wabisuke' and 'Wabisuke', carried the same, most common standard aceto-orcein-stained karyotype, if the presence of satellites was not considered, and 16 metacentric, 8 submetacentric, and 6 subtelocentric chromosomes. Actually, *C. japonica* L. var. *Macrocarpa* Masamune had satellites on four submetacentric chromosomes, and the other accessions had satellites on two submetacentric chromosomes (Kondo and Parks 1980). Prior to this, it was shown that the C-banding method can be applied to the somatic mid-metaphase chromosomes in *Camellia* taxa (Kondo and Parks 1979). These differentially stained bands in somatic midmetaphase chromosomes permitted the identification of individual chromosomes and made it possible to match the homologous pairs of chromosomes more precisely. Karyotypic variability and divergence among the seven accessions of *C. japonica* L. *sensu lato* with the same aceto-orcein-stained karyotype were revealed by C-banding method (Kondo and Parks 1981). In this way, the cytological marker was used to sort and classify a vast number of cultivars. However, due to the development of more sensitive DNA-based markers, attention was shifted towards the search of molecular markers.

6.8 DNA-Based Markers

A limited number of loci of morphological, biochemical as well as cytological markers showed lesser polymorphism, but with the advancement of sequencing techniques, such efforts were shifted towards the study of various molecular markers (Table 6.2).

Table 6.2 Molecular markers used in tea and other *Camellia* species (Mukhopadhyay et al. 2014)

Country	Marker	Objectives of the study	Population size	Reference
India	RAPD	Genetic fidelity	17 micropropagated plants	Balasaravanan et al. (2002)
	AFLP	Genetic diversity	49 genotypes	Balasaravanan et al. (2003)
	SSR	Development of SSR	21 genotypes	Bali et al. (2013)
	RAPD	Genetic diversity	14 genotypes	Bera and Saikia (2002)
	RAPD	Genetic fidelity	7 micropropagated plants	Borchetia et al. (2009)
	RFLP	Detection of adulteration in made tea	10 different tea samples	Dhiman and Singh (2003)
	AFLP, SSR, and RAPD	QTL identification	42 genotypes	Kamunya et al. (2010)
	AFLP	Genetic diversity	14 genotypes	Karthigeyan et al. (2008)
	AFLP	Genetic diversity	29 genotypes	Mishra and Sen-Mandi (2001)
	AFLP	Marker for drought tolerant	29 genotypes	Mishra and Sen-Mandi (2004)
	AFLP	Genetic diversity	29 genotypes	Mishra et al. (2009)
	RAPD	Genetic diversity	27 genotypes	Mondal (2000)
	ISSR	Genetic diversity	25 genotypes	Mondal (2002)
	RAPD	Development of DNA isolation technique	28 genotypes	Mondal et al. (2000)
	RAPD	Genetic fidelity	18 micropropagated plants	Mondal and Chand (2002)
	AFLP	Genetic diversity	32 genotypes	Paul et al. (1997)
	AFLP	Genetic diversity	32 genotypes	Rajasekaran (1997)
	AFLP	Genetic diversity	1,644 genotypes	Raina et al. (2012)
	RAPD and ISSR	Genetic diversity	10 genotypes	Roy and Chakraborty (2007)
	RAPD and ISSR	Genetic diversity	21 genotyeps	Roy and Chakraborty (2009)
	SSR	Development of SSR marker	32 genotypes of tea and 2 genotypes of *C. japonica*	Sharma et al. (2009)
	AFLP	Genetic diversity	123 genotype	Sharma et al. (2010)
	SSR	Cross-species validation of SSR	18 genotypes	Sharma et al. (2011)
	5S rDNA	Phylogenetic relationship	28 genotypes	Singh and Ahuja (2006)
	RAPD	Authentication of made tea sample	11 genotypes	Singh et al. (1999)
	RAPD	Genetic fidelity	16 vegetative propagated plants	Singh et al. (2004)
	RFLP	Genetic organization	1 tea and 11 non-tea species	Singh et al. (2011)
	ISSR	Genetic fidelity of somaclonal variant	20 genotypes	Thomas et al. (2006)
JAPAN	SSR	Identification of cultivars	41 genotypes	Kato et al. (2008)
	*mat*K	Genetic diversity	118 genotypes	Katoh et al. (2003)
	SSR	Genetic diversity	24 genotypes	Kaundaun and Matsumoto (2002)
	STS-RFLP	Authentication of made tea sample	46 samples	Kaundan and Matsumoto (2003a)
	CAPS	Species-specific probe	50 genotypes	Kaundan and Matsumoto (2003b)

Table 6.2 (continued)

Country	Marker	Objectives of the study	Population size	Reference
	SSR	Maternal inheritance of cp genome	6 genotypes	Kaundun and Matsumoto (2011)
	RFLP	Genetic analysis	30 genotypes	Matsumoto et al. (1994)
	RFLP	Genetic diversity	45 genotypes	Matsumoto et al. (2000)
	RFLP	Genetic analysis of *pal* gene	472 genotypes	Matsumoto et al. (2002)
	RFLP	Genetic diversity of *pal* gene	297 genotypes	Matsumoto et al. (2004)
	SSR	Development and validation of SSR	157 genotypes	Ohsako et al. (2008)
	e-RAPD	Development of DNA isolation	16 genotypes	Tanaka and Taniguchi (2002)
	RAPD	Identification of true crosses	2 parents and 38 F_1 progenies	Tanaka and Yamaguchi (1996)
	RAPD	Maternal inheritance	14 genotypes of F_1 population	Tanaka et al. (2006)
	Gene-specific PCR-RFLP	Phylogenic relationship	162 genotypes of 49 *Camellia* species	Tanikawa et al. (2008)
	SSR	Development of SSR	4 different species	Ueno et al. (1999)
	SSR	Genetic diversity	518 genotypes of *C. japonica*	Ueno et al. (2000)
	SSR, ALP	Genetic diversity	22 genotypes of *C. japonica*	Ueno and Tsumura (2009)
	CAPS	Cultivar identification	63 genotypes	Ujihara et al. (2011)
China	RAPD	Genetic analysis	23 species	Chen and Yamaguchi (2002)
	RAPD	Discrimination of genotypes	20 species	Chen and Yamaguchi (2005)
	RAPD	Standardization of DNA isolation protocol	15 genotypes	Chen et al. (1997b)
	RAPD and RFLP	Development of technique	7 genotypes	Chen et al. (1997a)
	RAPD	Genetic analysis	7 genotypes	Chen et al. (1998a)
	RAPD	Genetic analysis	15 genotypes	Chen et al. (1998b)
	RAPD	Genetic analysis	5 genotypes	Chen et al. (1999)
	RAPD	Genetic analysis	24 genotypes	Chen et al. (2002b)
	RAPD	Molecular phylogeny	24 species	Chen et al. (2002b)
	RAPD	Genetic polymorphism	24 species	Chen et al. (2002c)
	RAPD	Genetic diversity	15 genotypes	Chen et al. (2005b)
	RAPD	Genetic diversity	45 genotypes	Fang et al. (2003)
	ISSR	Genetic diversity	36 genotypes	Fan et al. (2010)
	SSR	Genetic diversity	185 genotypes	Fang et al. (2012)
	ISSR	Genetic diversity	14 genotypes	Hou et al. (2007)
	AFLP	Linkage map	69 F_1 genotypes	Huang et al. (2005)
	AFLP	Genetic diversity	40 genotypes	Huang et al. (2006b)
	RAPD and ISSR	Genetic mapping	94 F_1 genotypes	Huang et al. (2006a)
	SNP	Identification of SNP	40 genotypes	Huang et al. (2007)
	RAPD	Genetic diversity	23 genotypes	Hui et al. (2004)
	SSR	Development of SSR markers	24 genotypes of tea and 2 genotypes of *C. japonica*	Hung et al. (2008)
	ISSR	Genetic diversity	10 genotypes	Ji et al. (2007)
	ISSR	Genetic diversity	181 genotypes	Ji et al. (2011)
	SSR	Development of SSR markers	10 genotypes	Jin et al. (2006)

Table 6.2 (continued)

Country	Marker	Objectives of the study	Population size	Reference
	SSR	Genetic diversity	42 genotypes	Jin et al. (2007)
	RAPD	Genetic diversity	2 genotypes	Li et al. (2003)
	RAPD	Genetic diversity	69 genotypes	Li et al. (2005)
	RAPD	Genetic diversity	23 genotypes	Li et al. 2007
	ISSR	Genetic diversity	10 genotypes	Li et al. (2012)
	ISSR	Genetic diversity of made tea	26 genotypes	Lin et al. (2010)
	SSR	Development of BAC library for tea genome sequence	1 genotypes	Lin et al. (2011)
	ISSR	Genetic diversity	25 genotypes	Liu et al. (2008)
	ISSR	Genetic diversity	134 genotypes	Liu et al. (2010)
	pal gene, *rpl32-trnL* spacer	Genetic diversity and erosion	21 genotypes of *C. taliensis*	Liu et al. (2012a)
	ISSR	Genetic diversity	134 genotypes	Liu et al. (2012b)
	RAPD	Genetic relationship	21 genotypes	Luo et al. (2002b)
	RAPD	Parentage identification	15 genotypes	Luo et al. (2002a)
	RAPD	Genetic diversity	71 genotypes	Luo et al. (2004)
	ISSR	Genetic diversity	60 genotypes of *C. changii*	Luo et al. (2007)
	SSR	Genetic diversity	45 genotypes of 6 different species	Ma et al. (2010)
	SSR	Validation of markers	21 genotypes	Ma et al. (2012)
	RAPD	Genetic diversity	15 genotypes	Shao et al. (2003a)
	RAPD	Genetic diversity	25 genotypes	Shao et al. (2003b)
	RAPD	Genetic diversity	240 genotypes	Shen et al. (2007)
	SRAP	Genetic diversity	25 genotypes	Shen et al. (2009)
	RAPD	Genetic polymorphism	27 genotypes	Shen et al. (2002)
	RAPD	Phylogenic relationship	11 different species	Shi et al. (1998)
	SSR	Development of SSR markers	12 genotypes	Shi et al. (2013)
	RAPD, AFLP	Genetic diversity	120 genotypes of *C. nitidissima*	Tang et al. (2006)
	IST	Phylogenic relationship	30 species of *Camellia*	Tian et al. (2008)
	RAPD, AFLP	Genetic analysis	48 genotypes	Wachira et al. (2001)
	ISSR	Genetic diversity	114 genotypes of *C. reticulata*	Wang and Ruan (2012)
	RAPD	Genetic diversity	36 genotypes	Wang et al. (2007)
	RAPD	Genetic diversity	15 genotypes	Wang et al. (2010)
	RAPD	Genetic diversity	15 genotypes	Wang et al. (2011)
	ISSR	Genetic diversity	84 genotypes of *C. euphlebia*	Wei et al. (2005)
	ISSR	Genetic diversity	250 genotypes of *C. nitidissima*	Wei et al. (2008)
	SSR	Development of SSR markers	25 genotypes of *C. nitidissima*	Wei et al. (2010)
	RAPD	Genetic diversity	27 genotypes	Wen et al. (2002)
	SSR	Development of SSR markers	150 genotypes of *C. chekiangoleosa*	Wen et al. (2012)
	RAPD	Genetic diversity	31 genotypes	Wu et al. (2002a)
	RAPD	Parentage identification	3 hybrids	Wu et al. (2002b)
	EST-SSR	Development of SSRs	20 genotypes	Wu et al. (2013)
	AFLP	Genetic diversity	34 genotypes	Xiao et al. (2007)

Table 6.2 (continued)

Country	Marker	Objectives of the study	Population size	Reference
	SSR	Development of SSR markers	24 genotypes of *C. taliensis*	Yang et al. (2009)
	RAPD	Germplasm evaluation	200 genotypes	Yang et al. (2003)
	ISSR	Parent selection	48 genotypes	Yao et al. (2008)
	SSR	Genetic diversity	450 genotypes	Yao et al. (2012)
	RAPD	Genetic classification	45 genotypes	Zhang et al. (2003a)
	RAPD	Genetic classification	45 genotypes	Zhang et al. (2003b)
	ISSR	Identification of clones	10 genotypes of *C. oleifera*	Zhang et al. (2007)
	AFLP	Genetic diversity	23 genotypes	Zhao et al. (2006)
	SSR	Genetic diversity	40 genotypes	Zhao et al. (2008)
Kenya	RAPD, AFLP	Genetic linkage map	90 genotypes of F1 progenies	Hackett et al. (2000)
	RAPD, ISSR	Assessment of mating system	180 progenies of 6 genotypes	Muoki et al. (2007)
	RAPD	Genetic analysis	38 genotypes	Wachira et al. (1995)
	RAPD	Diagnosis of gene introgression	28 genotypes	Wachira et al. (1997)
	RAPD and AFLP	Genetic diversity	24 genotypes	Wachira et al. (2001)
Italy	SSR	Genetic relationships	132 accessions of *Camellia* species	Caser et al. (2010)
Taiwan	RAPD, ISSR	Genetic analysis	37 genotypes	Lai et al. (2001)
	RAPD	Genetic diversity	96 genotypes	Lou et al. (2004)
	Intronic sequence of *RPB2* gene	Barcoding of Species	25 genotypes of 4 species	Su et al. (2009)
	ITS of nrDNA	Phylogenic relationship	7 species	Vijayan and Tsou (2008)
	Nuclear ITS	Molecular taxonomy	112 species of *Camellia*	Vijayan et al. (2009)
Turkey	AFLP	Genetic diversity	32 genotypes	Kafkas et al. (2009)
SRI Lanka	RAPD	Genetic Diversity	39 genotypes	Mewan et al. (2005)
	SSR	Genetic map	3 genotypes	Mewan et al. (2007)
	SSR	Genetic diversity	27 genotypes	Ariyaratne et al. (2009)
	RAPD	Genetic diversity	46 Genotypes	Goonetilleke et al. (2009)
South Africa	RAPD	Genetic diversity	5 genotypes	Wright et al. (1996)
	RAPD	Development of trait-specific marker	18 genotypes	Mphangwe et al. (2013)
South Korea	RAPD	Genetic analysis	6 genotypes	Kaundan and Park (2002)
	RAPD	Phylogenic relationship	27 genotypes	Kaundan et al. (2000)
	AFLP	Genetic analysis	37 genotypes	Lee at al. (2003)
	RAPD	Genetic diversity	20 genotypes	Park et al. (2002)
Malaysia	RAM	Genetic variation	6 genotypes	Latip et (2010)
Pakistan	RAPD	Genetic diversity	75 genotypes	Afridi et al. (2011)
	RAPD	Genetic diversity	24 genotypes	Gul et al. (2007)
Portuguese	RAPD	Genetic analysis	71 genotypes	Jorge et al. (2003)
The USA	SBA	Evolutionary studies	30 species	Prince and Parks (1997)
	SBA	Phylogenic relationship	19 species	Prince and Parks (2000)
	SBA	Evolutionary studies	35 species	Prince and Parks (2001)
The UK	SSR	Development of SSR markers	15 genotypes	Freeman et al. (2004)

SBA sequence based analysis, *ALP* amplicon length polymorphism, *RAM* random amplified microsatellite markers

6.8.1 Random Amplified Polymorphic DNA (RAPD)

Since their discovery, random amplified polymorphic DNA (RAPD) markers have been used in a number of areas in population genetics. Initially, it was the most preferred DNA marker due to its greater speed, ease of performance, and non-requirement of radioactive materials (Williams et al. 1990). In tea and other species of *Camellia,* a considerable amount of work has been carried out which is summarized below.

6.8.1.1 Germplasm Characterization and Genetic Diversity

Present day tea plantation is developed largely from the selected genotypes based on the performance of yield, quality, and tolerance to the various biotic and abiotic stresses. As a consequence, widespread cultivation of clonal tea diminished the genetic diversity. Therefore, germplasm characterization at the molecular level of tea will help: (1) in varietal improvement of tea for agronomically important characters, (2) to preserve the intellectual property right of tea breeders, (3) to identify an individual tea cultivar by making a molecular passport, (4) to prevent the duplicate entry of different genotypes in tea gene pool, (5) in efficient selection of the varieties for hybridization programme, graft compatibility in composite plant production, etc., and (6) in taxonomic classification of tea genotypes on the basis of molecular markers.

Wachira et al. (1995) were the first to characterize 38 different cultivars of Kenyan tea. A total of 23 primers were used, which could generate 157 polymorphic bands. The maximum polymorphism of 20 bands was detected by primer SC10–56. The similarity matrix among the clones ranged from a minimum of 43 % to a maximum of 96 %, respectively. Based on average linkage cluster analysis, a dendrogram was constructed which clearly discriminated different genotypes of Assam, Cambod, and China tea. Further, to examine the evolutionary relationships, PCA was undertaken which was able to classify three varietal types of tea in a manner that was consistent with both the present taxonomy of tea and with the known pedigrees of some clones. In the same year, Tanaka et al. (1995) used several 10-mer and 12-mer primers to detect variation among Korean, Japanese, Chinese, Indian, and Vietnamese tea. Among all primers, OPF-2 was found to be the most polymorphic. Their study concluded that after introduction from China, Korean tea underwent little genetic diversification. On the contrary, Japanese tea showed a closer relationship with their Chinese and Indian counterparts, which revealed the fact that tea in Japan might have been brought from China and India. Twenty-five Indian tea cultivars and two ornamental species were characterized using RAPD markers (Mondal 2000). Out of the 40 random 10-mer primers, 11 generated polymorphic loci which were ranged from 7 to 21 per genotype. Out of the total 154 alleles, 138 were polymorphic resulting in a high genetic variability of 95.2 %. Shannon's index of diversity was used to partition the total phenotypic variations into intra and interpopulation components. On an average, 57 % within and 43 % between populations variability was revealed. A dendrogram was constructed on the basis of band sharing which separated the population into three clusters, i.e. China, Assam, and Ornamental type. The PCA revealed that the chinary clones were more dispersed than Assam clones, indicating a greater genetic diversity among chinary clones. The results showed that RAPD markers could be used effectively to distinguish and characterize Indian tea germplasm. Subsequently, several workers across the tea growing countries used RAPD markers to study the genetic diversity as well as to make the fingerprints of their own clones. This was primarily due to the simplicity and easy-to-perform nature of RAPD markers (Table 6.2).

6.8.1.2 Detection of Genetic Fidelity Among *In Vitro* Raised Plants

The most important part of any *in vitro* propagation system is mass multiplication of plantlets which are phenotypically uniform and genetically akin to the mother/donor plant; otherwise, the advantage of having desirable characters of elite supreme clones will not be achieved. Several approaches have been applied for identifying variants among micropropagated plants. These are phenotypic variation (Vuylsteke et al. 1988), karyotypic analysis of metaphase chromosomes (Jha and Sen 1992), and biochemical analysis (Damasco et al. 1996). Importantly, a major disadvantage of these techniques was the limited number of informative markers and the influence of environmental conditions or developmental processes (Rani et al. 1995). Due to these limitations, the above approaches were not fully suitable for detecting DNA sequence polymorphisms of *in vitro* raised plants. On the other hand, RAPD had been used very advantageously for a number of crop species to detect genetic diversity among micropropagated plants (Isabel et al. 1993; Rani et al. 1995; Damasco et al. 1996). Mondal and Chand (2002) did RAPD analysis with 50 primers to investigate genetic variability of *in vitro* raised tea plants of cultivar T-78. Among them, 39 primers developed 197 monomorphic bands in all the concerned plants. The primer SC 10–12 produced maximum eight bands, whereas primer SC 10–57 produced single monomorphic band. Twenty-four out of 221 bands produced by the remaining 11 primers (SC 10–15, SC 10–56, SC 10–71, SC 10–83, OPA-12, OPA-15, OPA-18, OPM-06, OPM-09, OPM-12, and OPM-18) were polymorphic for four micropropagated tea plants. These 24 loci were monomorphic for the remaining plants. The polymorphic amplification products of 11 primers found to be present in the four plants were compared to the remaining 14 plants by conducting a second DNA extraction from respective plants. In a separate experiment, survey with these 11 primers revealed monomorphism of the amplified DNA fragments among five vegetatively propagated plants of the same cultivar (T-78). Therefore, it was concluded that variation occurred as a result of mutation during micropropagation. Interestingly, the marker profiles among these four plants were identical with all 11 primers suggesting a complete homogeneity among them. Various studies indicated that *in vitro* culture produced abundant cytological anomalies affecting both structural and numerical chromosome constitutions, which can be correlated with many phenotypic abnormalities (Evans and Sharp 1986). Subsequently, several workers used RAPD markers to detect the genetic fidelity of *in vitro* as well as vegetatively derived tea plants (Table 6.2). Various reasons had been assigned to the occurrence of variability among *in vitro* raised plants. Smith (1998) reviewed the factors contributing to this variation and divided them into two, intrinsic factors, which largely depend on the genetic stability of the explants, and extrinsic factors such as culture media, particular growth regulators, etc. Different types of cultured tissue produced different genetic instability. They had been ranked in the order of low to high from micropropagation of isolated shoot tips and meristems, adventitious shoots, somatic embryos, organogenesis from callus, and protoplast-derived cells (Scowcroft 1984). Such variants at times had a negative effect, affecting the high commercial value of the crop such as tea. Importantly, Hammerschlag (1992) suggested that somaclonal variations among tissue-cultured plants could be either eliminated or minimized if special efforts are made to distinguish between axillary and adventitious shoots that are produced during *in vitro* propagation and roughing out of the adventitious shoots. However, it is not always possible to monitor these operations fully in commercial tissue culture laboratory where wages of the laboratory workers are given on the basis of the number of plants subcultured per day. Therefore, polymorphic RAPD primers can be used for testing the genetic fidelity of micropropagated tea plant.

6.8.1.3 Cultivar Identification

RAPD markers were also employed for identification of the true-crossing progenies in tea-breeding programme and to determine the relationship between parents and their hybrids. For example, two Japanese tea cultivars, Yutakamidori and Meiryoku, for which parentage identification for registration documents was identified using this marker (Tanaka and Yamaguchi 1996). Wright et al. (1996) used the same technique to characterize five different South African tea cultivars, namely SFS 150, SFS 204, PC1, PC81, and MFS87. Of the 20 arbitrary primers tested, only one (ABI-17) yielded a unique set of fingerprints for each cultivar, which allowed cultivar discrimination. Singh et al. (1999) isolated DNA from ten different processed dried commercial black and green tea samples. The isolated DNA was subjected to polymerase chain reaction (PCR) amplification by random primers. They demonstrated that this method had tremendous potential for testing the originality of commercial tea and for the identification of cultivars used by a tea manufacturer for a particular brand. Mondal et al. (2000) described a simple method of DNA isolation from eight different polyphenol-rich genus as well as from 20 commercially important tea cultivars. The method did not require liquid nitrogen or phenol purification step. The DNA was successfully used as a template for PCR amplification, which indicated the wide applicability of the marker. Liang et al. (2000) investigated the possibility of classification and identification of tea as well as closely related species. The results showed that the RAPD markers could specifically discriminate between species and varieties. While both Assam and China tea had a specific band, Japanese tea was closer to Chinese tea than others. Some of the tea varieties from Vietnam were the hybrids of Assam and China. Tanaka et al. (2001) used RAPDs to identify the pollen parent of popular Japanese green tea cultivar 'Sayamakaori'. They had screened the female parent 'Yabukita' along with 78 putative male plants, most of which were introduced from China, and concluded that pollen parent of 'Sayamakaori' was not present among the putative population. The genetic diversity and molecular phylogeny of 24 ornamental *Camellia* species and varieties were investigated by RAPD analysis. Fifteen decamer oligonucleotide primers were selected from the 61 screened, which generated a total of 95.3 % polymorphism of the amplified bands. The molecular phylogenetic dendrogram of 24 species was constructed using UPGMA that generated two groups, corresponding to 3- and 5-locular ovary in morphology. The genetic relationship and the molecular phylogeny among section *Thea* were discussed by Chen and Yamaguchi (2002). Maternal inheritance of chloroplast DNA (cpDNA) in some cross progenies between *C. vernalis* and *C. japonica* was investigated using the polymorphism of *atpH-atpI* region by RAPDs. The cpDNAs of all *C. vernalis* cultivars showed the same type as those of *C. sasanqua,* and all the progenies from *C. vernalis,* either open-pollinated or crossed, had the same cpDNA type as their maternal plants (Tateishi et al. 2007).

Internal transcribed spacer (ITS) of nuclear ribosomal DNA (nrDNA) had been widely employed for reconstructing phylogenetic relationships in plants, especially at the species level. In order to assess the efficacy of nrITS in elucidating the interspecific relationships of *Camellia,* Vijayan and Tsou (2008) conducted an experiment with seven closely or distantly related species. Extensive study of *Camellia,* based on Pfu polymerase-amplified ITS sequences, showed well-resolved interspecies relationships. Thus, the potential of nrITS in deducing the phylogenetic relationships in *Camellia* was demonstrated. RAPD markers were used for identification of *C. japonica* and related species as well as their hybrids. A wide range of markers such as random 10-mer and chloroplast-specific sequences were used and checked with the previously published monogram on *Camellia* for phylogenetic relationship. Even, the taxonomic classification as mentioned in the Chang's manual for different *Camellia* species was confirmed using RAPD markers (Prince and Parks 1997, 2000; Thakor 1997; Tiao and Parks 1997, 2003; Yoshikawa and Parks 2001; George and Adam 2006; Orel et al. 2007). However, this being a dominant marker with limited degree of polymorphism, attention was shifted to alternative advance markers.

6.8.2 Inter-Simple Sequence Repeat (ISSR)

Inter-simple sequence repeat (ISSR) markers had been used for genetic characterization of various plant species (Tsumara et al. 1996). Because of greater length of the ISSR primers, they show greater repeatability and stability of map position in the genome comparing genotypes of closely related individual (Zietkiewicz et al. 1994).

Twenty-five diverse tea cultivars were analyzed using the simple sequence repeat anchored PCR (SSR-anchored PCR) or Inter SSR-PCR (ISSR). Based upon the polymorphism, 12 out of 45 primers were chosen for final study. These 12 primers amplified a total of 130 bands, of which 108 (84 %) were polymorphic. A dendrogram was constructed using UPGMA method which revealed three distinct clusters of Cambod, Assam, and China type, which concurred with the known taxonomical classification of tea. These results suggested that the ISSR-PCR method can be used potentially for genetic fingerprinting and taxonomic classification of tea genotypes (Mondal 2002). Subsequently, it had been noticed that ISSR markers had been used for a number of applications such as studying the genetic diversity of tea germplasm as well as different species of *Camellia,* genetic fidelity of tissue-culture-raised plants and for formulating the conservation strategies (Table 6.2).

6.8.3 Restriction Fragment Length Polymorphism (RFLP)

Restriction fragment length polymorphism (RFLP) markers had been used to investigate genetic diversity of cultivated plants and wild relatives (Tanksley et al. 1989). In tea, Matsumoto et al. (1994) cloned the phenylalanine ammonia-lyase (*pal*) gene using the rice cDNA as heterologous probe and used for studying the genetic variations of Japanese green tea using RFLP marker. On the basis of the numbers and molecular weight of the detected fragments, they classified Japanese green tea cultivars into five groups that had different routes of origin. All the Assam hybrids used in their study could be distinguished from Japanese green tea cultivars on the basis of the grouping. Furthermore, the inheritance of *pal* gene in tea was investigated. It was concluded that *pal* gene was a single copy per haploid genome and was inherited as a single gene according to the Mendelian ratio of 1 : 2 : 1. The work was further extended to the Japanese tea with green tea cultivars using *pal* as DNA marker. The DNA fragments detected by RFLP markers were named as A, B, and D which were inherited as multiple allelic genes at one locus. They concluded that most of the cultivars belonging to the AA genotype group had been selected from local tea plants about 50 years ago. On the other hand, all the cultivars in the BD genotype group were either from cultivar Yabukita or its hybrids. The other genotypes (AD, AB, and DD) included cultivars selected from local tea plants and their hybrids. To investigate the reason for the absence of any BB genotypes, allelic frequency of all genotypes were studied and found that B was the least among the three with a frequency of 0.08. Hence, the chance of occurrence of BB genotype was low, i.e. 0.0064 (0.08×0.08) (Matsumoto et al. 2002). RFLP technique was also applied to identify the processed tea sample. In Japan, tea produced from 'Yabukita' fetched better realisation in the market, and hence there was a trade tendency to adulterate the low-grade tea with Yabukita, which was difficult to detect either visually or through tester tong. To solve the problem, Kaundun and Matsumoto (2003b) employed the sequence tag site (STS)-RFLP using the sequence information of three genes, namely, *pal*, chalcone synthase, and dihydroflavonol 4-reductase. The restriction digestion of the specific amplification authenticated 46 tea samples.

6.8.4 Simple Sequence Repeat (SSRs)

Simple sequence repeat (SSRs) markers, also known as microsatellites, are tandemly repeated DNA sequence motifs (usually 2–5 bp long) that are highly polymorphic in plant genomes (Wu and Tanksley 1993). Due to their hypervariability, relative ease of scoring by PCR, codominant nature, and high reproducibility, they are considered to be one of the most reliable genetic markers.

Ueno et al. (1999) were pioneers to develop the SSRs from *C. japonica,* a closely related species of tea. Out of the total 339 amplifications by RAPD markers, 21 were found to contain microsatellite repeats. Finally, four primer pairs were designed, which yielded single locus polymorphic amplification products. Using these primer pairs, 53 *C. japonica* ecotypes were genotyped, and population genetic parameters were calculated. The following year, Ueno et al. (2000) investigated the spatial genetic structure of *C. japonica* using four of these microsatellite primers. Spatial distribution of individuals was also assessed to obtain an insight into spatial relationships between individuals and alleles. Morisita's index of dispersion plotted 518 individuals of *C. japonica* in a single clump, and Moran's I spatial autocorrelation coefficient revealed weak genetic structure, indicating a low level of allele clustering among the individuals. Subsequently, SSR markers had been used by several workers for cultivar identifications, study of genetic diversity of tea as well as wild *Camellia* species (Table 6.2). At certain occasions, SSRs had been mined *in silico* and validated either on diverse genotypes of tea or tested for their cross transferability on different species of *Camellia* (Yang et al. 2009).

6.8.5 Amplified Fragment Length Polymorphism (AFLP)

Amplified fragment length polymorphism (AFLP) markers being a reliable and the most robust DNA marker (Vos et al. 1995) can detect more number of polymorphisms than RFLPs or RAPDs. Thus, AFLP markers offer an opportunity to perform detailed genetic studies in closely related population (Meksen et al. 1995). In tea, Paul et al. (1997) were the first to employ AFLP markers to detect diversity and genetic differentiation of 32 tea clones comprising Indian and Kenyan origin. Five enzyme-primer combinations revealed a total of 73 unambiguous polymorphic bands. The size of polymorphic bands ranged from 106 to 218 bp. Genetic diversity within population showed that the chinary types were more variable than Assam or Cambod type. The similarity matrix coefficient varied from 35 to 96 %. The dendrogram constructed on the basis of shared fragments into three known types, i.e. Assam, China, and Cambod, was generally consistent with the existing knowledge on the biosystematics of tea. According to the PCA, Assam clones from India and Kenya clustered closely indicating a common ancestry. In the same year, Rajasekaran (1997) reported the AFLP analysis of 42 tea clones that comprised 23 UPASI, 17 popular South Indian estate clones, and 2 Kenyan tea clones. Therefore, it was concluded that 90 % of the UPASI clones were inbred and thus unsuitable for commercial cultivation. Importantly, three clones, namely SMP-1, UPASI-15 as well as TRI-2025 clustered together, all of which were resistant to blister blight disease of tea whereas highly susceptible clones of the same diseases such as UPASI-4, UPASI-7, and UAPASI-12 grouped together. Therefore, he concluded that AFLP could be utilized successfully for developing the markers linked to blister blight disease tolerance. Later, several workers in several countries used AFLP markers to study the genetic diversity of tea as well as other *Camellia* species (Table 6.2). Interestingly, it had been also applied to identify the markers linked with drought tolerance in tea (Mishra and Sen-Mandi 2004).

6.8.6 Single Nucleotide Polymorphism (SNP)

Detection of allelic differences or variations in the plant growth regulators (PGRs) is an important application of genomic resources which can be achieved by highly robust DNA-based markers such as SNP or its haplotype (i.e. groups of SNPs that are linked to a particular trait). Due to higher availability and stability during inheritance as compared to other markers, such as SSRs, SNPs provide enhanced possibilities for studying PGR management in several ways such as cultivar identification, construction of genetic maps, detection of genetic diversity, detection of genotype vs. phenotype associations, and marker-assisted breeding. Although large-scale SNPs had been discovered in several crops using various sources of sequences, in tea, not many SNPs have been identified. Huang et al. (2007) identified the SNPs present in the coding region of polyphenol oxidase from different genotypes of tea. Thus, there was a tremendous scope to identify the large scale discovery of SNP markers.

6.8.7 Sequence-Tagged Microsatellite Site (STMS)

Recently, Caser et al. (2010) used sequence-tagged microsatellite site (STMS) markers to study the genetic variations with 132 accessions of *Camellia spp.,* which included 24 accessions representing 22 different species as well as 63 cultivars of *C. japonica,* 33 cultivars of *C. sasanqua,* 7 cultivars of *C. vernalis,* 3 cultivars of *C. hiemalis,* and 2 cultivars of *Camellia hybrida.* Cross transferability was obtained, which fruitfully amplified polymorphic alleles in all analyzed species, and 96 alleles were scored. PCA could differentiate the *C. japonica* cultivars from the winter camellias. The distribution of genetic variation, attributed by analysis of molecular variance (AMOVA), particularly highlighted genetic overlap among *C. sasanqua* cultivars and those belonging to *C. vernalis, C. hiemalis,* and *C. hybrida.*

Similarly, an initial study of STMS markers was undertaken by Matteo et al. (2010) with 132 accessions of *Camellia* species, which included 24 accessions representing 22 different species or varieties as well as 63 cultivars of *C. japonica,* 33 cultivars of *C. sasanqua,* 7 cultivars of *C. vernalis,* 3 cultivars of *C. hiemalis,* and 2 cultivars of *C. hybrida.* The four primer sets used (MSCJAF37, MSCJAH46, MSCJAF25, and MSCJAH38) successfully amplified polymorphic alleles in all the species analyzed, showing cross transferability. Overall, 96 alleles were scored. MSCJAH38 primer produced the highest number of alleles (30), while MSCJAH46 primer yielded the lowest number of alleles (15). Again, the PCA plot revealed a main differentiation between the *C. japonica* cultivars and the winter camellias. The distribution of the genetic variation, attributed by AMOVA, particularly highlighted genetic overlap among *C. sasanqua* cultivars and the cultivars belonging to *C. vernalis, C. hiemalis,* and *C. hybrida.* Thus, both the studies demonstrated that STMS markers offered a suitable method for detection of genetic variability and molecular study of *Camellia* genotypes.

6.8.8 Single-Strand Conformation Polymorphism (SSCP)

Wachira et al. (1997) analysed species introgression into cultivated gene pool of tea using 5 different organelle-specific primers in 19 species as well as 9 tea cultivars. Out of the five, three noncoding chloroplast regions as well as one mitochondrial region that amplified with universal primers did not reveal any polymorphisms. The remaining one cpDNA-specific PCR product revealed a single-strand conformation polymorphism (SSCP). This SSCP in the intergenic spacer between the *trnL* (UAA) 3′ exon and *trnF* (GAA) indicated that four species namely *C. furfuracea, C. assimilis, C. nokoensis,* and *C. tsaii* shared a common haplotype. Thus, the study indicated a possible hybridization between species of the sections involved.

6.8.9 Cleaved Amplified Polymorphic Sequence (CAPS)

Cleaved amplified polymorphic sequences (CAPSs) are alternative molecular markers, which combine both PCR and RFLP techniques. The techniques require minute amount of DNA and simple electrophoresis system to reveal polymorphisms. CAPS markers were successfully applied to a number of crops and forest species for which extensive nucleotide information is available. The advantage of these markers is the development of mapped cDNA clones that represent expressed genes. The genetic diversity of tea in both taxa was investigated based on CAPS analysis of *pal*, chalcone synthase 2 (*CHS2*), and dihydroflavonol 4-reductase (*DFR*) genes (Kaundun and Matsumoto 2003a). These three genes involve in catechin and tannin biosynthesis, and hence are directly responsible for tea taste and aroma. The critical factor in the development of CAPS markers is the right choice of PCR primers targeting a single locus at a time, in order to ensure reliable results. This was facilitated here because in tea both *pal* and *DFR* exist in one and two copies, respectively. The *CHS* gene is present in three copies but with sufficient sequence difference among them, allowing specific amplification of *CHS2* only. Based on their analysis, it was revealed that China type was more diverse than Assam type and that a higher proportion of overall diversity resided within genotypes as compared to between genotypes. Even though no specific DNA profile was found for either tea varieties following any single PCR-RFLP analysis, a factorial correspondence analysis was carried out on all genotypes and markers separating the tea samples into two distinct groups according to their origin. This reflects the larger differences between the two taxa in their polyphenolic profiles. Thus, STS-based markers developed will be very useful in future mapping, population genetics, and fingerprinting studies of this important species and other *Camellia* species. This work also opens the way for correlative analysis between newly established tea genotypes based on phenylpropanoid genes and chemotypes, mainly polyphenolics. Similarly, Ujihara et al. (2011) developed EST-based CAPS markers and used them to identify different cultivar.

6.9 Organelle DNA Analysis

Because of the relative resistance to evolutionary changes of organelle DNA than nuclear DNA, cpDNA and mitochondrial DNA (mtDNA) sequences had been widely used to investigate interspecific relationships (Jorgensen and Cluster 1989; Waugh et al. 1990). cpDNA encodes for many agronomically important genes including large subunit of RUBP carboxylase oxygenase, a 32 kd thylakoid membrane protein (Bedrock and Kolodner 1979), and some other coupling factors (Nelson et al. 1980). Noncoding regions displayed higher rates of evolution than coding regions; hence the former were desirable targets for phylogenic studies. The resolutions of many such noncoding regions had been amplified by the universal PCR primers (Taberlet et al. 1991). However, the relatively high frequency of insertions/deletions may even, in some cases, make it possible to use the size of PCR product as a genetic marker. The choice of cpDNA and mtDNA sequences that maximized phylogenic information, however, depended upon the evolutionary time scale of the plant system. Borthakur et al. (1998) made the first attempt to study the cpDNA of tea. Good quality of cpDNA using a discontinuous sucrose gradient was isolated from ten different tea cultivars. To overcome the phenolic problem, 0.1 % polyvinylpyrrolidone along with 0.1 % BSA was used during homogenization. The clear greenish yellow bands of intact chloroplasts were collected from the interface of 20–45 % sucrose gradient from where cpDNA was isolated. Thakor (1997) re-examined taxonomic relationship among the genus *Camellia* with the help of 5 cp DNA sequences from 25 *Camellia* species covering 4 subgenera of tea family. Out of which, 4 cpDNA sequences showed a low degree of variability (less than 2 %). The remaining one revealed a much higher degree of variability (3.8–20 %). Interestingly, the phylogenic analysis using parsimony (PAUP) of these sequences contradicted the sectional or subgeneric grouping of either Sealy's or Chang's monograph. Prince and Parks (1997) analysed the evolutionary relationship in tea subfamily Theoideae based upon two cpDNA regions, namely *rbcL* and *matK* sequence, for 4 species of subfamily Ternstroemioideae and 24 species from Theoideae. Later on, the same workers also examined the same cpDNA region (*rbcL* and *matK*) to confirm the family Theaceae, a natural group, as well as to evaluate the validity of circumscription of tribes and genus of its subfamily Theoideae (Prince and Parks 2000). The nucleotide sequences of *rbcL* gene in chloroplast DNA were determined on the native tea varieties of Japan, Korea, China, South East Asia, Sri Lanka, and India. Direct sequencing of the amplified cpDNA products were carried out. Alignments were obtained by assuming two substitutions, at nucleotide position 40 (adenine in China tea) and 948 (guanine in Assam tea). The nucleotide sequences of the *rbcL* gene in China and Assam were 99.8 % similar. On the other hand, the 1,370 nucleotide sequences of *rbcL* gene among *C. irrawadiensis*, *C. taliensis*, and Assam tea were the same, except a different base at position 627. At this position in *C. irrawadiensis* and *C. taliensis*, thymine and adenine were observed, respectively, as specific bases (Kato 2001). The two major molecular phylogenetic investigations of the genus *Camellia* (one by Vijayan et al. (2009) with *nrITS* sequence and the other with *nDNA RPB2* sequence by Xiao and Parks (2003) had provided considerable insight into the interspecies relationships, which could not be provided by many previous attempts with the use of cpDNA sequences (Yang et al. 2006). These two molecular phylogenetic investigations shared many important findings. Both studies revealed the need to revise the existing classifications and supported the monophyly of sections *Thea* and *Furfuracea*. Also, they suggested that the species *C. hongkongensis* should be shifted from section *Camellia* to section *Furfuracea*, and both studies revealed that sections *Eriandra* and *Theopsis* were closely related, not separable, and that species of sections *Tuberculata*, *Chrysantha*, as well as *C. szechuanensis* from section *Pseudocamellia*, were closely related. Finally, results of both studies equally supported the section *Camellia* as polyphyletic. The species from the section *Camellia* are grouped on the basis of geographical origin and distribution, and species in this section distributed in the southeastern and eastern China, Korea, and Japan were well separated from those in southern and southwestern China. Nevertheless, both studies disagreed on many points. The most notable disagreement was the monophyly of the section *Paracamellia* defined by Ming (2000) and Sealy (1958), which was bifurcated of the section *Paracamellia* (Xiao and Parks 2003), and supported Chang's (1981) creation of the section *Oleifera* from the section *Paracamellia*. A recent study of leaf anatomical characters also supported the separation of *Oleifera* from the section *Paracamellia* (Lin et al. 2008). Other important differences were: First, species of the sections *Eriandra* and *Theopsis* formed a monophyletic clade in our tree, but mixed together with species from the section *Camellia* and divided into two well-separated clades in the study by Xiao and Parks (2003). Second, the positions of some species in small sections and isolates differed; for example, *C. amplexicaulis* of section *Longipedicellata* was isolated and was a sister to the clade of *Eriandra* and *Theopsis*, but was associated with clades of species in sections *Camellia*, *Oleifera*, and *Paracamellia* in the dendrogram tree of Xiao and Parks (2003). Also, *C. yunnanensis* of section *Stereocarpus* was embedded in the clade consisted of sections *Chrysantha* and *Tuberculata*, but was allied to section *Furfuracea* in the dendrogram tree of Xiao and Parks (2003). These types of conflicts in results are not uncommon in molecular phylogeny and can arise from both analytical and biological factors (Rokas et al. 2003a). Analytical factors that generally affect phylogenetic reconstruction were choice of optimality criterion (Huelsenbeck 1995), data availability (Cummings et al. 1995), taxon sampling (Graybeal 1998), and specific assumptions in the modelling of sequence evolution (Yang et al. 1994). The major biological factor that affects phylogenetic reconstruction was the evolutionary dynamics that may cause the history of the genes under analysis to obscure the history of the species (Rokas et al. 2003b).

6.10 Genetic Linkage Map

Construction of genetic linkage map is very important to know about the genome structure of an organism. It has several uses in plant breeding. However, the prerequisite for any such work is a mapping population from where segregation pattern of the marker can be tagged. This is difficult in a woody species like tea due to the several problems discussed earlier (Chap. 2).

A preliminary linkage map for tea plant was constructed with RAPD markers along with phenotypic characters such as theanine content, date of bud sprouting, resistance to anthracnose, and tolerance to cold (Tanaka 1996). However, in tea, a population thought to be derived from two known non-inbred parents was scored for RAPD and AFLP markers in order to develop a linkage map (Hackett et al. 2000). A very high proportion of the markers exhibited unexpected segregation ratio in the light of their configurations in parents and an exploratory statistical analysis revealed patterns in the marker scores, which could most easily be explained by the hypothesis of three male parents contributing pollen to this cross. The map had 15 linkage groups with three or more markers, agreeing with the haploid chromosome number of tea. The statistical methods that revealed the subpopulation were easy to apply routinely and may prove to be useful diagnostic tools for the analysis of non-inbred mapping populations. There were 126 markers, covering 1,349.7 cM, with an average distance of 11.7 cM between loci on the map. Recently, an AFLP linkage map for tea plant was also constructed in China. The map of a female parent included 17 linkage groups and contained 208 markers, covering a total length of 2,457.7 cM. The average distance between markers was 11.9 cM. A map from a male parent included 16 linkage groups and located 200 markers, covering a total length of 2,545.3 cM, and the average distance between markers was 12.8 cM (Huang et al. 2005). Most recently, a partial genetic map of backcross F_1 generation between 'Zhenong 129' (selected from the open pollination of 'Fuding Dabaicha'×'Yunnna Dayecha' parents) and 'Fuding Dabaicha' was also generated using RAPD and ISSR markers (Huang et al. 2006b). However, in previous studies, the number of individuals for mapping was limited to 94 only, and the density was not high enough to meet the demand of precise mapping. These constructed maps were still limited to locate QTLs linked with some important traits due to the low distribution of molecular markers.

Recently, a population of 54 F_1 clones derived from reciprocal crosses between 'Sayamakaori' and 'Kana-Ck17' was used for the linkage analysis. Maps of both parents were constructed from the F_1 population that was taken for BC_1 population. The order of most of the dominant markers in the parental maps was consistent. Thus, a core map was constructed by merging the linkage data for markers that detected polymorphisms in both the parents. The core map contained 15 linkage groups, which correspond to the basic chromosome number of tea. The total length of the core map was 1,218 cM. Thus, the reference map as a central core map is sandwiched between the parental maps for each linkage group; the combined maps containing 441 SSR, 7 CAPS, 2 STS, and 674 RAPD markers were developed. This newly constructed linkage map can be used as a basic reference linkage map of tea (Taniguchi et al. 2007, 2012).

Another integrated genetic map of tea using a segregating F_1 population derived from a cross between two commercial cultivars ('TTES 19' and 'TTES 8') was developed. Of the 505 mapped markers, there were 265 paternal markers (52.5%), 163 maternal markers (32.3%), 65 doubly heterozygous dominant markers (12.9%), and 12 codominant markers (2.4%). The codominant markers and doubly heterozygous dominant markers were used as bridge loci for the integration of the paternal and maternal maps. The integrated map comprised 367 linked markers, including 36 SSR, 3 CAPS, 1 STS, 250 AFLP, 13 ISSR, and 64 RAPD markers, that were assigned to 18 linkage groups. The linkage groups represented a total map length of 4,482.9 cM with a map density of 12.2 cM. The average density of the integrated genetic map constructed in this study was 12.2 cM (based on Haldane function) or 10.4 cM (based on Kosambi function), which was similar to the result of Huang et al. (2005) who used 208 and 200 markers to construct maternal and paternal maps with the density of 11.9 and 12.8 cM, respectively. This genetic map had the highest genetic coverage so far, which could be applied to comparative mapping, QTL mapping, and marker assisted selection in the future (Hu et al. 2013).

6.11 Genomic Resources

In absence of whole genome sequence, genomic resources are essential especially for woody plants where sequence-driven researches such as marker assisted breeding, association mapping, cloning of genes, mapping of QTLs are extremely important to develop the elite genotypes within a short period of time. In *Camellia,* 48,672

Table 6.3 Different genomic resources of tea

Name	Quantity	Reference
Popular cultivars	600	Mondal et al. (2004)
Different species of *Camellia*	325	Mondal et al. (2004)
BAC clones	401,280	Lin et al. (2011)
QTLs	23	Kamunya et al. (2010)
Cloned genes	55	NCBI
miRNAs	16	Das and Mondal (2010); Prabu and Mandal (2010); Mohanpuria and Yadav (2012)
ESTs	48,672	NCBI
SSR markers	250 for tea; 5,844 for *C. chekiangoleosa*	Kato et al. (2008); Sharma et al. (2009); Shi et al. (2013)
SNPs	5	Huang et al. (2007)

expressed sequence tags (as on 30 July 2013) are available in NCBI which are mainly associated with quality and few with abiotic and biotic stresses (Table 6.3). Lin et al. (2011) constructed bacterial artificial chromosome library of tea and generated 401,280 clones with an average insert size of 135 kb. Among the other genomic resources, while Kamunya et al. (2010) identified 23 minor QTLs, approximately 250 polymorphic SSR markers were validated by various groups (Kato et al. 2008, Sharma et al. 2009). Apart from that, a few SNPs (Huang et al. 2007) and microRNAs of tea (Das and Mondal 2010; Prabu and Mandal 2010; Mohanpuria and Yadav 2012; Zhu and Luo 2013) had also been identified.

C. reticulata was found as diploid, tetraploid as well as hexaploid and hence is expected to have a complex genome. To resolve the polyploid complexity, physical maps of the 18S–26S rDNA ribosomal RNA genes (rRNA) were generated by fluorescent in situ hybridization for *C. reticulata,* including three types of ploidy of *C. reticulata* with its related species such as *C. japonica, C. yunnanensis, C. pitardii,* and *C. saluenensis*. An advanced method was used for preparing chromosome spreads. Eight, twelve and eighteen rDNA sites were observed on the genomes of diploid, tetraploid, and hexaploid *C. reticulata,* respectively. Eight, four, five, and four rDNA loci were located on the chromosomes of *C. pitardii, C. japonica, C. saluenensis,* and *C. yunnanensis,* respectively. The number and position of rDNA sites in these species were compared for analysis. The results supported some of the earlier phylogenetic speculations about this complex genome and suggested the occurrence of some structural rearrangements in chromosome (Gu and Xiao 2003).

6.12 Conclusion

In general, genetic improvement of tea was mainly contributed by conventional breeding; however, in the last two decades great attempts had been made to intervene some of the problems of conventional breeding using modern techniques such as molecular markers. Till now, several DNA markers have been used to make fingerprints, which need to be documented systematically and should be made available for public use to preserve the intellectual property rights of tea breeders. Priority in the area of molecular breeding should be given: (1) to undertake a massive characterization of tea germplasm across the world through a common 'Tea germplasm characterization consortium', which already exists for several similar crops; (2) to identify DNA markers to do early selection at nursery stage for various biotic (such as flower blight) and abiotic stresses (such as cold hardiness), which will revolutionize *Camellia* breeding where works suffer due to the lack of selection criteria and long gestation periods, (3) to develop the molecular markers for hybrid identification, (4) to generate large scale SSRs and SNPs which have high potential for marker assisted breeding, (5) to develop the trait-specific 'core' germplasm, and (6) to mine alleles for important genes and their utilization either for developing markers or for transgenic plant production. Finally, being a woody perennial, which highly outbreeds in nature, tea possesses a good potential for association mapping than the simple biparental mapping which is totally unexplored in this genus.

References

Ackerman WL (1971) Genetic and cytological studies with *Camellia* and related genera. Technical Bull No 1427. USDA/U.S. Govt. Print Office, Washington DC, p 115

Afridi SG, Ahmad H, Alam M, Khan IA, Hassan M (2011) DNA landmarks for genetic diversity assessment in tea genotypes using RAPD markers. Afr J Biotech 10:15477–15482

Alcázar A, Ballesteros O, Jurado JM, Pablos F, Martín MJ, Vilches JL, Navalón A (2007) Differentiation of green, white, black, Oolong, and Pu-erh teas according to their free amino acids content. J Agric Food Chem 55:5960–5965

Anderson S (1994) Isozyme analysis to differentiate between tea clones. Inligtingsbulletin Institut vir Tropiese en Subtropiese Gewasse 266:15

Ariyaratne PNK, Mewan KM, Goonetilleke WASNST, Attanayake DPSTG (2009) Study of genetic relationships of tea (*Camellia sinensis* (L.) O. Kuntze) using SSR markers and pedigree analysis. Proc 9th Agr Res Symp 1:293–298

Balasaravanan T, Pius PK, Kumar RR (2002) Assessment of genetic fidelity among the *in vitro* propagated culture lines of *Camellia sinensis* (L.) O kuntze using RAPD markers. In: Proceedings of the 15th Plantation Crops Symposium PLACROSYM XV Mysore, 10–13 Dec, pp 181–184

Balasaravanan T, Pius PK, Kumar RR, Muraleedharan N, Shasany AK (2003) Genetic diversity among south Indian tea germplasm (*Camellia sinensis*, *C. assamica* and *C. assamica* spp. Lasiocalyx) using AFLP markers. Plant Sci 165:365–372

Bali S, Raina SN, Bhat V, Aggarwal RK, Goel S (2013) Development of a set of genomic microsatellite markers in tea (*Camellia* L.) (Camelliaceae). Mol Breeding 32:735–741. doi: 10.1007/s11032-013-9902-4

Banerjee B (1987) Can leaf aspect affect herbivory? A case study with tea. Ecol 68:839–834

Banerjee B (1992) Botanical classification of tea. In: Wilson KC, Clifford MN (eds) Tea cultivation to consumption. Chapman and Hall, London, pp 25–51

Barua DN (1958) Leaf sclereids in the taxonomy of the *Thea camellias*. I. Wilson's and related *Camellias*. Phytomorphology 8:257–264

Barua DN, Dutta AC (1959) Leaf sclereids in taxonomy of *Thea camellias* II. *Camellia sinensis* L. Phytomorphology 9:372–382

Barua DN, Dutta KN (1971) Distribution of shoots on the plucking surface of a tea bush and its relation to spacing. Two Bud 18:8–11

Barua PK (1963) Classification of tea plant. Two Bud 10:3–11

Bedrook JR, Kolodner R (1979) The structure of choloroplast DNA. Annu Rev Plant Physiol 30:593–620

Bera B, Saikia H (2002) Randomly amplified polymorphic DNA (RAPD) marker analysis in tea (*Camellia sinensis* L) generative clones. In Proceedings of the 15th plantation crops symposium PLACROSYM XV Mysore 10–13 Dec, pp 235–238

Bezbaruah HP (1971) Cytological investigation in the family theaceae—I. Chromosome numbers in some *Camellia* species and allied genera. Caryologia 24:421–426

Borchetia S, Das SC, Handique PJ, Das S (2009) High multiplication frequency and genetic stability for commercialization of the three varieties of micropropagated tea (*Camellia* spp.). Sci Hort 120:544–550

Borse BB, Rao LJM, Nagalakshmi S, Krishnamurthy N (2002) Fingerprint of black teas from India: identification of the regio-specific characteristics. Food Chem 79:419–424

Borthakur S, Mondal TK, Borthakur A, Deka PC (1995) Variation in peroxidase and esterase isoenzymes in tea leaves. Two Bud 42:20–23

Borthakur S, Mondal TK, Parveen SS, Guha A, Sen P, Borthakur A, Deka PC (1998) Isolation of chloroplast DNA from tea, *Camellia* sp. Ind J Exp Biol 36:1165–1167

Caser M, Torello Marinoni D, Scariot V (2010) Microsatellite-based genetic relationships in the genus *Camellia*: potential for improving cultivars. Genome 53:384–99

Chang HT (1981) A taxonomy of the genus *Camellia*. Acta Scientiarum Naturalium Universitatis Sunyatseni, Monograph series 1:1–180

Chen C (1996) Analysis on the isozymes of tea plants F_1 hybrids. J Tea Sci 16:31–33

Chen J, Wang PS, Xia YM, Xu M, Pei SJ (2005a) Genetic diversity and differentiation of *Camellia sinensis* L. (cultivated tea) and its wild relatives in Yunnan province of China, revealed by morphology, biochemistry and allozyme studies. Genetic Res Crop Evol 52:41–52

Chen L, Chen D, Gao Q, Yang Y, Yu F (1997a) Isolation and appraisal of genomic DNA from tea plant (*Camellia sinensis* (L.) O. Kuntze. J Tea Sci 17:177–181

Chen L, Gao Q, Yang Y, Yu F, Chen D (1998a) Optimum amplification procedure and reaction system for RAPD analysis of tea plant (*Cameliia sinensis* (L.) O. Kuntze. J Tea Sci 18:16–20

Chen L, Gao QK, Chen DM, Xu CJ (2005b) The use of RAPD markers for detecting genetic diversity, relationship and molecular identification of Chinese elite tea genetic resources [*Camellia sinensis* (L.) O. Kuntze] preserved in a tea germplasm repository. Biodiver Conser 14:1433–1444

Chen L, Qiqing T, Qikang G, Jilin S, Fulian Y (1997b) Observation on pollen morphology of 8 species and 1 variety in genus *Camellia*. J Tea Sci 17:183–188

Chen L, Tong Q, Zhuang W (1992) Studies on pollen morphology and fuzzy clustering analysis of tea. Acta Agri Uni Zhejiang (Chinese) 18:29–36

Chen L, Wang PS, Yamaguchi S (2002a) Discrimination of wild tea germplasm resources (*Camellia* sp.) using RAPD markers. Agri Sci China 1:1105–1110

Chen L, Wang PS, Yamaguchi S (2002b) Discrimination of wild tea germplasm resources (*Camellia* sp.) using RAPD markers. Agri Sci China 1:1105–1110

Chen L, Zhou Z (2005) Variations of main quality components of tea genetic resources [*Camellia sinensis* (L.) O. Kuntze] preserved in the China National Germplasm Tea Repository. Plant Food Hum Nut 60:31–35

Chen L, Yamaguchi L (2002) Genetic diversity and phylogeny of tea plant (*Camellia sinensis*) and its related species and varieties in the section Thea genus *Camellia* determined by randomly amplified polymorphic DNA analysis. J Hort Sci Bio 77:729–732

Chen L, Yamaguchi S (2005) RAPD markers for discriminating tea germplasms at the inter-specific level in China. Plant Breed 124:404–409

Chen L, Yamaguchi S, Wang PS, Xu M, Song WX, Tong QQ (2002b) Genetic polymorphism and molecular phylogeny analysis of section Thea based on RAPD markers. J Tea Sci 22:19–24

Chen L, Yang Y, Yu F, Gao Q, Chen D (1998b) Genetic diversity of 15 tea (*Camellia sisnensis* (L.) O.Kuntze) cultivars using RAPD markers. J Tea Sci 18:21–27

Chen L, Yu F, Yang Y, Gao Q, Chen D, Xu C (1999) A study on genetic stability of excellent germplasm (*Camellia sisnensis* (L.) O. Kuntze) using RAPD markers. J Tea Sci 19:13–16

Chen Q, Zhao J, Liu M, Cai J (2008) Nondestructive identification of tea (*Camellia sinensis* L.) varieties using FT-NIR spectroscopy and pattern recognition. Czech J Food Sci 26:360–367

Chen Y, Yu M, Xu J, Chen X, Shi J (2009) Differentiation of eight tea (*Camellia sinensis*) cultivars in China by elemental fingerprint of their leaves. J Sci Food Agric 89:2350–2355

Chengyin L, Weihua L, Mingjun (1992) Relationship between the evolutionary relatives and the variation of esterase isozymes in tea plant. J Tea Sci 12:15–20

Chung MY, Epperson BK, Chung MG (2003) Genetic structure of age classes in *Camellia japonica* (THEACEAE). Evol 57:62–73

Clark JY, Warwick K (1998) Artificial keys for botanical identification using a multilayer perception neural network (MLP). Artif Intell Rev 12:95–115

Cummings MP, Otto SP, Wakeley J (1995) Sampling properties of DNA sequence data in phylogenetic analysis. Mol Biol Evol 12:814–822

Damasco OP, Godwin ID, Smith MK, Adkins SW (1996) Gibberellic acid detection of dwarf off-types in micropropagated Cavendish bananas. Aus J Expt Agric 36:237–241

Das A, Mondal TK (2010) In silico analysis of miRNA and their targets in tea. Amer J Plant Sci 1:77–86

Deng WW, Ogita S, Ashihara H (2010) Distribution and biosynthesis of theanine in Theaceae plants. Plant Physiol Biochem 48:70–72

Dhiman B, Singh M (2003) Molecular detection of cashew husk (*Anacardium occidentale*) adulteration in market samples of dry tea (*Camellia sinensis*). Planta Med 69:882–884

Eden T (1976) Tea. Longman, London, p 236

Evans DA, Sharp WR (1986) Somaclonal and gametoclonal variation. In: Evans DA, Sharp WR, Ammirato PV (eds) Technique and Applications. Handbook of Plant Cell Culture, vol 42. Macmillan Publishing Company, New York, pp 97–132

Fan K, Hong Y-C, Ding Z-T, Wang Y (2010) Analysis of genetic diversity among natural hybrid progenies of *Camellia sinensis* 'Huangshanzhong'. Acta Hort Sinica 37:1357–1362

Fang SW, Hua PR, Sheng WP, Mei X, Xing DH, Ping ZY, Hua LJ, Shao WF, Pang RH, Wang PS, Xu M, Duan HX, Zhang YP, Li JH (2003) RAPD analysis of tea trees in Yunnan. Sci Agrire Sinica 36:1582–1587

Fang W, Cheng H, Duan Y, Jiang X, Li X (2012) Genetic diversity and relationship of clonal tea (*Camellia sinensis*) cultivars in China as revealed by SSR markers. Plant Sys Evol 298:469–483

Ferguson JM, Grabe DF (1986) Identification of cultivars of perennial rye grass by SDS-PAGE of seed proteins. Crop Sci 26:170–176

Fernández-Cáceres PL, Martín MJ, Pablos F, González AG (2001) Differentiation of tea (*Camellia sinensis*) varieties and their geographical origin according to their metal content. J Agric Food Chem 49:4775–4779

Freeman SJ, West CJ, Lea V, Mayes S (2004) Isolation and characterization of highly polymorphic microsatellites in tea (*Camellia sinensis*). Mol Ecol Notes 4:324–326

Fukushima E, Iwasa S, Endo N, Yoshinari T (1966) Cytogenetics studies in Camellia. I. Chromosome survey in some *Camellia* species. Jap J Hort 35:413–421

George O, Adam M (2006). Investigation into the evolutionary origins of Theaceae and genus *Camellia*. Int Camellia J 38:78–89

Ghosh-Hazra N (2001) Advances in selection and breeding of tea—a review. J Plant Crop 29:1–17

Goonetilleke WASNST, Priyantha PGC, Mewan KM, Gunasekare MTK (2009) Assessment of genetic diversity of tea (*Camellia sinensis* L.O. Kuntze) as revealed by RAPD-PCR markers. J Nat Sci Found Sri Lanka 37:147–150

Graybeal A (1998) Is it better to add taxa or characters to a difficult phylogenetic problem? Syst Biol 47:9–17

Gu Z, Xiao H (2003) Physical mapping of the 18S-26S rDNA by fluorescent in situ hybridization (FISH) in *Camellia reticulata* polyploid complex (Theaceae). Plant Sci 164:279–285

Gul S, Ahmad H, Khan IA, Alam M (2007) Assessment of diversity of tea genotypes through RAPD markers. Pak J Biol Sci 10:2609–2611

Gulati A, Rajkumar S, Karthigeyan S, Sud RK, Vijayan D, Thomas J, Rajkumar R, Das SC, Tamuly P, Hazarika M, Ahuja PS (2009) Catechin and catechin fractions as biochemical markers to study the diversity of Indian tea (*Camellia sinensis* (L.) O. Kuntze) germplasm. Chem Biodivers 6:1042–1052

Gunasekara MTK, Arachchige JDK, Mudalige AK, Peiris TUS (2001) Morphological diversity of tea (Camellia *sinensis* L.) genotypes in Sri Lanka. In: Proceedings of the 57th annual session of Sri Lanka Association for the Advancement of Science (SLAAS) Part I. Colombo, p 83

Hackett CA, Wachira FN, Paul S, Powell W, Waugh R (2000) Construction of a genetic linkage map for *Camellia sinensis* (tea). Heredity (Edinb.) 85:346–355

Hairong X, Qiqing T, Wanfanz Z (1987) Studies on the genetic tendency of tea plant hybrid generation using isozyme technique. In: Proceedings of the international symposium on tea quality and human health. China, 4–9 November 1987, p 21

Hammerschlag FA (1992) Somaclonal variation. In: Hammerschlag FA, Lit RE (eds) Biotechnology of perennial fruit crops. CAB International, Wallingford, pp 35–55

Hazarika M, Mahanta PK (1984) Composition changes in chlorophylls and carotenoids during the four flushes of tea in North-East India. J Sci Food Agri 35:298–303

Hirai M, Kozaki I (1986) Isozymes of citrus leaves. In: Kitaura K, Akihama T, Kukimura H, Nakajima H, Horie M, Kozaki I (eds) Development of new technology for identification and classification of tree crops and ornamentals. Fruit Tree Research Station, Ministry of Agriculture, Forestry and Fisheries, Japan, pp 73–76

Hou YJ, He Q, Li ZL, Li PW, Liang GL, Xu J (2007) ISSR applied to the germplasms identification of *Camellia sinensis*. Southwest China J Agri Sci 26:1272–1276

Hu CY, Lee TC, Tsai HT, Tsai YZ, Lin SF (2013) Construction of an integrated genetic map based on maternal and paternal lineages of tea (*Camellia sinensis*). Euphytica 191(1):141–152. doi: 10.1007/s10681-013-0908-0

Huang FP, Liang YR, Lu JL, Chen RB (2006a) Genetic mapping of first generation of backcross in tea by RAPD and ISSR markers. J Tea Sci 26:171–176

Huang JA, Huang YH, Luo JW, Li JX, Gong ZH, Liu ZH (2007) Identification of single nucleotide polymorphism in polyphenol oxidase gene in tea plant (*Camellia sinensis*). J Hunan Agri Uni 33:454–458

Huang JA, Li JX, Huang YH, Luo JW, Gong ZH, Liu ZH (2005). Construction of AFLP molecular markers linkage map in tea plant. J Tea Sci 25:7–15

Huang JN, Li J, Huang Y, Luo J, Zong Z, Liu Z (2006b) Genetic diversity of tea [*Camellia sinensis* (L.) O. Kuntze] cultivars revealed by AFLP Analysis. Acta Hort Sinica 33:317–322

Huelsenbeck JP (1995) Performance of phylogenetic methods in simulation. Syst Biol 44:17–48

Hui LX, Lin LC, Peng SZ, Wu LJ, Wen SC, Hua GZ, Xuan C, Li XH, Liu CL, Shi ZP, Luo JW, Shen CW, Gong ZH, Chen X (2004) Analysis of genetic relationships among "Rucheng Baimao Cha" plants with RAPD method. J Tea Sci 24:33–36

Hung CY, Wang KH, Huang CC, Gong X, Ge XJ, Chiang TY (2008) Isolation and characterization of 11 microsatellite loci from *Camellia sinensis* in Taiwan using PCR-based isolation of microsatellite arrays (PIMA). Conserv Genet 9:779–781

Ikeda N, Kawada M, Takeda Y (1991) Isozymic analysis of *Camellia sinensis* and its interspecific hybrids. In: Proceedings of the international symposium of tea science, Shizouka, Japan, 26–28 Aug 1991, p 98

IPGRI (1997) Descriptors for tea (*Camellia sinensis*). International Plant Genetic Resources Institute, Rome

Isabel NL, Tremblay MM, Tremblay FM, Bousquet J (1993) RAPD as an aid to evaluate the genetic integrity of somatic embryogenesis derived population of *Picea mariana* (Mill) B.S.P. Theor Appl Genet 86:81–87

Jeyaramraja PR, Jayakumar D, Kumar RR, Pius PK (2002) Peroxidase isozyme—a versatile tool for screening of drought tolerant clones. Newsl UPASI Res Found 12:1–7

Jha TB, Sen SK (1992) Micropropagation of an elite Darjeeling tea clone. Plant Cell Rep 11:101–104

Ji PZ, Li H, Gao LZ, Zhang J, Chen GZQ, Huang XQ (2011) ISSR diversity and genetic differentiation of ancient tea (*Camellia sinensis* var. *assamica*) plantations from China: implications for precious tea germplasm conservation. Pak J Bot 43:281–291

Ji PZ, Zhang J, Wang PS, Huang XQ, Xu M, Tang YC, Liang MZ (2007) Genetic diversity of ancient tea plant in Yunnan province of Chian revealed by inter-simple sequence repeat (ISSR) polymerase chain reaction. J Tea Sci 27:271–279

Jin JQ, Cui HR, Chen WY, Lu MZ, Yao YL, Xin Y, Gong XC (2006) Data mining for SSR in ESTs and development of EST-SSR marker in tea plant (*Camellia sinensis*). J Tea Sci 26:17–23

Jin JQ, Cui HR, Gong XC, Chen WY, Xin Y (2007) Studies on tea plants (*Camellia sinensis*) germplasms using EST-SSR marker. Yi Chuan 29:103–108

Jorge S, Pedroso MC, Neale DB, Brown G (2003) Genetic differentiation of Portuguese tea plant using RAPD markers. Hort Sci 38:1191–1197

Jorgensen RA, Cluster PD (1989) Modes and tempos in the evolution of nuclear ribosomal DNA: new character for evolutionary studies and new markers for genetic and population studies. Ann Mo Bot Gard 75:1238–1247

Kabir SE, Ghosh-Hajra N, Chaudhuri TC (1991) Performance of certain clones under the agroclimatic conditions of Darjeeling. Tea Board India Tech Bull 5:1–8

Kafkas S, Ercisxli S, Doğan Y, Ertürk Y, Haznedar A, Sekban R (2009) Polymorphism and genetic relationships among tea genotypes from Turkey revealed by amplified fragment length polymorphism markers. J Amer Soc Hort Sci 134:428–434

Kamunya SM, Wachira FN, Pathak RS, Korir R, Sharma V, Kumar R, Bhardwaj P, Chalo R, Ahuja PS, Sharma

RK (2010) Genome mapping and testing for quantitative trait loci in tea (*Camellia sinensis* (L.) O. Kuntze). Tree Genet Genom 6:915–929

Karthigeyan S, Rajkumar S, Sharma RK, Gulati A, Sud RK, Ahuja PS (2008) High level of genetic diversity among the selected accessions of tea (*Camellia sinensis*) from abandoned tea gardens in Western Himalaya. Biochem Genet 46:810–819

Kato F, Taniguchi F, Monobe M, Ema K, Hirono H, Maeda-Yamamoto M (2008) Identification of Japanese tea (*Camellia sinensis*) cultivars using SSR marker. J Jap Soc Food Sci Tech 55:49–55

Kato M (2001) Analysis of differentiation of tea using DNA markers in evergreens forest. Int Conf on O-Cha (tea) Cult Science, Shizuoka, Japan, p 18

Katoh Y, Katoh M, Takeda Y, Omori M (2003) Genetic diversity within cultivated teas based on nucleotide sequence comparison of ribosomal RNA maturase in chloroplast DNA. Euphytica 134:287–295

Kaundun SS, Matsumoto S (2002) Heterologous nuclear and chloroplast microsatellite amplification and variation in tea, *Camellia sinensis*. Genome 45:1041–1048

Kaundun SS, Matsumoto S (2003a) Development of CAPS markers based on three key genes of the phenylpropanoid pathway in tea, *Camellia sinensis* (L.) O. Kuntze, and differentiation between *assamica* and *sinensis* varieties. Theor Appl Genet 106:375–383

Kaundun SS, Matsumoto S (2003b) Identification of processed Japanese green tea based on polymorphism generated by STS-RFLP analysis. J Food Chem 51:1765–1770

Kaundun SS, Matsumoto S (2011) Molecular evidence for maternal inheritance of the chloroplast genome in tea, *Camellia sinensis* (L.) O. Kuntze. J Sci Food Agric 91:2660–2663

Kaundun SS, Park YG (2002) Genetic structure of six Korean tea populations revealed by RAPD-PCR markers. Crop Sci 42:594–601

Kaundun SS, Zhyvoloup A, Park YG (2000) Evaluation of genetic diversity among elite tea (*Camellia sinensis var. sinensis*) accessions using RAPD markers. Euphytica 115:7–16

Kerio LC, Wachira FN, Wanyoko JK, Rotich MK (2012). Characterization of anthocyanins in Kenyan teas: Extraction and identification. Food Chem 131:31–38

Kondo K (1975) Cytological studies in cultivated species of *Camellia*. PhD thesis, University of North Carolina Chapel Hill, p 260

Kondo K (1977) Chromosome number in the genus *Camellia*. Biotrapica 9:86–94

Kondo K (1978a) Cytological studies in cultivated species of *Camellia*. In: Encyclopedia of *Camellia* Japan. Camellia Soc, vol 2. Kodansha Publ Co., Tokyo, p 456

Kondo K (1978b) Cytological studies in cultivated species of *Camellia*. Shi-Kaki 99:41–53

Kondo K, Parks CR (1979) Giemsa C-banding and karyotype of *Camellia* (-banded karyotypes can tell more detail on inter and intra-specific relationships in *Camellia*). Am Camellia Yearb 34:42–47

Kondo K, Parks CR (1980) Giemsa C-banding and karyotype of *Camellia*. Proc Inter Camellia Cong Kyoto, pp 55–57

Kondo K, Parks CR (1981) Cytological studies in cultivated species of *Camellia*. VI. Giemsa C-banded karyotypes of seven accessions of *Camellia japonica* L. *sensu lato*. Jap J Breed 31:25–34

Kulasegaram S (1980) Technical developments in tea production. Tea Q 49:157–183

Lai JA, Yang WC, Hsiao JY (2001) An assessment of genetic relationships in cultivated tea clones and native wild tea in Taiwan using RAPD and ISSR markers. Bot Bull Acad Sin 42:93–100

Latip, SNH, Muhamad R, Manjeri G, Tan SG (2010) Genetic variation of selected *Camellia sinensis* (cultivated tea) varieties in Malaysia based on random amplifed microsatellite (RAMs) markers. Pertanika J Trop Agric Sci 33:259–267

Lee S, Kim J, Sano J, Ozaki Y, Okubo H (2003) Phylogenic relationships among tea cultivars based on AFLP analysis. J Food Agri 47:289–299

Li B, Yin Y, Zhou Y, Deng PQ, Yang HW (2003) Genetic diversity of two sexual tea cultivars detected by RAPD markers. J Tea Sci 23:46–150

Li J, Jiang CJ, Wang ZX (2005) RAPD analysis on genetic diversity of the preconcentrated core germplasms of *Camellia sinensis* in China. Yi Chuan 27:765–771

Li JB, Hashimoto F, Shimizu K, Sakata Y (2013) Chemical taxonomy of red-flowered wild Camellia species based on floral anthocyanins. Phytochemistry 85:99–106

Li JH, Nesumi A, Shimizu K, Sakata Y, Liang MZ, He QY, Zhou HJ, Hashimoto F (2010) Chemosystematics of tea trees based on tea leaf polyphenols as phenetic markers. Phytochemistry 71:1342–1349

Li L, Zhong YH, Ji YZ, Zhi YZ, Sui N (2012) Analysis on genetic diversity of ten insular populations of *Camellia japonica*. Acta Hort Sinica 39:1531–1539

Li XH, Zhang CZ, Liu CL, Shi ZP, Luo JW, Chen X (2007) RAPD analysis of the genetic diversity in Chinese tea germplasm. Acta Hort Sinica 34:507–508

Liang YR, Tanaka JC, Takeda YY (2000) Study on diversity of tea germplasm by RAPD method. J Zhejiang For Coll 17:215–218

Lin J, Kudrna D, Wing RA. (2011) Construction, characterization, and preliminary BAC-end sequence analysis of a bacterial artificial chromosome library of the tea plant (*Camellia sinensis*). J Biomed Biotech 2011:476723–476731

Lin XY, Peng QF, Tang X, Hu ZH (2008) Leaf anatomy of Camellia sect. Oleifera and sect. Paracamellia (Theaceae) with reference to their taxonomic significance. J Syst Evol 46:183–190

Lin YP, Hu CY, Tsai YZ, Lin SF (2010) Studies on the fast molecular identification technologies of made tea (*Camellia sinensis*) varieties and their applications. Crop Envir Bioinform 7:37–51

Liu B, Sun X, Wang Y, Li Y, Cheng H, Xiong C, Wang P (2012b) Genetic diversity and molecular discrimination of wild tea plants from Yunnan Province based on inter-simple sequence repeats (ISSR) markers. Afr J Biotech 11:11566–11574

Liu BY, Li YY, Tang YC, Wang LY, Cheng H, Wang PS (2010) Assessment of genetic diversity and relationship of tea germplasm in Yunnan as revealed by ISSR markers. Acta Agro Sinica 36:391–400

Liu BY, Wang PS, Ji PZ, Xu M, Cheng H (2008) Study on genetic diversity of peculiar sect. *Thea* (L.) Dye in Yunnan by ISSR markers. J Yunn Agri Uni 23:302–308

Liu Y, Yang SX, Ji PZ, Gao LZ (2012a) Phylogeography of *Camellia taliensis* (Theaceae) inferred from chloroplast and nuclear DNA: insights into evolutionary history and conservation. BMC Evol Biol 12:92–117

Lou ZC, Shin YL, Ying LC (2004) Genetic relationship of Taiwan tea varieties. In: Proceedings of the 2004 international conference on O-CHA culture and science, 4–6 November, Shizuoka, Japan, pp. 226–228

Lu H, Jiang W, Ghiassi M, Lee S, Nitin M (2012) Classification of *Camellia* (Theaceae) species using leaf architecture variations and pattern recognition techniques. PLoS One 7:29704–29722

Luo JW, Shi ZP, Li JX, Shen CW, Huang YH, Gong ZH (2002a) Study on the application of RAPD techniques to parentage identification of tea plant. J Hunan Agric Univ 3:3–6

Luo JW, Shi ZP, Shen CW, Liu CL, Gong ZH, Huang YH (2002b) Studies on genetic relationships of tea cultivars (*Camellia sinensis (L.) O.* Kubtze) by RAPD analysis. J Tea Sci 22:140–146

Luo JW, Shi ZP, Shen CW, Liu CL, Gong ZH, Huang YH, Luo JW, Shi ZP, Shen CW, Liu CL, Gong ZH, Huang YH (2004) The genetic diversity of tea germplasms [*Camellia sinensis (L.) O.* Kuntze] by RAPD analysis. Acta Agro Sin 30:266–269

Luo XY, Zhuang XY, Yang YS (2007) Genetic diversity of *Camellia Changii* Ye (Theaceae) using ISSR marker. J Tea Sci 15(2):1005–1012

Ma JQ, Zhou YH, Ma CL, Yao MZ, Jin JQ, Wang XC, Chen L (2010) Identification and characterization of 74 novel polymorphic EST-SSR markers in the tea plant, *Camellia sinensis* (Theaceae). Ame J Bot 97:153–156

Ma, JQ, Ma CL, Yao MZ, Jin JQ, Wang ZL, Wang XC, Chen L (2012) Microsatellite markers from tea plant expressed sequence tags (ESTs) and their applicability for cross-species/genera amplification and genetic mapping. Sci Hort 134:167–175

Magambo MJS, Cannell MGR (1981) Dry matter production and partition in relation to yield of tea. Exper Agri 17: 33–38

Magoma GN, Wachira FN, Imbuga MO, Agong SG (2003) Biochemical differentiation in *Camellia sinensis* and its wild relatives as revealed by isozyme and catechin patterns. Biochem System Ecol 31:995–1010

Magoma GN, Wachira FN, Obanda M, Imbuga M, Agong SG (2000) The use of catechins as biochemical markers in diversity studies of tea (*Camellia sinensis*). Genetic Resour Crop Evol 47:107–114

Mariya JKM, Sasikumar R, Balasubramanian M, Saravanan M, Rajkumar R (2003) Influence of light on catechin biosynthesis in tea. Tea 24:80–86

Matsumoto S, Kiriwa Y, Yamaguchi S (2004) The Korean tea plant (*Camellia sinensis*): RFLP analysis of genetic diversity and relationship to Japanese tea. Breed Sci 54:231–237

Matsumoto S, Kiriwa Y, Takeda Y (2000) Analysis of genetic diversity in Chinese tea (*Camellia sinensis*) using RFLP and detection of difference on tea from Japanese. Breed Sci 2:209

Matsumoto S, Kiriwa Y, Takeda Y (2002) Differentiation of Japanese green tea as revealed by RFLP analysis of phenyl-alanine ammonia-lyase DNA. Theo Appl Genet 104:998–1002

Matsumoto S, Takeuchi A, Hayastsu M, Kondo S (1994) Molecular cloning of phenylalanine ammonia-lyase cDNA and classification of varieties and cultivars of tea plants (*Camellia sinensis*) using the tea PAL cDNA probes. Theor Appl Genet 89:671–675

Matteo C, Marinoni T, Daniela VS (2010) Microsatellite-based genetic relationships in the genus Camellia: potential for improving cultivars. Genome 53:384–399

McKenzie JS, Jurado JM, Pablos FD (2010) Characterization of tea leaves according to their total mineral content by means of probabilistic neural networks. Food Chem 123(3):859–864

Meksen K, Leister D, Peleman J, Zabeau M, Salamini F, Gebhardt C (1995) A high resolution map of the vicinity of the R1 locus on chromosome V of potato based on RFLP and AFLP markers. Mol Gen Genet 249:74–81

Mewan KM, Liyanage AC, Everard JMDT, Gunasekara MTK, Karunanayake EH (2005) Studying genetic relationships among the tea (*Camellia sinensis*) cultivars in Sri Lanka using RAPD Markers. Sri Lanka J Tea Sci 70(1):42–53

Mewan KM, Saha MC, Konstatin C, Pang Y, Abeysinghe ISB, Dixon RA (2007) Construction of genomic and EST-SSR based genetic linkage map of tea (*Camellia sinensis*). In: Proceedings of the 4th international conference on o-cha (tea) culture and science, 2–4 November 2007, Shizuoka, Japan, p 52

Ming TL (2000) Monograph of the genus *Camellia*. Kunming Institute of Botany, Chinese Academy of Sciences, Yunnan Science and Technology Press, Kunming

Mishra RK, Chaudhary S, Ahmad A, Pradhan M, Siddiqi TO (2009) Molecular analysis of tea clones (*Camellia sinensis*) using AFLP markers. Int J Integr Boil 5:130–135

Mishra RK, Sen-Mandi S (2001) DNA fingerprinting and genetic relationship study of tea plants using amplified fragment length polymorphism (AFLP) technique. Ind J Plant Genet Resour 14:148–149

Mishra RK, Sen-Mandi S (2004) Molecular profiling and development of DNA marker associated with drought tolerance in tea clones growing in Darjeeling. Curr Sci 87:60–66

Mohanpuria P, Yadav SK (2012) Characterization of novel small RNAs from tea (*Camellia sinensis* L.). Mol Biol Rep 39:3977–986

Mondal TK (2000) Studies on RAPD marker for detection of genetic diversity, *in vitro* regeneration and Agrobacterium-mediated genetic transformation of tea (*Camellia sinensis*). PhD thesis, Utkal University, India

Mondal TK (2002) Detection of genetic diversity among the Indian tea (*Camellia sinensis*) germplasm by inter-simple sequence repeats (ISSR). Euphytica 128:307–315

Mondal TK, Bhattacharya A, Laxmikumaran M, Ahuja PS (2004) Recent advance in tea Biotechnology. Plant Cell Tiss Org 75:795–856

Mondal TK, Chand PK (2002) Detection of genetic instability among the micropropagated tea (*Camellia sinensis*) plants. *In Vitro* Cell Dev-Pl 37:1–5

Mondal TK, Singh HP, Ahuja PS (2000) Isolation of genomic DNA from tea and other phenolic rich plants. J Plantation Crops 28:30–34

Mphangwe NIK, Vorster J, Steyn JM, Nyirenda HE, Taylor NJ, Apostolides Z (2013) Screening of tea (*Camellia sinensis*) for trait-associated molecular markers. Appl Biochem Biotechnol 171(2):437–449. doi: 10.1007/s12010-013-0370-4

Mugnai S, Pandolfi C, Azzarello E, Masi E, Mancuso S (2008) *Camellia japonica* L. genotypes identified by an artificial neural network based on phyllometric and fractal parameters. Plant Syst Evol 270:95–108

Mukhopadhyay M., Sarkar B, Mondal T.K (2014) Omics advances in tea (*Camellia sinensis*). In: Omics applications in Crop Sciences (Ed. Barh D) CRC press. New York. pp 467–490

Muoki RC, Wachira FN, Pathak RS, Kamunya SM (2007) Assessment of the mating system of *Camellia sinensis* in biclonal seed orchards based on PCR markers. J Hort Sci Biotech 82:733–738

Nagarajah S, Ratnasurya R (1981) Clonal variability in root growth and drought resistance in tea (*Camellia sinensis*). Plant Soil 60:153–155

Nelson N, Melson H, Schatz G (1980) Biosynthesis and assembly of the protein-translocating adenosine triphosphete complex from chloroplasts. Proc Natl Acad Sci 77:1361–1364

Neog B, Yadav RNS, Singh ID (2004) Peroxidase, polyphenol oxidase and acid phosphatase activities in the stigma-style tissue of *Camellia sinensis* (L) O. Kuntze following compatible and incompatible pollination. J Indian Inst Sci 84:47–52

Ng Etich WK, Wachira FN (2003) Variations in leaf anatomy and gas exchange in tea clones with different ploidy. J Hort Sci Biotech 78:173–176

Ohsako T, Ohgushi T, Motosugi H, Oka K (2008) Microsatellite variability within and among local landrace populations of tea, *Camellia sinensis* (L.) O. Kuntze, in Kyoto, Japan. Genet Resour Crop Evol 55:1047–1053

Orel G, Marchant AD, Wei CF, Curry AS (2007) Molecular Investigation and Assessment of *C. azalea* (syn. *C. changii* Ye 1985) as potential breeding material. Int Camellia J 39:64–75

Otaghvari AM, Garehyazie B, Hassanpur M, Sehgal D (2010) Isozyme banding patterns in Iranian tea clones. Bioinfolet 7:7–12

Owuor PO (1989) Differentiation of teas by the variations of linalools and geranol contents. Bull Chem Soc Ethip 3:31–35

Owuor PO, Obanda M (1998) The use of chemical parameters as criteria for selecting for quality in clonasl black tea in Kenya: achievements, problems and prospects. A Rev Tea 19:49–58

Owuor PO, Reeves SG, Wanyoko JK (1986) Co-relation of flavins content and valuation of Kenyan black teas. J Sci Food Agri 37: 507–513

Pandolfi C, Mugnai S, Azzarello E, Bergamasco S, Masi E, Mancuso S (2009) Artificial neural networks as a tool for plant identification: a case study on Vietnamese tea accessions, Euphytica 166:411–421

Park YG, Kaundun SS, Zhyvoloup A (2002) Use of the bulked genomic DNA-based RAPD methodology to assess the genetic diversity among abandoned Korean tea plantations. Genet Res Crop Evol 49:159–165

Parks CR, Case KF (1968) Chromatographic evidence for the genetic contamination of Camellia saluenesis in cultivation. Am Camellia Yearb 23:124–134

Paul S, Wachira FN, Powell W, Waugh R (1997) Diversity and genetic differentiation among population of Indian and Kenyan tea (*Camellia sinensis* (L.) O. Kuntze) revealed by AFLP markers. Theor Appl Genet 94:255–263

Pedro L, Caceres F, Martin MJ, Pablos F, Gonzalez G (2001) Differentiation of tea (*Camellia sinensis*) varieties and their geographical origin according to their metal content. J Agri Food Chem 49:4775–4779

Pi E, Peng Q, Lu H, Shen J, Du Y, Huang F, Hu H (2009) Leaf morphology and anatomy of Camellia section *Camellia* (Theaceae). Bot J Linn Soc 159:456–476

Piyasundara JHN, Gunasekare MTK, Wikremasinghe IP (2008) Identification of discriminating morphological descriptors for characterization of tea (*Camellia sinensis* L.) germplasm in Sri Lanka. Trop Agric Res 20:193–199

Piyasundara, JHN, Gunasekare MTK, Peiris TUS, Wickramasinghe IP (2006) Phenotypic diversity of Sri Lankan tea (*Camellia sinensis L*) germplasm based on morphological descriptors. Trop Agri Res 18:237–243

Prabu GR, Mandal AK (2010) Computational identification of miRNAs and their target genes from expressed sequence tags of tea (*Camellia sinensis*). Genom Proteom Bioinform 8:113–121

Prince LM, Parks CR (1997) Evolutionary relationships in the tea subfamily Theoideae based on DNA sequence data. Int Camellia J 29:135–144

Prince LM, Parks CR (2000) Estimation on Relationships of Theoideae (Theaceae) infreed from DNA data. Int Camellia J 32:79–84

Prince ML, Parks CR (2001) Phylogenic relationships of Theaceae inferred from chloroplast DNA sequence data. Amer J Bot 88:2309–2320

Raina SN, Ahuja PS, Sharma RK, Das SC, Bhardwaj P, Negi R, Sharma V, Singh SS, Sud RK, Kalia RK,

Pandey V, Banik J, Razdan V, Sehgal D, Dar TH, Kumar A, Bali S, Bhat V, Sharma S, Prasanna BM, Goel S, Negi MS, Vijayan P, Tripathi SB, Bera B, Hazarika M, Mandal AKA, Kumar RR, Vijayan D, Ramkumar S, Chowdhury BR, Mandi SS (2012) Genetic structure and diversity of India hybrid tea. Genet Resour Crop Evol 59:1527–1541

Rajanna L, Ramakrishnan M (2010) Isozyme studies on some selected *Camellia* clones. Int J Eng Sci Tech 2:6918–6921

Rajanna L, Ramakrishnan M, Simon L (2011) Evaluation of morphological diversity in south Indian tea clones using statistical methods. Maejo Int J Sci Technol 5(1):1–12

Rajasekaran P (1997) Development of molecular markers using AFLP in tea. In: Varghese JP (ed) Molecular approaches to crop improvement. Proceedings of the national seminar on molecular approaches to crop improvement 29–31 Dec, Kottayam, Kerala, India, pp 54–58

Rani V, Ajay P, Raina SN (1995) Random amplified polymorphic DNA (RAPD) markers for genetic analysis in micropropagated plants of *Populus deltoides* Marsh. Plant Cell Rep 14:459–462

Roberts EAH, Wight W, Wood DJ (1958) Paper chromatography as an aid to the identification of *Thea camellias*. New Phytol 57:211–225

Rokas A, King N, Finnerty J, Carroll SB (2003a) Conflicting phylogenetic signals at the base of the metazoan tree. Evol Dev 5:346–359

Rokas A, Williams BL, King N, Carroll SB (2003b) Genome scale approaches to resolving incongruence in molecular phylogenies. Nature 425:798–804

Roy SC, Chakraborty BN (2007) Evaluation of genetic diversity in tea of the Darjeeling foot hills, India using RAPD and ISSR markers. J Hill Res 20:13–19

Roy SC, Chakraborty BN (2009) Genetic diversity and relationship among tea (*Camellia sinensis*) cultivars as revealed by RAPD and ISSR based fingerprinting. Indian J Biotech 8:370–376

Sabhapondit S, Karak T, Bhuyan LP, Goswami BC, Hazarika M (2012) Diversity of catechin in Northeast Indian tea cultivars. Sci World J. doi:10.1100/2012/485193

Sanderson GW (1964) The chemical composition of fresh tea flush as affected by clone and climate. Tea Quart 35:101–109

Saravanan M, Maria John KM, Raj Kumar R, Pius PK, Sasikumar R (2005) Genetic diversity of UPASI tea clones (*Camellia sinensis* (L.) O. Kuntze) on the basis of total catechins and their fractions. Phytochem 66:561–565

Satyanarayan N, Sharma VS (1982) Biometric basis for yield prediction in tea clonal selection. Proceedings of the 14th Plantation Crops Symposium PLACROSYM IV, 3–5 Dec 1981, Mysore, India, pp 237–243

Satyanarayan N, Sharma VS (1986) Tea (*Camellia* L. spp) germplasm in South India. In: Srivastava HC, Vatsya B, Menon KKG (eds) Plantation crops: opportunity and constraints. Oxford IBH Publishing Co, New Delhi, pp 173–179

Scowcroft WR (1984) Genetic variability in tissue culture: impact on germplasm conservation and utilization. Report on the IBPGR, Rome, p 152

Sealy JR (1958) A revision of the genus *Camellia*. Royal Horticultural Society, London

Sen P, Bora U, Roy BK, Deka PC (2000) Isozyme characterization in *Camellia* spp. Crop Res 19:519–524

Seurei P (1996) Tea improvement in Kenya: a review. Tea 17:76–81

Shao WF, Pang RH, Duan HX, Wang PS, Xu M, Zhang YP, Li JH (2003a) RAPD analysis of tea in Yunnan. Sci Agric Sinica 36:1582–1587

Shao WF, Pang RH, Duan HX, Wang PS, Xu M, Zhang YP, Li JH (2003b) Use of RAPD analysis to classify tea trees in Yunnan. Agric Sci China 2:1290–1296

Sharma P, Deka PC (2002) Identification of tea hybrids by leaf and pollen isozyme analysis. Res Crop 3:411–415

Sharma RK, Bhardwaj P, Negi R, Mohapatra T, Ahuja PS (2009) Identification, characterization and utilization of unigene derived microsatellite markers in tea (*Camellia sinensis* L.). BMC Plant Biol 9:53–77

Sharma RK, Negi MS, Sharma S, Bhardwaj P, Kumar R, Bhattachrya E, Tripathi SB, Vijayan D, Baruah AR, Das SC, Bera B, Rajkumar R, Thomas J, Sud RK, Muraleedharan N, Hazarika M, Lakshmikumaran M, Raina SN, Ahuja PS (2010) AFLP-based genetic diversity assessment of commercially important tea germplasm in India. Biochem Genet 48:549–564

Sharma RKK, Negi MS, Sharma S, Bhardwaj P, Kumar R, Bhattachrya E, Tripathi SB, Vijayan D, Baruah AR, Das SC, Bera B, Rajkumar R, Thomas J, Sud RK, Muraleedharan N, Hazarika M, Sharma H, Kumar R, Sharma V, Kumar V, Bhardwaj P, Ahuja PS, Sharma RK (2011) Identification and cross-species transferability of 112 novel unigene-derived microsatellite markers in tea (*Camellia sinensis*). Amer J Bot 98:133–138

Shen CW, Huang YH, Huang JA, Luo JW, Liu CL, Liu DH (2007) RAPD analysis for genetic diversity of typical tea populations in Hunan Province. J Agri Biotech 15:855–860

Shen CW, Luo JW, Shi ZP, Gong ZH, Tang HP, Liu FZ, Huang YH (2002) Study on genetic polymorphism of tea plants in Anhua Yuntaihan population by RAPD. J Hunan Agric Univ 28:320–325

Shen CW, Ning ZX, Huang JA, Chen D, Li JX. (2009) Genetic diversity of *Camellia sinensis* germplasm in Guangdong Province based on morphological parameters and SRAP markers. Ying Yong Sheng Tai Xue Bao 20:1551–1558

Shi J, Dai X, Chen Y, Chen J, Shi J, Yin T (2013) Discovery and experimental analysis of microsatellites in an oil woody plant *Camellia chekiangoleosa*. Plant Syst Evol 299:1387–1393

Shi SH, Tang SQ, Chen YQ, Qu LH, Chang HT (1998) Phylogenic relationship among yellow flowered *Camellia* species based on random polymorphic DNA. Acta Phytotaxo Sinica 36:317–322

Shu J, Chen L (1996) Study on the evolution route of tea pollen morphology. J Tea Sci 16:115–118

Singh D, Ahuja PS (2006) 5S rDNA gene diversity in tea (*Camellia sinensis* (L.) O. Kuntze) and its use for variety identification. Genome 49:91–96

Singh HP, Ravindranath SD (1994) Occurrence and distribution of PPO activity in floral organs of some standard and local cultivars of tea. J Sci Food Agric 64:117–120

Singh ID (1980) Nonconventional approaches to the breeding of tea in north-east India. Two Bud 27:3–6

Singh ID (1999) Plant improvement. In: Jain NK (ed) Global advances in tea. Aravali Book International (P) Ltd., Delhi, pp 427–448

Singh M, Bandana, Ahuja PS (1999) Isolation and PCR amplification of genomic DNA from market samples of dry tea. Plant Mol Biol Rep 17:171–178

Singh M, Dhiman B, Sharma C (2011) Characterization of a highly repetitive DNA sequence in *Camellia sinensis* (L.) O. Kuntze genome. J Biotech Res 3:78–83

Singh M, Saroop J, Dhiman B (2004) Detection of intraclonal genetic variability in vegetatively propagated tea using RAPD markers. Biologia Plant 48:113–115

Singh S, Sud RK, Gulati A, Joshi R, Yadav AK, Sharma RK (2013) Germplasm appraisal of western Himalayan tea: a breeding strategy for yield and quality improvement. Genet Resour Crop Evol 60:1501–1513

Smith MK (1998) A review of factors influencing the genetic stability of micropropagated banana fruits. Aust J Expt Agric 43:219–223

Staub JE, Kuhns LJ, May B, Grun P (1982) Stability of potato tuber isozymes under different storage regimes. J Am Soc Hort Sci 107:405–408

Su MH, Hsieh CF, Tsou CH (2009) The confirmation of *Camellia formosensis* (Theaceae) as an independent species based on DNA sequence analyses. Bot Stud 50:477–485

Su MH, Tsou CH, Hsieh CF (2007) Morphological comparisons of Taiwan native wild tea plant (*Camellia sinensis* (L.) O. Kuntze forma formosensis Kitamura) and two closely related taxa using numerical methods. Taiwania 52:70–83

Taberlet P, Gielly L, Pauton G, Bouvet J (1991) Universal primers for amplification of three non-coding regions of choloroplast DNA. Plant Mol Biol 17:1105–1109

Takeo T (1981) Variations in amounts of linalool and geraniol produced in tea shoots by mechanical injury. Phytochemistry 30:2149–2151

Takeo T (1983) Variation in the aroma compound content of semi-fermented tea and black tea. Nippon Nogei Kagaku Kaishi 57:457–459

Tanaka J (1996) RAPD linkage map of tea plant and the possibility of application in tea genetics and breeding. Tea Res J 84:44–45

Tanaka J (2006) Study on the utilization of DNA markers in tea breeding. Bull Nat Inst Veg Tea Sci 5:113–155

Tanaka J, Taniguchi F (2002) Emphasized-RAPD (e-RAPD): a simple and efficient technique to make RAPD Bands clearer. Breed Sci 52:225–229

Tanaka J, Yamaguchi N, Nakamura Y (2001) Pollen parent of tea cultivar Sayamakaori with insect and cold resistance may not exist. Breed Res 3:43–48

Tanaka JI, Sawai Y, Yamaguchi S (1995) Genetic analysis of RAPD markers in tea. J Jap Breed 45:198–199

Tanaka JI, Yamaguchi S (1996) Use of RAPD markers for the identification of parentage of tea cultivars. Bull Nat Res Inst Veg Orna Plant Tea 9:31–36

Tang S, Bin X, Wang L, Zhong Y (2006) Genetic diversity and population structure of yellow Camellia (*Camellia nitidissima*) in China as revealed by RAPD and AFLP markers. Biochem Genet 44:449–461

Taniguchi F, Furukawa K, Ota-Metoku S, Yamaguchi N, Ujihara T, Kono I, Fukuoka H, Tanaka J (2012) Construction of a high-density reference linkage map of tea (*Camellia sinensis*). Breed Sci 62: 263–273

Taniguchi F, Tanaka J, Kono I (2007) Construction of genetic linkage map of tea using SSR markers. In: Proceedings of the 3rd international conference on o-cha (tea) culture and science (ICOS), Shizuoka, Japan

Tanikawa N, Onozaki T, Nakayama M, Shibata M (2008) PCR-RFLP analysis of chloroplast DNA variations in the atpI-atpH spacer region of the Genus *Camellia*. J Jap Soc Hort Sci 77:408–417

Tanksley SD, Yong ND, Paterson AH, Bonierbals MW (1989) RFLP mapping in plant breeding: new tools for an old science. Biotech 7: 257–264

Tateishi N, Ozaki Y, Okubo H (2007) Occurrence of ploidy variation in *Camellia vernalis*. J Fac Agric Kyushu Univ 52:11–15

Thakor BH (1997) A re-examination of the phylogenetic relationships within the genus Camellia. Int Camellia J 29:130–134

Thomas J, Vijayan D, Joshi SD, Joseph Lopez S, Raj Kumar R (2006) Genetic integrity of somaclonal variants in tea (*Camellia sinensis* (L.) O Kuntze) as revealed by inter simple sequence repeats. J Biotech 123:149–154

Tian M, LI JY, NI S Li XL (2008) Phylogenetic study on section *Camellia* based on ITS sequences data. J Acta Hort Sinica 35:1685–1688

Tiao JX, Parks CR (1997) Identification of closely related *Camellia* hybrid and mutant using molecular markers. Int Camellia J 29:111–116

Tiao JX, Parks CR (2003) Research for a new classification system for the genus *Camellia*. Int Camellia J 33:109–112

Toyao T, Takeda Y (1999) Studies on geographical diversity of floral morphology of tea plant (*Camellia sinensis* (L.) O. Kuntze) using the method of numerical taxonomy. Tea Res J 87:39–57

Tsumura Y, Ohba K, Strauss SH (1996) Diversity and inheritance of Inter-simple sequence repeat polymorphism in Douglas fir (*Pseudotsuga menziessi*) and sugi (*Cryptomeria japonica*). Theor Appl Genet 92:40–45

Ueno S, Tomaru N, Yoshimaru H, Manabe T, Yamamoto S (2000) Genetic structure of *Camellia japonica* L. in an old-growth evergreen forest, Tsushima, Japan. Mol Eco 9:647–656

Ueno S, Tsumura Y (2009) Development of microsatellite and amplicon length polymorphism markers for *Camellia japonica* L. from tea plant (*Camellia sinensis*) expressed sequence tags. Mol Ecol Resour 9:814–816

Ueno S, Yoshimaru H, Tomaru N, Yamamoto S (1999) Development and characterization of microsatellite markers in *Camellia japonica* L. Mole Eco 8:335–336
Ujihara T, Taniguchi F, Tanaka J, Hayashi N (2011). Development of expressed sequence tag (EST)-based Cleaved Amplified Polymorphic Sequence (CAPS) markers of tea plant and their application to cultivar identification. J Agric Food Chem 59:1557–1564
Vijayan K, Tsou CH (2008) Technical report on the molecular phylogeny of *Camellia* with nrITS: the need for high quality DNA and PCR amplification with *Pfu*-DNA polymerase. Bot Stud 49:177–188
Vijayan K, Zhang WJ, Tsou CH (2009) Molecular taxonomy of *Camellia* (Theaceae) inferred from NRITS sequences. Amer J Bot 96:1348–1360
Vo TD (2006) Assessing genetic diversity in Vietnam tea [Camellia sinensis (L.) O. Kuntze] using morphology, inter-simple sequence repeat (ISSR) and microsatellite (SSR) markers. PhD Thesis, Georg-August Gottingen University, Germany
Vos P, Hogers R, Bleeker M, Reijans M, van de Lee T, Hornes M, Freijters A, Pot J, Peleman J, Kuiper M, Zabeau M (1995) AFLP: a new techniques for DNA fingerprinting. Nucleic Acid Res 23:4407–4414
Vuylsteke D, Swennen R, Wilson GF, Langhe ED (1988) Phenotypic variation among *in vitro* propagated plantain (*Muse* sp. Cultivar 'AAB'). Sci Hort 36:79–80
Wachira F, Tanaka J, Takeda Y (2001) Genetic variation and differentiation in tea *(Camellia sisnensis)* germplasm revealed by RAPD and AFLP variation. J Hort Sci Biotech 76:557–563
Wachira FN (1990) Desirable tea plants: an overview of a search for markers. Tea 11:42–48
Wachira FN, Powell W, Waugh R (1997) An assessment of genetic diversity among *Camellia sinensis* L. (cultivated tea) and its wild relatives based on randomly amplified polymorphic DNA and organelle specific STS. Heredity 78:603–611
Wachira FN, Waugh R, Hackett CA, Powell W (1995) Detection of genetic diversity in tea (*Camellia sinensis*) using RAPD markers. Genome 38:201–210
Wang BY, Ruan ZY (2012) Genetic diversity and differentiation in *Camellia reticulata* (Theaceae) polyploid complex revealed by ISSR and ploidy. Genet Mol Res 11:503–511
Wang XF, Zheng HY, Zheng WH, Ao CQ, Jin HY, Zhao LH, Li N, Jia LR (2011) RAPD-based genetic diversities and correlation with morphological traits in *Camellia* (Theaceae) cultivars in China. Genet Mol Res 10:849–859
Wang XF, Zheng WH, Zheng HX, Xie QQ, Zheng HY, Tang H, Tao YL (2010) Optimization of RAPD-PCR reaction system for genetic relationships analysis of 15 *Camellia* cultivars. Afr J Biotech 9:798–804
Wang XP, Ma Bt, Qi GN, Tian H, Fang CU, Zhang ZC, Yin XM (2007) RAPD analysis on the genetic relationsh ips of tea cultivars grown in Sichuan. J Acta Hort Sinica 34:242–244
Waugh R, Vande Ven WTG, Phillips MS, Powell W (1990) Chloroplasts DNA diversity in the genus *Rubus* (Rosaceae) revealed by southern hybridization. Plant Syst Evol 172:65–75
Wei JQ, Chen ZY, Wang ZF, Tang H, Jiang YS, Wei X, Li XY, Qi XX (2010) Isolation and characterization of polymorphic microsatellite loci in *Camellia nitidissima chi* (Theaceae). Amer J Bot 97:89–90
Wei K, Wang L, Zhou J, He W, Zeng J, Jiang Y, Cheng H (2011) Catechin contents in tea (*Camellia sinensis*) as affected by cultivar and environment and their relation to chlorophyll contents. Food Chem 125:44–48
Wei X, Cao H-L, JIang YS, Ye WH, Ge XJ, LI F (2008) Population genetic structure of *Camellia nitidissima* (Theaceae) and conservation implications. Bot Stud 49:147–153
Wei X, Wei JQ, Cao HL, Li F, Ye WH (2005) Genetic diversity and differentiation of *Camellia euphlebia* (Theaceae) in Guangxi, China. Ann Bot Fennici 42:365–370
Wen Q, Xu L, Gu Y, Huang M, Xu L (2012) Development of polymorphic microsatellite markers in *Camellia chekiangoleosa* (Theaceae) using 454-ESTs. Am J Bot 99:203–205
Wen SC, Wu LJ, Peng SJ, Hua GZ, Ping TH, Zhi LF, Huan HY, Shen CW, Luo JW, Shi ZP, Gong ZH, Tang HP, Liu FZ, Huang YH (2002) Study on genetic polymorphism of tea plants in Anhua Yuntaishan population by RAPD. J Hunan Agri Univ 28:320–325
Wendel JF, Parks CR (1982) Genetic control of isozyme variation in *Camellia japonica* L. J Hered 73:197–204
Wendel JF, Parks CR (1983) Cultivar identification in *Camellia japonica* L. using allozyme polymorphisms. J Amer Soc Hort Sci 108:290–295
Wendel JF, Parks CR (1984) Distorted segregation and linkage of alcohol dehydrogenase genes in *Camellia japonica* L. (Theaceae). Biochem Genet 22:739–748
Wendel JF, Parks CR (1985) Genetic diversity and population structure in *Camellia japonica* L. (Theaceae). Amer J Bot 72:52–65
Wickramaratna MRT (1981) Variations in some leaf characteristics in tea (*Camellia sinensis* L.) and their use in identification of clones. Tea Quart 50:183–198
Wight W (1954) Morphological basis of quality in tea. Nature 173:630–631
Wight W (1958) The agrotype concept in tea taxonomy. Nature 181:893–895
Wight W, Barua DN (1954) Morphological basis of quality in tea. Nature 173:630–631
Williams JGK, Kubelik AR, Livak KJ, Rafaliski JA, Tingey SV (1990) DNA polymorphism amplified by arbitory primers are useful as genetic markers. Nucl Acid Res 18:6531–6535
Wright LP, Apostolides Z, Louw AI (1996) DNA fingerprinting of tea clones. In: Whittle AM and Khumalo FRB (eds). Proceedings of the 1st Regional Tea Research Seminar. Blantyre, Malawi 22–23 March 1995, pp. 44–50
Wright LP, Mphangwe NIK, Nyirenda HE, Apostolides Z (2000) Analysis of caffeine and flavan-3-ol composition in the fresh leaf of *Camellia sinesis* for predicting the quality of the black tea produced in Central and Southern Africa. J Sci Food Agric 80:1823–1830

Wright LP, Mphangwe NIK, Nyirenda HE, Apostolides Z (2002) Analysis of the theaflavin composition in black tea (*Camellia sinensis*) for predicting the quality of tea produced in Central and Southern Africa. J Sci Food Agric 82:517–525

Wu H, Chen D, Li J, Yu B, Qiao X, Huang H, He Y (2013). *De novo* characterization of leaf transcriptome using 454 sequencing and development of EST-SSR markers in tea (*Camellia sinensis*) Plant Mol Biol Rep 31:524–538

Wu KS, Tanksley SD (1993) Abundance, polymorphism and genetic mapping of microsatellite in rice. Mol Genet 241:225–235

Wu LJ, Peng SJ, Wen SC, Lin LC, Hua GZ, Huan HY, Luo JW Shi ZP, Shen CW, Liu CL, Gong ZH, Huang YH (2002a) Studies on genetic relationships of tea cultivars [*Camellia sinensis* (L.) O. Kuntze] by RAPD analysis. J Tea Sci 22:140–146

Wu LJ, Peng SZ, Xian Li J, Wen SC, Huan HY, Hua GZ, Luo JW, Shi ZP, Li JX, Shen CW, Huang YH, Gong ZH (2002b) Studies on genetic diversity of tea cultivars [*Camellia sinensis* (L.) O. Kuntze] by RAPD analysis. J Tea Sci 12:121–127

Xiao LZ, Yan CY, Li JX, Luo JW, He YM, Zhao CY (2007) AFLP analysis on genetic diversity of Fenghuang-Dancong tea plant germplasm. J Tea Sci 27:280–285

Xiao TJ, Parks CR (2003) Molecular analysis of the genus *Camellia*. Int Camellia J 35:57–65

Xu H, Ton Q, Zhuang W (1987) Studies on genetic tendency of tea plant hybrid generation using isozyme technique, Proceedings of the International Tea Quality and Health Symposium, pp 21–25

Xu ZX, Liu ZH, Wang KB, Liu F, Shi L, Gao DZ, Xu ZX, Liu ZH, Wang KB, Liu F, Shi L, Gao DZ (2004) On relationship between green tea polyphenols and catechins and green tea raw materials. J Hunan Agri Univ 30:257–260

Yamaguchi S (2001) Summarized remarks on the origin of Japanese tea by the genetic resource study in Eastern Asia and Japan. Proc 2001 Intl Conf O-Cha Cult Science, Shizuoka, Japan, 5–8 Oct

Yang JB, Li HT, Yang SX, Li DZ, Yang YY (2006) The application of four DNA sequences to studying molecular phylogeny of *Camellia* (Theaceae). Acta Bot Yunn 28:108–114

Yang JB, Yang J, Li HT, Zhao Y, Yang SX (2009) Isolation and characterization of 15 microsatellite markers from wild tea plant (*Camellia taliensis*) using FIASCO method. Conserv Genet 10:1621–1623

Yang Y, Sun T (1994) Study on the esterase isoenzyme in tea mutagenic breeding. China Tea 16:4–9

Yang YJ, Yu FL, Chen L, Zeng JM, Yang SJ, Li SF, Shu AM, Zhang ZF, Wang YS, Wang HS, Wang PS, Xu M, Song WX, Guo JC, Yang RX, Zhang WJ, Chen ZH, Yang YJ, Yu FL, Chen L, Zeng JM, Yang SJ, Li SF, Shu AM, Zhang ZF, Wang YS, Wang HS, Wang PS, Xu M, Song WX, Guo JC, Yang RX, Zhang WJ, Chen ZH (2003) Elite germplasm evaluation and genetic stability of tea plants. J Tea Sci 23:1–8

Yang Z, Goldman N, Friday A (1994) Comparison of models for nucleotide substitution used in maximum-likelihood phylogenetic estimation. Mol Biol Evol 11:316–324

Yao MZ, Chen L, Liang YR (2008) Genetic diversity among tea cultivars from China, Japan and Kenya revealed by ISSR markers and its implication for parental selection in tea breeding prog. Plant Breed 127:166–172

Yao MZ, Ma CL, Qiao TT, Jin JQ, Chen L (2012) Diversity distribution and population structure of tea germplasms in China revealed by EST-SSR markers. Tree Genet Genom 8:205–220

Yoshikawa N, Parks CR (2001) Systematic studies of *Camellia japonica* and closely related species. Int Camellia J 33:117–121

Zhang GW, Zhong WB, Wu Y, Tan XF, Du TZ (2007) Identification of oil tea (*Camellia oleifera*) superior clones by ISSR molecular marker. Forest Res 20:278–282

Zhang YP, Li JH (2003a) RAPD analysis of tea trees in Yunnan. Sci Agri Sin 36:1582–1587

Zhang YP, Li JH (2003b). Use RAPD analysis to classify tea trees in Yunnan. Agri Sci China 2:1290–1296

Zhao CY, Zhou LH, Luo JW, Huang JA, Tan HP (2006) AFLP analysis of genetic diversity of tea plant germplasm in Guangdong province. J Tea Sci 26:249–252

Zhao LP, Liu Z, Chen L, Yao MZ, Wang XC (2008) Generation and characterization of 24 novel EST derived microsatellites from tea plant (*Camellia sinensis*) and cross-species amplification in its closely related species and varieties. Conserv Genet 9:1327–1331

Zhu QW, Luo YP (2013) Identification of microRNA and their targets in tea (*Camellia sinensis*) J Zhejian Uni Sci-B (BioMed and Biotech) 5:1–10

Zietkiewicz E, Rafalski A, Labuda D (1994) Genome fingerprinting by simple-sequence repeat (SSR) anchored polymerase chain reaction amplification. Genomics 20:176–183

7 Stress Physiology

7.1 Introduction

Being perennial, tea plants experience multiple abiotic and biotic stresses (ABstreses) that adversely influence and reduce the growth of the plants and yield as well (Mondal 2003, 2009). While biotic stresses typically reduced 20 % yield, abiotic stresses are estimated to reduce 65 % of yield in tea (Waheed et al. 2012). Stress evaluation and exploration of alleviatory actions are the two most active and kinetic research fields of plant science. Tea plants encounter stresses in several ways due to their unique growth habits: (1) Being perennial, with a life span of more than 100 years, bushes experience all season stresses throughout their life; (2) the plants are maintained to a height of 60–80 cm by continuous pruning that leads to mechanical injury throughout their lives; (3) the young leaves are continually harvested which leads to wounding stress; (4) usually grown in higher altitude, thus experience UV radiation stress; (5) the necessary shade trees grown in tea plantation provide low light intensity stress; (6) being cultivated as monoculture, they provide complimentary circumstances for diverse pests; (7) near equator, they encounter winter dormancy and (8) are always grown in low soil pH or acidic soils.

7.2 Abiotic Stress

Therefore, tea bushes need to combat an array of stresses throughout their whole life to facilitate adaptation as well as survival through the counteractive mechanisms to endure stresses. Nevertheless, excessive stresses beyond tolerance level will unavoidably culminate in oxidative damage attributable to intensive production of reactive oxygen species (ROS) (Smirnoff 1993). Oxidative stress is a phenomenon which surfaces as a consequence of unevenness involving the ROS production (Apel and Hirt 2004). In this chapter, progress in the research of abiotic and biotic stresses of tea plants has been appraised.

7.2.1 Moisture Stress

The physiochemical aspects of tea plant due to low moisture stress had been studied in detail. Water stress is a major limitation for plant survival and growth (Kefei et al. 1997). Tea suffers both moisture deficit and inundation (Mokgalaka et al. 2004). Tea plants under low moisture stress showed reduced water potential (ψ) and sluggish growth rate (Yang et al. 1987). Tender leaves of tea exhibited higher diffusion resistance through stomatal regulation but tolerant clones possessed higher relative water content (RWC) than their drought-susceptible counterparts (Sandanam et al. 1981). However, clones that minimized water loss and possessed efficient osmotic adjustments could endure moisture deficiency. Soil moisture stress reduced RWC, ψ, photosynthetic rate (*Pn*); severely affected PS-II reaction centres and increased diffusive resistance (Wijeratne et al. 1998). A colossal 258 % increase in *Pn* had been observed in fully irrigated tea compared to

T. K. Mondal, *Breeding and Biotechnology of Tea and its Wild Species*,
DOI 10.1007/978-81-322-1704-6_7, © Springer India 2014

drought-stressed tea (Smith et al. 1994). Experimental evidences suggested that, critical levels of leaf ψ and transpiration rate (E) of drought-tolerant tea clones were relatively higher than the susceptible genotypes (Anandacoomaraswamy et al. 2002). Similarly, maximal fluorescence (Fm), variable fluorescence (Fv) and their ratio (Fv/Fm) as well as activity potential of PS-II (Fv/Fo) also shrunk, leading to non-photochemical quenching and reduction of Pn (Guo et al. 2009). It had been detected that unavailability of moisture drastically reduced shoot: root ratio, leaf area, stem diameter, stomatal conductance (gs), net CO_2 assimilation and adjusted the dry matter partitioning to roots which ultimately decreased yield (Bore 2008). Soil water status influenced gs that, in turn, affected Pn, E and played a key role in shaping the water status in tea leaves. Decreased gs in response to increasing irradiance, leaf temperature (T_L) and air vapour pressure were noticed in tea (Smith et al. 1993, 1994; De Costa et al. 2007). Further, it was found that total chlorophyll (chl) content increased during rainy season and decreased during summer indicating that chl synthesis was also affected by low moisture. Studies indicated that high humidity generated maximum Pn whereas low temperature along with low soil moisture eased Pn. Under excessive water along with high humidity, photosynthetic photon flux density (PPFD) and Pn were reduced (Hajra and Kumar 2002) whereas low moisture stress increased electrical conductivity (Yang et al. 1987).

Several studies indicated that low moisture induced the activity of antioxidant enzymes such as peroxidase (POD), superoxide dismutase (SOD), catalase (CAT), glutathione reductase (GR) polyphenol oxidase (PPO), phenylalanine ammonia-lyase (PAL), etc. along with antioxidants such as polyphenol and osmolites such as proline. The increments were rapid in tolerant cultivars compared to susceptible cultivars, which might have acted as a defense mechanism to combat low moisture stress (Yang et al. 1987; Wu and Pan 1995; Jeyaramraja et al. 2002a; Chakraborty et al. 2002; Upadhyay and Panda 2004; Upadhyaya et al. 2008). Tea seeds under desiccation accrued hydrogen peroxide (H_2O_2), increased ascorbate peroxidase (APX) and SOD activities (Chen et al. 2011b) whereas B deficiency and water stress together decreased dry weight (DW) as well as photochemical parameters but increment in proline and antioxidant activity was also evident (Hajiboland and Bastani 2012). In an another study, low moisture stress reduced PAL activity in UPASI-2, -8 and -9 tea clones along with lowering the synthesis of epigallocatechin gallate (EGCG) and epicatechin gallate (ECG) due to molecular rearrangement with elevated leaf temperature. Similarly, flooding (high moisture stress) deteriorated made tea quality, because gallic acid and caffeine concentrations declined leading to reduced synthesis of epi-theaflavic acid, epitheaflavic acid-3′-gallate and theaflavic acid (Jeyaramraja et al. 2003a). Drought-tolerant clones had been found to translocate more photosynthates to shoots than susceptible ones (Jeyaramraja et al. 2003b). Drought along with high temperature influenced dry matter partitioning and more dry matter was diverted to roots than to shoots during drought. Under drought, the amount of dry matter partitioned to leaves, stems and harvested shoots was decreased by 80–95 % (Burgess and Carr 1996).

It was found that low soil moisture stress reduced the accumulation of polyphenols in leaves of tea but again the reduction was much faster in susceptible cultivar than tolerant cultivars. Additionally, tolerant clones maintained a high content of polyphenols even at low soil moisture content. Thus, polyphenols may be used as biochemical indicators for selecting drought-tolerant cultivars (Cheruiyot et al. 2007).

The underlying mechanism of post-drought stress recovery in tea had been studied. Application of $CaCl_2$ thwarts drought-induced oxidative stress by increasing DW, prolines, phenols, SOD, POD, CAT and GR activities and plummeting H_2O_2 and MDA (Upadhyaya et al. 2011). Accumulation of potassium (K), calcium (Ca), manganese (Mn) and B accelerated during post-drought recovery because they positively modulated enzymes like, SOD, CAT, POX, PPO and GR (Upadhyaya et al. 2012). Under drought, continuous loss of soil moisture, by evapotranspiration, led to water deficiency in plants that eventually became fatal. Hence, application of antitranspirants may be a rewarding choice. Phenyl mercuric acetate (PMA) was the most useful (Nagarajah and Ratnasooriya 1977) due to its efficient translocation and obliteration within 20 days of application (Nagarajah and Ratnasooriya 1977). Alongside abscisic acid (ABA), Raliidhan and Antistress (antitransparent) were crucial as they increased stomatal diffusion resistance, ψ and relative turgidity, and reduced E (Handique and Manivel 1990). A linear relationship between total DW and the ratio between E and mean saturation vapour pressure deficit in clonal tea plants had been established that manifested discernible significance (Anandacoomaraswamy et al. 2000).

Several reports indicated that water deficiency regulated the catechin metabolism in tea plants. Longer rainy season (sufficient moisture stress) along with high humidity altered the quality constituents, such as EGC, epicatechin, ECG, EGCG, TC and caffeine (Wang et al. 2011). Interestingly, EC quinone (ECQ) and EGCG quinone (EGCGQ), the oxidation products of EC and EGCG, increased up to 100- and 30-fold, respectively, in tea plants exposed to 19 days of water deficit. Oxidation of EC and EGCG preceded proanthocyanidins (PAs) accumulation in leaves, which increased from 35 to 53 mg/g DW after 26 days of water deficit. Under low moisture stress, formation of ECQ and EGCGQ negatively correlated with the extent of lipid peroxidation in leaves, thus supporting a protective role for these compounds in drought-stressed plants (Hernandez et al. 2006). While EGCG and ECG showed no significant response to changes in soil water content, the shoot contents of EC and EGC in the six tea clones showed varied responses, with a distinct pattern in the water-stress tolerant clones (TRFK 6/8 and TRFK 31/30). The results suggested a potential use for EC and EGC as indicators to identify drought-tolerant cultivars in tea (Cheruiyot et al. 2008).

7.2.2 Temperature Stress

Low temperature is one of the key environmental stresses that tea bushes necessitate to cope with their entire life cycle. Tea plants growing beyond 16° N or S latitudes encounter winter dormancy with corresponding variations in day and night temperatures (Barua 1989; Nandi and Palni 1993). Environmental cues including low temperature forced plants to undergo oxidative stress, which was accompanied by enhanced production of ROS. These toxic ROS were scavenged using an array of enzymes. Among them, SOD played an important role which was isolated and characterized from tea leaf (Vyas and Kumar 2005a).

Barua (1969) induced dormant tea buds by application of GA_3 and suggested that the balance of growth promoting and inhibiting substances might determine the state of dormancy in tea shoots. High levels of free ABA and phenols were observed in dormant shoots of tea and it was suggested that they could possibly play a major role in dormancy (Nagar 1996). The possible involvement of PAs (Kakkar and Nagar 1997), GA_3 (Nagar and Kumar 2000) and brassinosteroids (Gupta et al. 2004) were established in tea shoots during dormancy. Conjugated indole-3-acetic acid (IAA) levels were increased with the onset of dormancy reaching its maximum value when free IAA levels were lowest. Nagar and Sood (2006) reported that with increase of free IAA levels, the conjugated IAA levels decreased in shoots prior to dormancy, suggesting of metabolic interconversion of IAA during these periods. Effects of low temperature on the response of two cultivars of tea with contrasting periods of winter dormancy were studied. The two cultivars were 'Teenali 17/154' (TNL) and 'Kangra Jat', where the former showed shorter periods of winter dormancy than later. Low temperature (5 °C) resulted in an increase of superoxide (O_2^-) ion and cellular damage (as measured by tetrazolium chloride reduction test) in both the clones; however, the increase was lesser in the case of TNL compared to KNJ. Activities of SOD, APX and GR increased in both the clones in response to low temperature, however, GR activity exhibited significant differences ($P<0.05$) between the two clones. Low temperature caused increase in the activity of various isozymes such as SOD, APX and GR. A new isozyme of SOD (Cu/Zn type) was induced in both the clones at low temperature. Significantly, higher GR activity in both the clones suggested a role of this enzyme in imparting better protection to tea at low temperature. Also, variations for GR isozyme were observed between the clones. Therefore, it was concluded that TNL, a clone with relatively lesser period of winter dormancy experienced lesser oxidative stress in response to low temperature compared to KNJ (Vyas and Kumar 2005b).

Frost during winter further injured young pluckable shoots which led to defoliation (Mondal 2003). During winter dormancy, low temperatures alone or in combination with low irradiation induced oxidative stress in plants. It had been found that genotypes having shorter dormancy period displayed elevated activities of antioxidant enzymes such as SOD, APX and GR due to accrual of ROS. Efficient scavenging of toxic oxygen species lowered down accumulations during winter months and thus it was coupled with reduced winter dormancy (Vyas et al. 2007). Photosynthesis can be a potential marker to categorize low-temperature stress effects on plants. In fact, low-temperature tolerant tea genotypes possessed higher MDA, proline and carotenoid (car) contents along with enhanced ROS scavenging machinery than the susceptible clones (Netto et al. 2005). Clones, for instance, TTL-1, TTL-4 and UPASI-9 were found to be frost tolerant compared to clones TTL-6, SM/OM/54 as well as SMP-1 and thus former clones were suitable for a region having recurrent frost occurrence because cultivation of these clones will ensure higher yield and quality (Netto et al. 2005). It was reported that low temperature induced albinism in tea but reverted back the albino phenomenon when the temperature was above 20 °C. In albino tea genotype, the development of chl from etioplast and the accumulation of chl a and b were blocked at 15 °C; thus culminating in albino leaf. Experimental evidences specified that chl content reduced with decrease in temperature (Du et al. 2008). However, external applications of proline and betaine on tea leaves decreased cold susceptibility, enhanced the content of methylglyoxal as well as MDA and increased GR and glutathione-*S*-transferase activity (Kumar and Yadav 2009). Tea growing under the shade tree (*Grevillea robusta*) had substantially lower transpiration rates than unshaded tea because shade tree reduced *E* principally by minimizing irradiance that strikes the canopy and by lowering the canopy temperature (Anandacoomaraswamy et al. 2000).

7.2.3 Ultraviolet (UV) Radiation Stress

It is well documented that UV-B is harmful to plants (Frederick et al. 1989). It reduced the synthesis of chl a/b-binding proteins as well as the D1 polypeptide of PS-II, thus inhibiting the overall growth (Stapleton 1992; A-H-Mackerness et al. 1997). In response to UV-B radiation, tea plants upregulated genes required for synthesis of protective phenolic compounds and catechins (Mamati et al. 2006). Short-term irradiation of UV-B stimulated the accumulation of total catechins. Conversely, excessive irradiation of UV-B suppressed catechins accumulation (Zheng et al. 2008). Under *in vitro* culture, UV-B radiation impeded number of callus-forming cells and promoted the accumulation of soluble phenols. Deposition of phenolic compounds on cell wall and intercellular spaces was increased (Zagoskina et al. 2003). White light supplemented with UV-B, induced formation of chl-bearing cells in callus culture and increased the phenols. PS-II activity was stimulated in the phototrophic cells of the callus culture in tea (Zagoskina et al. 2005). Tea plants were accustomed to broad range of irradiance that regulated carbon (C) assimilation, and experimental evidences revealed that 1200 μmol m^{-2} s^{-1} PPFD generated maximum *Pn* (Barman et al. 2008).

7.2.4 Low Light-Induced Stress

Under shaded conditions, tea plants possess photosynthetic apparatus customized to function, with utmost capacity (De Costa et al. 2007). Shade can influence *Pn* indirectly by regulating leaf temperature. Experimental evidences indicated that photosynthetic apparatus was tailored to function with maximum capacity under shade

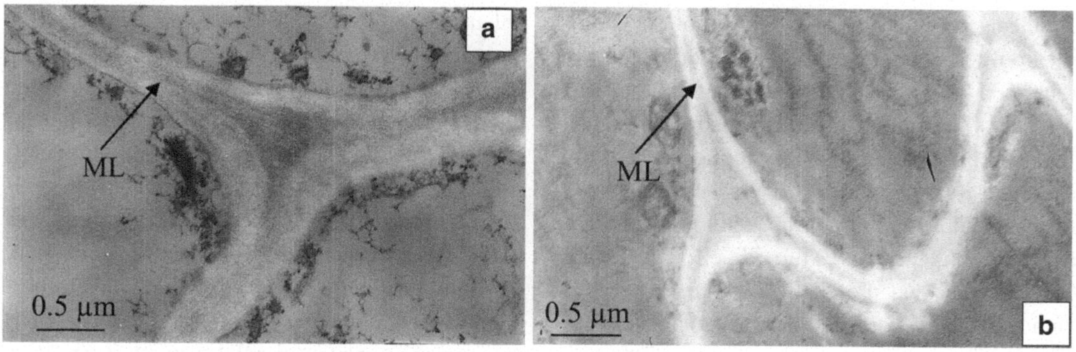

Fig. 7.1 Ultrastructural changes of tea leaf under B deficiency and optimum B concentration. **a** Cell wall of B deficient (0 mg/l) palisade cell with thickened (*arrow*) middle lamella (*ML*) (1 cm = 0.5 μm) without excessive thickening (*arrow*) in palisade cell of tea plantlet treated with B sufficiency (0.15 mg/l). **b** *Bars* indicate the measurements as mentioned. (Aadopted from Mukhopadhyay et al. 2013a)

(Mohotti and Lawlor 2002). The rate of *Pn*, *gs* and apparent quantum yield of seedlings under shade-grown tea remained consistently higher as compared to unshaded conditions (Mohotti et al. 2000). It had also been observed that light intensity beyond 1,400–1,500 μmol $m^{-2}s^{-1}$ resulted in photoinhibition that, in turn, declined *Pn* (Mohotti and Lawlor 2002). However, shade reduced photoinhibition by increasing *gs* (De Costa et al. 2007).

Overshade may reduce yield. It had been emphasized that PS-II, in tea, offered lower efficiency in capturing excitation energy and transport of electrons, apart from having low rates of RuBP carboxylation and RUBISCO activity under the overshade conditions. Besides, stomatal resistance of CO_2 diffusion towards carboxylation sites made CO_2 unavailable for carboxylation; thus reducing yield (De Costa et al. 2007).

The effect of environmental factors on contents of tea leaves had received meager attention and some invaluable breakthroughs had eventually surfaced. Under overshaded conditions, accumulation of epicatechins, ECG, EGC and EGCG decreased in newly developing buds because the conversions of glucose, shikimic acid and phenylalanine were depressed alongside the activity of PAL (Saijo 1980). At the outset, low light intensity inhibited synthesis of flavanols, causing decline in gallocatechin and leucoanthocyanin concentrations. The flavogen content of darkened stems, especially in seedlings, was much less decreased than that of the leaves (Forrest 1969). *C. sinensis* var. Yabukita, under shading treatments, appreciably gained phenylpropanoids/benzenoids and phenylalanine whereas upstream metabolites such as shikimic acid, prephenic acid and phenylpyruvic acid were reduced (Yang et al. 2012).

7.2.5 Elemental Stress

Tea encounters typical predicament of element deficiency due to extended life span, unremitting harvest without proper replenishment of nutrients and deterioration of soil fertility due to continuous uptake. Apart from deficiency, heavy metal contaminations due to relentless handling and industrial emission have become a critical alarm to the tea plantation worldwide. Enormous improvements have been made to date in unveiling a handful of elements that wreak havoc resulting in an exponential amount of crop loss. However, fortified with swift accumulation of information along with accelerated breakthrough has enabled scientists to foray into this particular field with full prospective returns. A summary of different elemental stresses on tea plants are depicted in Fig. 7.1.

7.2.5.1 Nitrogen (N)

Tea plants require huge quantity of N not only for yield (green leaf) but also for amino acid biosynthesis, as well as for synthesis of secondary metabolites (Chamuah 1988). Although tea was reckoned low-pH tolerant and favoured ammonium (NH_4^+) because of quicker absorption, it also took nitrate which noticeably reduced glutamine synthetase (GS) activity, decreased concentrations of total N, free amino acids and glucose, as a consequence, leaves turned yellow reflecting N deficiency (Ruan et al. 2007). Further, N deficiency affected water relations by increasing stomatal resistance and transpiration rate (*E*) but did not alter stomatal opening, leaf water potential and root resistance probably (Nagarajah 1981). The macroelement increased *Pn* but reduced photoinhibition and hence promoted production (Mohotti and Lawlor 2002).

7.2.5.2 Boron (B)

B deficiency is prevalent in the tea cultivating soils. B deficiency reduced chl a/b ratio, net assimilation, *E*, and *gs* of the young leaves, whereas increased car in the older leaves parallel with non-photochemical quenching. Under B deficiency, NO_3^- concentrations diminished while NO_2^- concentrations increased in the young leaves and among the antioxidative defensive enzymes, APX and peroxidase (POD) increased in older leaves and roots respectively in tea (Hajiboland et al. 2011a). B-deficient tea plants mitigated the reduction of photosynthetic energy conversion under intermediate and highly illuminated circumstances. The best possible light intensity for CO_2 assimilation was achieved under intermediate light for the young leaves and high for the old leaves. Activity of APX, SOD and concentration of proline remained lower under intermediate conditions. In the young leaves, photochemical events were protected additionally under excess light and low B supply. B deficiency-induced activation of antioxidant defense system alleviated the effect of high light stress (Hajiboland et al. 2011b, 2013). B deficiency culminated in an overall inhibition of plant growth as indicated by reduction of *Pn,* stomatal conductance (*gs*), *E*, pigment concentrations, dry weight (DW) and fresh weight (FW) of shoot and root, but activity of all antioxidant enzymes, total sugar, reducing sugar, starch, ascorbic acid (ASA) and phenolics enhanced, although protein contents decimated under stress. Amount of superoxide anions (O_2^-), MDA, electrolyte leakage and H_2O_2 augmented under deficiency. Furthermore, B deficiency increased the transcript of antioxidant enzymes, such as POD, SOD, APX and transcripts of 14-3-3 gene indicating that oxidative metabolism was the prime defense mechanism under B deficiency of young tea plants that brought a cascade of biochemical changes and implied the essence of B in upbringing of tea plantlets. In addition, closing of stomata, increasing amount of ABA and proline content in deficient plants indicated that B deficiency induced low moisture stress (Mukhopadhyay et al. 2013a). Transmission electron micrograph (TEM) of cell walls elucidated stress obligation by the presence of thickened middle lamellae of leaf cells (Fig. 7.2), which was a key marker of B deficiency, since the B-adequate cells did not possess such structural modifications. Under deficiency, the B-rhamnogalacturonan-II complex formed due to increase in monomers concurrently solidify the middle lamella. Thus, damage of cell-wall ultrastructure was a corroboration of B-induced stress to tea plants (Mukhopadhyay et al. 2013a).

7.2.5.3 Zinc (Zn)

Zn is an essential micronutrient, deficiency of which is quite common in tea cultivation. Apart from other factors, applications of excessive phosphate fertilizer in soil induced Zn deficiency (Nelson 2006). Deficiency and sufficiency of Zn led to growth inhibition, shoot as well as root FW and DW reduction, reduction of *Pn, E* and *gs* in conjunction with reduction of both chl a and b, O_2^-, MDA and H_2O_2. Zn stresses both deficiency as well as sufficiency increased electrolyte leakage, phenol concentration and the activities of APX, CAT, SOD and POD enzyme in tea suggesting that antioxidative mechanism was the prime defense system under Zn deficiency. Transcripts of the antioxidative enzymes were upregulated under stress corroborating with the biochemical assays. Largely the results implied that Zn stress increased ROS production and activities of the enzymes but the defensive system en bloc did not afford adequate fortification from

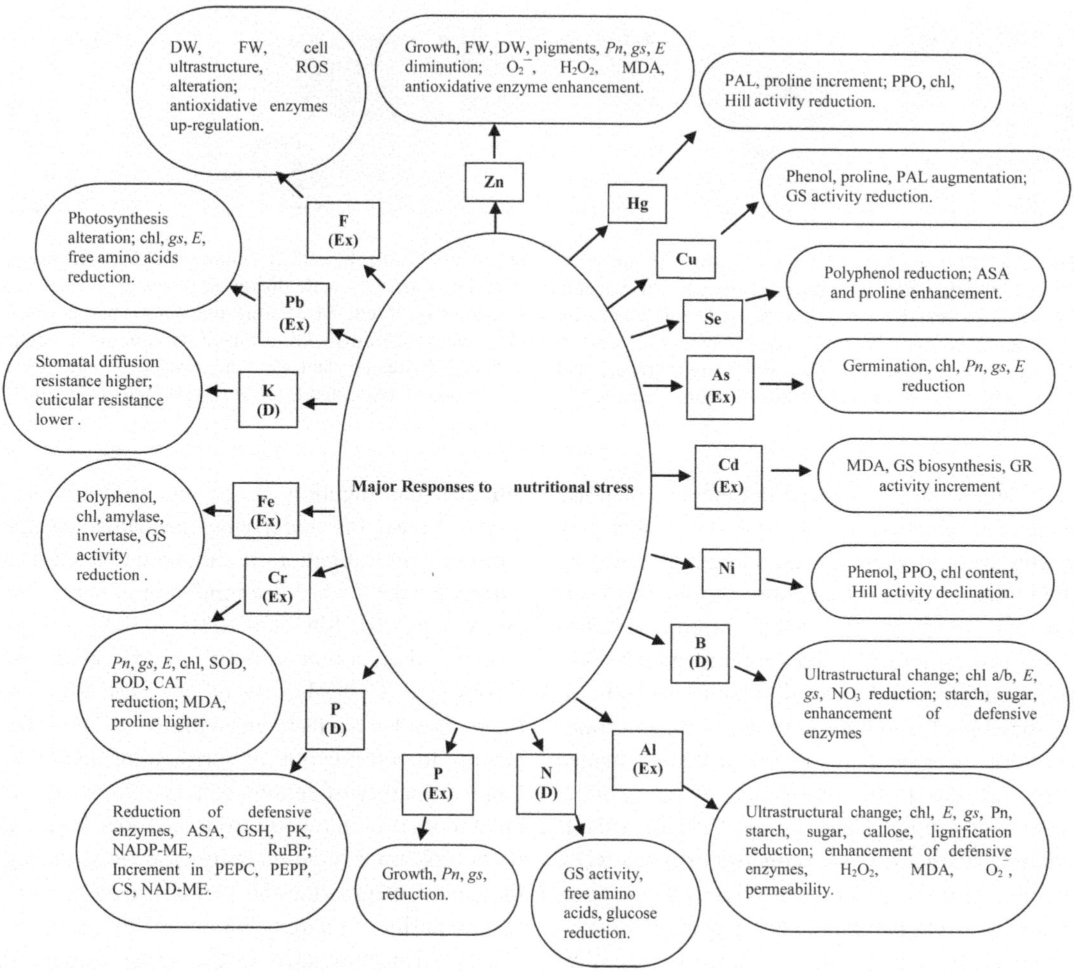

Fig. 7.2 Schematic representation of effects of elemental stresses. *Ex excess, D deficiency*

oxidative damage (Mukhopadhyay et al. 2013b). Typical toxicity symptom of Zn included prominent browning of leaves although severe toxicity led to the death of the plants. UPASI-9 cultivars, under Zn toxicity of 2,000 mg/kg died within 15 days, while 1,000 mg/kg extended the death by 36 days. Elemental interaction had also been studied in tea plants. It had been found that excessive Zn stimulated Mn translocation, whereas antagonized phosphorous (P), magnesium (Mg) and iron (Fe) absorption (Venkatesan et al. 2006). Combination of Zn and selenium deficiency increased PPO activity (Qian et al. 2009). The contents of polyphenols, catechins, amino acids and soluble sugars increased with the increased Zn application (Huiqun and Yuchen 1987) whereas Zn deficiency enhanced the accumulation of amino acids, glucose, activity of SOD but decreased MDA, PPO activity (Jinghua et al. 1997) and chl (Owuor 2001). Zn enrichment enhanced photosynthesis as well as protein synthesis (Huiqun and Yuchen 1987) but reduced the uptake of P and K, besides the reduction of the contents of Fe, Mn, Ca and Mg in tissues (Wu and Fang 1994). Under Zn deficiency, the youngest leaves became narrow, strap-like, erect, and outward appearance became a rosette like at the stem summit. Leaf blades became yellow and severely affected leaves became stunted with inward-curling margins; apical growth and branch extension retarded (Nelson 2006) along with dwindled DNA contents in tea seedlings (Wu

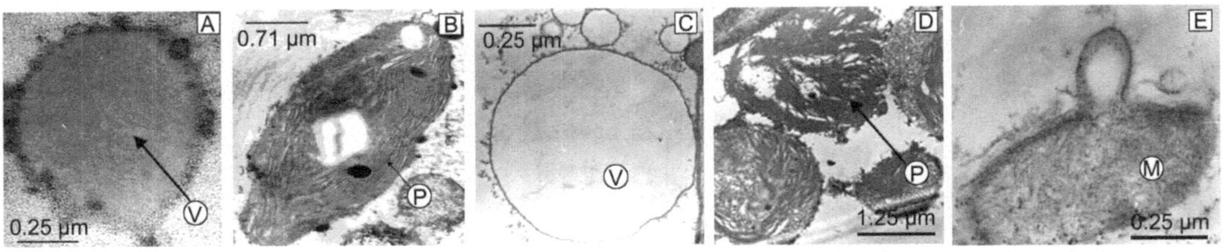

Fig. 7.3 Ultrastructural changes of tea leaf and root under Zn deficiency as well as Zn-excess. **a** Vacuole (*V*) of Zn-deficient (without Zn) palisade cell filled with electron dense phenolic compounds (1 cm =0.25 µm). **b** Nnormal chloroplast (*P*) of Zn (0.05 mg/l) treated leaf (1 cm =0.71 m) and **c** normal ultrastructure vacuole (*V*) (1 cm =0.25 µm) in leaf Zn-optimum (0.5 mg/l) plants. **d** Chloroplast (*P*) with distorted grana at 5 mg/l Zn treatment (1.25 cm =1.25 µm) and **e** mitochondria (*M*) (1.25 cm =0.25 µm) with localized swelling in leaf cells of 5 mg/l Zn treated plant. *Bars* indicate the measurements as mentioned. (Adopted from Mukhopadhyay et al. 2012)

and Fang 1994). Under Zn deficiency, vacuoles contained electron dense material resembling phenolic compounds that could be reaffirmed by the higher content of phenolic compounds. Over Zn dose produced deformed chloroplast ultrastructures as indicated by ruptured membranes, and disintegrated thylakoid, which adversely influenced photosynthetic activity. Similarly, mitochondria were swelled and cristae were damaged (Fig. 7.3) due to the imposition of stress under redundant supply. These imply that antioxidants did not satisfactorily scavenge superfluous ROS, to safeguard the tissues from free radical injury under Zn stress (Mukhopadhyay et al. 2013b).

Zn modulated drought-mediated biochemical damages in tea plants. It was demonstrated that drought induced the decrease of RWC, dry mass of leaf and antioxidants such as ascorbate and glutathione in the tested tea clones (TV-1, TV-17 and TV-29) by the application of zinc sulphate ($ZnSO_4$) before water withholding for 7 days. Increase of phenolic content with decrease in H_2O_2 and lipid peroxidation and differential activities of enzymes such as SOD, CAT, POX, PPO, GR and APX with concomitant increased Zn uptake in leaf were noticed under drought treatment (Upadhyaya et al. 2013).

7.2.5.4 Aluminum (Al)

Tea plants and its wild species are Al hyperaccumulator and can accumulate as high as 30,000 mg/kg Al in mature leaves on DW basis (Matsumoto et al. 1976; Zeng et al. 2011), a concentration that is toxic for other plants. Thus, it attracted the attention of the scientific community to reveal the underlying mechanism. Contemporary investigations highlighted that age and variety determined Al accumulation in tea (Ruan and Wong 2001; Shu et al. 2003) and Al absorption varied on soil conditions and extractable Al (Wong et al. 1998). Uptake of Al^{+++} and NH_4^+ by the roots of tea plants induced proton (H^+) release leading to acidification of surrounding medium. Higher quantity of cations concentrations stimulated more proton release from the roots, because roots took up a huge quantity of cations during the course of growth, and then released them to restore parity of charge (Wan et al. 2012). In the leaves, Al accumulated in the upper epidermal cell walls of mature leaves (Carr et al. 2003). In tea leaves, Al was mainly present in two dominant forms, i.e. as Al-catechin complex (Nagata et al. 1992) and Al-citrate (1:1) complex in the xylem sap (Morita et al. 2004) and accumulated more in Chinary types plant than Assam types (Fung et al. 2003). The tolerance mechanism rested on the fact that Al retention took place in epidermal leaf apoplast and not in the symplast (Tolrà et al. 2011). Furthermore, N fertilization augmented extractable Al levels but decreased soil pH and exchangeable base cations, but Al quantity in shoots was reduced by enormous N fertilizer application (Ruan et al. 2006).

While Al induced the uptake of Ca, Mg, K and Mn, it reduced the uptake of Fe, copper (Cu) and Zn. Moreover, total phenols, catechins in tea plant tissues increased with increasing availability of Al in soil (Chen et al. 2011a). In hydroponic culture, Al at lower concentration (2 mg/l), improved plant growth, but at higher concentrations, it induced oxidative damage. In accordance with Al application, Al contents increased in the grown-up tissues than their younger complements and it had been observed that accumulation started after 6 h of exposure of the young root tips to Al solution. Pigments content, *Pn, E* and *gs* were reduced due to Al toxicity. Activated oxygen metabolism was evidenced by increasing MDA, membrane injury, evolution of O_2^- and H_2O_2 under Al toxicity. Phenol contents initially increased but later reduced at advanced levels whereas total sugar and starch concentrations decimated beyond 2 mg/l of Al. Activities of antioxidative enzymes such as APX, GR, POD, CAT and SOD, increased with the increase of Al. Expression of *citrate synthase* (*CS*) transcript seemed to be upregulated in the mature leaves, young and old roots with increased Al content in those parts, perhaps due to Al-citrate complex formation. Simultaneously citrate and malate production also increased due to Al toxicity. These results considerately indicated that Al at lower levels promoted growth but became a stressor at elevated concentrations (Mukhopadyay et al. 2012). Apart from all these defense mechanisms, callose formation was a frequent rejoinder to stresses and lower doses of Al stimulated root elongation and callose reduction but elevated doses stimulated its accumulation (Lian et al. 1998). Similarly, TEM analysis of Al-treated leaves, revealed significant membrane damage at 70.67 mg/l of Al because at this level the equilibrium between formation and detoxification of ROS was not much effective (Li et al. 2011a). Interestingly, Al-toxicity tolerance depends upon the genotype of tea. For example, Pingyangtezao and Wuniuzao, two tea cultivars, under Al toxicity performed in a different way as indicated by different levels of chl content with differential activity of SOD, CAT and POD enzymes among the two cultivars (Yu et al. 2012).

It had also been reported that the content of proline and activities of POD, CAT and SOD enzymes reduced under high Al: F but increased with the increase of fluorine (F) alone in tea plants (Wang et al. 2009a). Upon application of Al, tea plants accumulated higher levels of ASA,

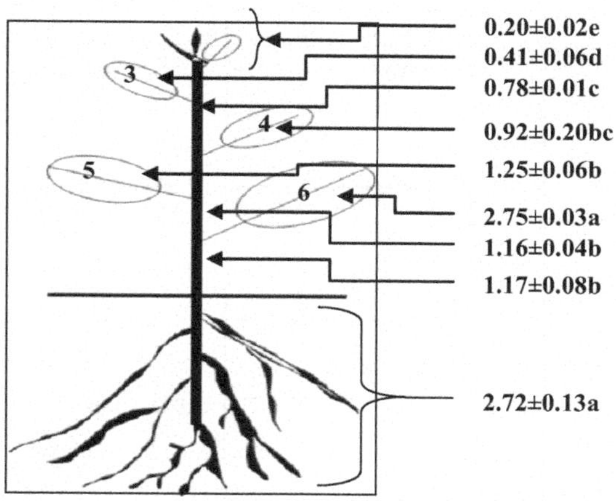

Fig. 7.4 Al content (g kg^{-1} DW) in tea plants treated with 53 g/l of $AlCl_3$ at pH 4.2 (leaves below the growing shoot apex are indicated as *3, 4, 5* and *6*). Each value represents the mean ± SE of three replicates (DW = dry weight). Different *letters* indicate significant difference at $P<0.05$. (Adopted from Mukhopadhyay et al. 2012)

chl, car, phenolics, proline as well as quality constituents such as gallocatechin, EGCG, gallocatechin gallate, ECG and catechin gallate indicating the fact that optimum dose of Al might induce the quality of made tea (Napaporn et al. 2012). Al-induced enhancement of antioxidant enzymes activities, membrane integrity and delayed lignifications afforded the stimulatory effects of Al on the growth of tea plant (Ghanati et al. 2005). Recent investigations revealed that Al was detoxified by exclusion mechanism forming Al-ligand complexes and thus suppressed Al uptake. It was also proposed that oxalate was a key Al-chelating compound in the tea roots (Morita et al. 2008). Oxalate exudation along with small quantity of malate, citrate and oxalic acid metabolisms were coupled to Al accumulation in roots (Morita et al. 2001). In tea cultivar TS-78, Al contents increased from young to mature leaves and from immature to full-grown stem suggesting that Al accumulation was specific to the developmental stage of tea plant (Fig. 7.4). Proportional variations of Al accumulation in the sixth mature leaf revealed that Al concentration steadily increased from the centre towards the margin of the leaf (Fig. 7.5). Maximum quantity of accumulation occurred at the margin of the leaf (Mukhopadyay et al. 2012). Interestingly, protein played important role to protect the tea plant from Al toxicity

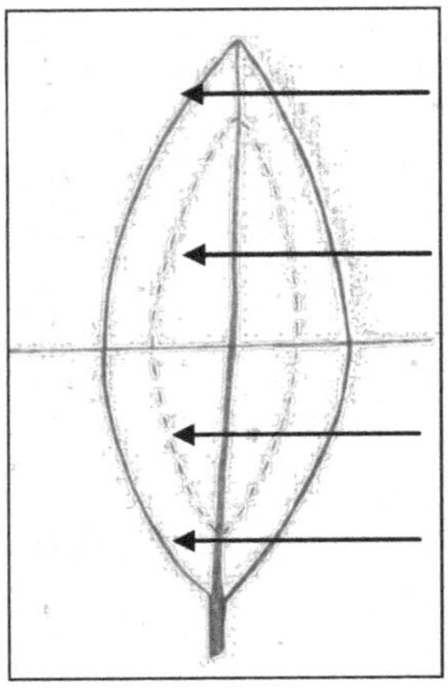

Fig. 7.5 Distribution Al (g kg^{-1} DW) in the 6th leaf of tea plant treated with 53 mg/l of $AlCl_3$ at pH 4.2. The leaf tissues were cut with scissor along with the line for Al analysis. Each value represents the mean ± SE of three replicates (DW = dry weight). Different *letters* indicate significant difference at $P<0.05$. (Adopted from Mukhopadhyay et al. 2012)

as it was found that poly-L-aspartic acid and poly-L-glutamic acid protected growth of tea pollen tube from toxic effects of Al. It was assumed that the polypeptides might prevent Al from either interfering with Ca^{++} transport or binding to calmodulin that was implicated in pollen tube growth (Konishi et al. 1988).

7.2.5.5 Phosphorus (P)

Tea is cultivated in humid and subhumid tropical, subtropical and temperate regions of the world, mainly on acidic soils where P deficiency is often experienced (Lin et al. 2009a). Hence, P fertilizers are used regularly in tea plantations to increase productivity (Salehi and Hajiboland 2008). Being an essential element, P played a pivotal role and P deficiency lowered the activities of SOD, APX, GR, CAT, monodehydroascorbate reductase (MDAR), dehydroascorbate reductase (DHAR) enzymes together with the contents of ASA and reduced glutathione (GSH). Nevertheless, in spite of all biochemical changes, MDA levels remained unchanged indicating the active participation of the antioxidant system to provide substantial safeguard against photooxidative injury (Lin et al. 2012). Increased quantity of P inhibited growth due to reduction of *Pn* as well as *gs* that affected the photochemistry of P-deficient leaves (Salehi and Hajiboland 2008). P deficiency induced malate and citrate discharge accompanied by increasing activity of PEP carboxylase (PEPC), PEP phosphatase (PEPP), CS and NAD-malic enzyme (NAD-ME) in leaf and decreased activity of pyruvate kinase (PK), NADP-ME and NADP-isocitrate dehydrogenase activities in roots. Compared to roots, malate accumulation was more in the leaves along with increased activities of NADP-ME, NAD-ME and PK (Lin et al. 2011). In tea, P deficiency reduced particularly P and Mg content in the tissue but increased C/N ratios under P-deficiency treatment (Lin et al. 2009b). P deficiency decreased electron transport due to impairment in the electron transport chain thus ultimately reducing ATP production terminating in a smaller amount RuBP regeneration and CO_2 assimilation as well. It was proposed that energy dissipation was increased to facilitate the protection of P deficiency (Lin et al. 2009b).

7.2.5.6 Lead (Pb)

Pb is a well-known pollutant having the ability to function as a stressor. In Pb-contaminated soil, phytotoxic symptoms appeared in plants together with reduction of growth, caffeine and amino acids but the enhancement of catechin content. Under such situation, maximum accumulation of Pb occurred in tea roots (Wu et al. 2011). Excess Pb decreased chl content, *Pn, gs* and along with *E* significantly (Tang et al. 2008a). Longjing 43 showed an enhancement of *Pn* when both Longjing 43 and Zhenong 117 were subjected to Pb stress. POD and SOD activities increased with higher Pb concentration in Zhenong 117 but in Longjing 43, POD and SOD activity increased initially and then decreased gradually. It was also observed that application of $CaCO_3$ significantly decreased Pb concentration in tea plants (Han et al. 2007).

In another study, Pb was found to increase the growth of Longjing 43, but reduced the growth of Zhenongl 17 at the same concentration (430 mg/kg). Conversely, at a lower level (160 mg/kg), it reduced the net *Pn* of Zhenong 117, while increased the *Pn* of Longjing43 indicating that different genotypes behaved differently under Pb. In Zhenong 117, POD and SOD activities increased with increasing Pb availability whereas in Longjing 43, POD activity increased initially and then decreased indicating that the effects of Pb varied on genotypes of tea (Ruan et al. 2012).

7.2.5.7 Fluorine (F)

Tea accumulates large amounts of F and transports them readily to the leaves. F concentration increased linearly with availability in soil. Interestingly, its uptake and translocation further increased in presence of Al (Ruan et al. 2003). Chinary tea accrued higher F contents compared to Assam variety (Fung et al. 2003). It was suggested that increasing F concentrations in soil decreased FW, DW, chl content and *Pn* whereas CAT and guaiacol peroxidase (GPX) activities increased significantly along with the increase of H_2O_2, malodialdehyde (MDA) and proline (Li et al. 2011b).

Ca application reduced F uptake, possibly due to the effect of Ca on the properties of cell wall or membrane permeability or may be due to alteration in F speciation (Ruan et al. 2004). Increased F application often damaged chl membrane, which ultimately reduced the *Pn* (Li et al. 2011b). Conversely, F also mitigated the adverse effect of Al and stimulated pollen tube growth either in combination or single-handedly (Konishi and Miyamoto 1983).

7.2.5.8 Selenium (Se)

Se increases chl, car, phenol, proline and ASA content. It had been reported that at 50 mg/l of Se offered best chemical quality of made tea and highest phenolic contents (Napaporn et al. 2012). Foliar applications of Se augmented yield and quality along with sweetness and aroma, but bitterness decreased significantly. Though polyphenol contents reduced at higher dose of Se, total amino acids and vitamin C content increased extensively by Se application (Hu et al. 2003). In another study, tea enriched with Se, by selenate application, exhibited the maximum inhibition of lipid oxidation, whereas R-tocopherol showed the lowest inhibition. Se-enriched green tea harvested during the early spring had been found to be much superior compared to regular tea indicating that along with other factors, Se might be responsible to improve the quality of green tea (Xu et al. 2003).

7.2.5.9 Arsenic (As) and Cadmium (Cd)

Enhanced As^{++} in soil decreased germination of tea seeds and subsequently the growth of tea seedlings, along with instantaneous reduction of chl content, *Pn, gs* and *E* in tea (Tang et al. 2008b). In As and Cd excess soil, feeder roots played the buffering role so that minimum quantity of As gets transported to the shoots. In fact, Cd persuaded growth, photosynthesis and chl synthesis when present in optimum quantity but excess quantity impaired the plants (Tang et al. 2008b). Similarly, Cd toxicity decreased chl as well as protein content, but enhanced MDA and upregulated the transcripts of GSH biosynthetic genes such as; γ-glutamylcysteine synthetase and glutathione synthetase, and GR, indicating the protective role of these enzymes under excess Cd stress (Mohanpuria et al. 2007). Negative correlations between Cd concentration with *Pn, E* and biomass production were found (Shi et al. 2008). Cd induced higher chl, MDA and soluble sugar in spring than summer spring tea (Xia and Lan 2008). Depending on the tissue of callus origin, Cd concentration influenced phenol metabolism under *in vitro* culture, indicating that the production of phenol under *in vitro* conditions was dependent on original tissue of callus origin. In leaf and stem origin calli, Cd decreased the flavans but uniqueness remained unchanged in calli originated from roots. Simultaneously, lignin content in the calli of root and stem origin increased, although it remained unaltered in the leaf-derived callus (Zagoskina et al. 2007).

7.2.5.10 Others (Chromium, Iron, Potassium, Mercury, Copper, Magnesium, Manganese and Nickel)

Despite the fact that nutrients play important role in plant growth, effects of certain elements on tea are less studied which are discussed here.

Chromium (Cr) is a widespread natural contaminant in soil and its accumulation in soil creates a disparaging effect on tea plants. It not only affects the yield and quality but also jeopardizes human health either by direct contact or through food chain. In tea, higher level of Cr diminished chl content, *Pn, gs* and *E* and ultimately reduced the number, length as well as weight of the young buds (Tang et al. 2008a). Under Cr toxicity, activity of antioxidant enzymes such as SOD, POD and CAT was reduced, while proline and MDA content were increased (Tang et al. 2012). Fe toxicity reduced chl content and polyphenols along with a decrease of activity of amylase, invertase, aspartate aminotransferase and glutamate synthase enzyme in tea (Hemalatha and Venkatesan 2011).

K-deficient tea plants demonstrated higher stomatal diffusive resistance along with reduction of night opening. Deficient plants also showed lower cuticular resistance (Nagarajah 1979).

Mercury (Hg) is a heavy metal and can induce oxidative stress in tea. Upon exposure, Hg decreased phenol contents, activity of PPO enzyme and chl content. On the contrary, PAL activity and proline content were increased (Basak et al. 2001). Tea plants exposed to high Hg showed reduction of *Pn,* root browning and chlorosis of leaf (Yadav and Mohanpuria 2009).

Cu ions can oxidize thiol bonds causing disruptions in protein structure and functions (Yadav and Mohanpuria 2009). While Cu toxicity stress increased phenols, proline content and PAL activity, but upon exposure, it reduced the activity of PPO, GS enzyme and chl content (Basak et al. 2001; Rana et al. 2008). It had been reported that Chinary cultivar was more tolerant than Assam, because Assam type accumulated more ROS and less proline (Yadav and Mohanpuria 2009). Cu-based fungicides are being used for decades to control fungal diseases in tea, which may lead to accumulation of Cu in the soil. Thus, in order to know the biochemical responses to increase concentrations of Cu, leaves of two cultivars of Darjeeling hills were investigated. Exposure to excess Cu resulted in increased lipid peroxidation, reduced chlorophyll content, higher level of phenolic compounds and an increase in PPO enzyme activity. Two peroxidase isozymes (POD1 and POD2) were detected in plants exposed to Cu. In addition, biochemical responses in two tested cultivars, TS-462 and TS-520 differed significantly. TS-520 was found to be more sensitive to Cu. From this study, it appears that Cu exposure led to the production of ROS in the leaves resulting in significant lipid peroxidation. Tea plants try to mitigate this oxidative damage through accumulation of phenolic compounds and induction of antioxidant enzymes (Saha et al. 2012). Report on nickel (Ni) interaction of tea plant is also scanty. Excessive Ni decreased phenols, chl content and PPO activity but increased PAL activity and proline content (Basak et al. 2001). Soil applied Mg improved free amino acids and quality of made tea (Ruan et al. 1999). Nevertheless, Mg inhibited polyphenols content but up to a certain extent increased chl content. Excess Mg induced coppery colour all over the leaf surface, however, it led to defoliation (Venkatesan and Jayaganesh 2010). Mn had significant influence on yield of tea. It was proven that all the chemical constituents of quality such as theaflavin, thearubigin, total colour and caffeine content increased with the enhancement of Mn level. Mn applications influenced uptake of leaf N, P and shoot Mn content significantly (Gohain et al. 2000). All these studies indicated that these elements under sufficiency or deficiency conditions become a stressor as indicated by increasing antioxidant enzymes as well as antioxidants.

7.3 Biotic Stress

Tea mosquito bug, mites such as red, pink as well as purple, thrips, termites, red slug, looper caterpillar, leaf-hopper, etc. are important pests of tea (Gurusubramanian and Borthakur 2005), although more than 1,000 pests are reported in tea (Das 1965). Significantly, pest infestation also changed important biochemical constituents of plants depending upon some factors including the genotypes (Chakraborty and Chakraborty 2005; Saikia et al. 2011). Blister blight (*Exobasidium vexans*), a biotrophic fungus, causes a havoc and crop loss in tea plantation. Premkumar et al. (2008) reported that upon infestation leaf area, FW and DW, *Pn, gs* and *E* were reduced together with total sugars, amino acids, total N, polyphenols and catechins in the more in susceptible genotype than the tolerance genotype. In general, it had been found that under severe blister blight infection, total phenols, catechins, N, amino acids, theaflavins, thearubigins, caffeine, as well as PPO activity along with aroma components, such as hexane-1-ols, phenyl acetaldehyde, linalool, methyl salicylate, geraniol, indoles, β-ionones and nerolidol were reduced, which led to production of a very inferior quality of made tea (Gulati et al. 1999). A simultaneous decrease of 2-phenylethanol content and prephenate dehydratase (PDT) activity was also noticed since it was assumed that loss of PDT activity may be due to palisade tissue injury along with epidermal layers containing the stomata. These ultimately yielded lower *Pn,* thus sinking the carbon flow through the shikimate pathway (Sharma et al. 2011). In susceptible tea, content of total phenols and ortho-dihydroxy phenols decreased upon infection of Sclerotial blight (*Sclerotium rolfsii*) of tea. Activity of PAL, POD increased specifying their pivotal roles in defense mechanism (Bhagat and Chakraborty 2010). Increased activity of DNase, chitinase, PAL, tyrosine ammonia lyase (TAL), PPO and POD was reported due to the infection of Bird's-eye spot disease (*Cercospora theae*) (Gnanamangai et al. 2011).

Similarly, the physiological and biochemical changes due to *Pestalotiopsis* infection were also studied in tea under greenhouse conditions. It had been found that while *Pn, E,* water use efficiency and chl fluorescence were reduced in the infected leaves, stomatal resistance was increased. Biochemical constituents were higher in healthy leaves compared to that of healthy tissue of infected leaves. There was no significant variation in pigments such as chl and car, but in the infected tissue, these constituents reduced significantly. Activity of both PAL and TAL were higher in the healthy zone of infected leaves than that of infected zone of the same leaves. On the other hand, activity of PPO and peroxidase enzymes was higher in the lesion region and healthy portion of infected leaves (Sanjay and Baby 2007).

7.3.1 Relevance of Microbes in Stress Alleviation

Microbial infections alter the biochemical changes as a consequence of stress. Foliar spray of *Bacillus licheniformis* lowered the caffeine content of tea without affecting flavour and aroma (Ramarethinam and Rajalakshmi 2004). Upon application of bioformulations of *Bacillus megaterium,* phenolics and pigment contents were increased noticeably (Chakraborty et al. 2012). *Trichoderma harzianum* and *Azospirillum brasilense* stimulated rooting and acclimatization of *in vitro* grown tea shoots, and plantlets due to superior nutrient uptake (Thomas et al. 2010). Application of K-solubilizing bacteria (*Pseudomonas putida*) in permutation with N, P and K increased chl, car, N, P and K contents of the shoots together with total polyphenols, catechins, amino acids and sugars (Bagyalakshmi et al. 2012). It had been found that *B. megaterium,* isolated from rhizosphere, inhibited brown rot disease (*Fomes lamaoensis*) of tea. Root colonization by *B. megaterium* and subsequent inoculation with *F. lamaoensis* increased the phenols and activity of enzymes such as POD, chitinase, β-1,3-glucanase, PAL and also produced IAA, siderophore and antifungal metabolites that promoted tea growth (Chakraborty et al. 2006). Arbuscular mycorrhizal fungi, upon inoculation, increased the content of amino acids and total proteins, respectively along with total polyphenols, content of caffeine as well as sugar (Singh et al. 2010). These works collectively suggested that beneficial microbes upon infection promoted growth of tea plants and simultaneously reduced the adverse effects of stress (Table 7.1).

7.3.2 Autotoxins

Autotoxicity, exists both in natural and agricultural ecosystem, is a process by which a species or its decomposing residues discharge phytotoxins to inhibit growth of the other plants (Eden 1940). Analysis of the effects of autotoxins designate that the chemicals possess both inhibitory

Table 7.1 Biotic stress-induced alteration of different biochemical parameters of tea plants

Causal organisms	Biochemicals amended	References
H. theivora	ABA ↑ GA_3 ↑ IAA ↓	Saikia et al. (2011)
H. theivora	POD ↑ PPO ↑ APX ↑ PAL ↑ Polyphenols ↓ Catechins ↓	Chakraborty and Chakraborty (2005)
E. vexans	Caffeine ↑ Theaflavins ↑ Thearubigins ↑ Linalool ↑ Geraniol ↑ β-ionone ↑ Methyl salicylate ↑ Total sugar ↓ Amino acids ↓ PPO ↓ Nitrogen ↓ Catechins ↓	Gulati et al. (1999)
E. vexans	Total sugar ↓ Amino acids ↓ Nitrogen ↓ Polyphenols ↓ Catechins ↓	Premkumar et al. (2008)
E. vexans	2-phenylethanol ↓ Prephenate dehydratase ↓	Sharma et al. (2011)
S. rolfsii	PAL ↑ POD ↑ Total phenol ↓ Ortho dihydroxy phenol ↓	Bhagat and Chakraborty (2010)
C. theae	DNase ↑ Chitinase ↑ PAL ↑ TAL ↑ POD ↑ PPO ↑	Gnanamangai et al. (2011)

and stimulatory effects on antioxidant activities in which stimulatory actions can be seen at lower concentrations, but an increment over time unveils obvious inhibitory effects. It had also been observed that adverse effects of autotoxins were counteracted by the enzymes of antioxidant defense system and resulted in higher MDA formation in the leaves of tea (Cao et al. 2011a).

7.3.3 Effect of Plant Growth Regulators (PGR)

PGRs such as 2,4-D and NAA determined the growth of *in vitro* culture of tea plants (Mondal et al. 1998). Under stress, external applications of SA induced expansion of leaf area suggesting that it acted as a phytohormone. Leaf TSS and invertase activity increased appreciably, thus making the way for potential plant resistance against environmental stresses (Kaveh et al. 2004). Brassinolides stimulated rapid healthy shoot growth by modifications in stomatal response during the period of low moisture stress (Ng'etich and Bore 2001). Ploidy levels of tea also influenced photosynthesis. It had been found that the net CO_2 assimilation and *gs* were lower in the tetraploids than in either diploids or triploids and were also lower in triploids than diploids (Ng'etich and Wachira 2003).

Effects of exogenous application of ABA on physiological and metabolic changes in tea were investigated. The highest improvement of yield was evident in response to regular application (a day after every harvest) of ASA at 600 mg/l followed by 400 mg/l. Foliar application of ASA (600 mg/l) at regular intervals enhanced chl a and b contents besides a significant increase of total polyphenols and catechins when compared to the untreated control. Exogenous application of ASA at alternate harvesting rounds increased the activity of PPO whereas APX remained unchanged. An increase of PAL activity was noticed with foliar application of ASA irrespective of its concentration. TFs and TRs of made tea were increased when ABA was applied with 400 and 600 mg/l at alternate dose. The scores of brewed tea liquor characteristics, viz. infusion, total colour, strength and briskness were higher even at the lower concentration of ASA treatment. Results suggested that foliar application of ASA (400 mg/l) proved to be a useful measure to improve the yield, physiological attributes and antioxidant properties of tea (Murugan et al. 2012).

7.4 Effects of Stress on Quality of Made Tea

Aroma is an indispensable aspect affecting the quality of tea and pivotal for its marketing and price realization. Many intrinsic and extrinsic factors interplay to achieve the most desirable feature of made tea. Quality of made tea depends upon the age of leaf, i.e. younger leaves produced better quality. In immature leaves, a relatively large proportion was used to produce catechins, a compound responsible for quality of black tea (Sanderson and Sivapalan 1966).

Adequate supply of Mg nutrient promoted the synthesis of theanine in roots and its accumulation in the young leaves of tea plants. Mg status in soil was an important factor influencing the mobility of amino acids and sugars via xylem and phloem especially when N and C reserves remobilized to support spring growth of young shoots (Ruan et al. 2012). Application of $MgSO_4$ followed by kieserite induced higher content of chl, car, quantity of TFs, TRs and polymerized substances (Jayaganesh et al. 2011). Activity of theanine synthetase, an enzyme involved in theanine biosynthesis, could be influenced by salt treatments. Salt treatments positively affect the levels of theanine, total free amino acids and the level of theanine product was corroborated with that of theanine synthatase transcript expression. (Deng et al. 2012). However, salinity is not a major problem for tea cultivation worldwide.

7.5 Emerging Physiological Stresses

Several physiological factors are related to the quality and yield of tea and any alteration of them becomes a stressor. A brief account of them is enumerated here (Table 7.2).

CO_2 assimilation capacity of tea plants decreased with the increase of temperature due to high irradiance while maximum *Pn* was found to be around 30–35 °C, but no net photosynthesis can be seen beyond 42 °C (Mohotti and Lawlor 2002). Respiration rate also influenced growth rate due to removal of the photoassimilates as substrate. Mature tea used almost 67 % of the photoassimilates for respiration (Tanton 1979). In tea, root respiration rate increased with increasing N supply, owing to increase respiratory cost for NO_3^- uptake (Anandacoomaraswamy et al. 2002). Tea plants produced lower biomass ascribed to continuous elimination of young shoots. It had been found that unplucked tea produced 36 % more total biomass and 64 % more woody tissue than plucked tea. It had also been established that biomass production rate and tea yield maintained inverse relationships (Magambo and Cannell 1981). Plucking of young shoots before maturity reduced the sink capacity of the tea bush and thus reduced biomass production (Tanton 1979). Photoperiod affected shoot growth of tea and the growth of tea shoots was repressed with 11 h photoperiod combined with night temperature around 20 °C. Additionally it had been proposed that tea shoots become dormant below a photoperiod of 11.16 h (De Costa et al. 2007).

7.6 Impact of Various Stresses on Wild Species of Tea

Like any other plants, different *Camellia* species too have different growth habits under different climatic conditions (He et al. 2012). A brief account of the effects of stresses on growth of some *Camellia* species is discussed here.

The relationship between the low temperature and the cell injury in eight species of *Camellia* showed that *C. oleifera* was found to be the best suitable species for high-temperature zones whereas *C. vietnamensis* was more suitable for temperate regions. *C. oleifera* required optimum concentrations of Al and P for its growth (Tan et al. 2011) but under Al toxicity and P inadequacy reduced growth and photosynthesis. Lime in combination with P led to increase in *Pn, gs* and a decrease in intercellular CO_2 concentration. Simultaneously content of pigments such as car, chl was reduced significantly, indicating its crucial role of car in photoprotection (He et al. 2010). Another study revealed that P deficiency led to higher activity of acid phosphatase in some tea varieties (Chen et al. 2011c). In fact, leaf age and leaf position were found to be two important

Table 7.2 Changes of physiochemical parameters of tea plants under various abiotic stresses

Stress	Physiological changes	Biochemical changes	References
Low moisture stress	Water potential ↑, Growth ↓	-	Yang et al. (1987)
	Diffusion resistance, RWC↑	-	Sandanam et al. (1981)
	Diffusion resistance↑ RWC, *Pn* ↓	-	Wijeratne et al. (1998)
	F_0, Thermal dissipation Fm, Fv, Fv/Fm, Fv/F_0↑, Photosynthesis ↓	-	Guo et al. (2009)
	-	Peroxidase ↑, POD ↑	Jeyaramraja et al. (2002b)
	E, *gs*, root: shoot, CO_2 assimilation ↓	chl ↑	Bore (2008)
	-	Proline, H_2O_2,MDA, O^{2-}, ↓ CAT, SOD chl, car ABA, GPX, GR ↑	Upadhyay and Panda (2004)
	-	MDA, O_2^-, POD, GR, CAT, Phenol, H_2O_2 ↑ ASA, Glutathione ↓	Upadhyaya et al. (2008)
	-	Electrical conductivity ↑ POD, PPO, *Pn* ↓	Yang et al. (1987)
	-	Proline ↑	Wu and Pan (1995)
	-	H_2O_2, APX, SOD ↑	Chen et al. (2011b)
	-	SOD, APX, GR ↑	Vyas et al. 2007
	-	MDA, Proline, car ↑	Netto et al. (2005)
	-	Number of chloroplasts ↓	Du et al. (2008)
Low temperature	-	-	Mamati et al. (2006)
	-	EGCG ↑	Zheng et al. (2008)
	Callus forming cell size ↓	Soluble phenols ↑	Zagoskina et al. (2003)
	PS-II ↑	Phenolic compunds ↑	Zagoskina et al. (2005)

Table 7.2 (Continued)

low light intensity	-	EGCG, Epicatechin, ECG, PAL, EGC ↓	Saijo (1980)
	-	Flavanols ↓	Forrest 1969
	-	Phenylpropanoids, ↑ Phenylalanine Shikimic ↓ acid, Prephenic acid, Phenylpyruvic acid	Yang et al. (2012)
2,4-D	Growth ↑	-	Nikolaeva et al. (2009)
NAA	Growth, Biomass ↓	Phenol, Flavan ↑	Nikolaeva et al. (2009)
ABA	-	TSS, Starch, Protein ↑	Sharma et al. (2004)
SA	Leaf area, FW, DW ↑	TSS, Invertase ↑	Kaveh et al. (2004)

influencing factors that determined the *Pn* in *C. oleifera* (Wang et al. 2009b). Density of planting also influenced *Pn* because lower planting yielded higher *Pn* than compact planting in *C. oleifera,* a fact which is well established (He and Lu 2008). Under high illumination, 6-benzylamino purine (BAP) reduced the accumulation of chl as well as protein content in *C. oleifera* (Hu 1998). Low Al content increased the content of chl a as well as b, *gs* and *Pn* but decreased when Al concentration increased. Simultaneously *Pn* also affected biomass accumulation (He et al. 2010). It had been found that under γ radiation (up to the dose of 1,000 rad), SOD activity was increased but decreased beyond 1,000 rad. Similarly, POD activity was also found to be influenced by intensity of the radiation (Huang et al. 2009). Leaf blight of *C. oleifera* is one of the major leaf diseases, which causes premature defoliation, hence reduction of growth of the plants. It had been detected that *Bacillus subtilis* offered antagonism against this disease (Li et al. 2011c). High-temperature stress increased the contents of soluble protein, sugar and free proline in *C. oleifera* leaves (Wang et al. 2012). Low moisture stress also affected the antioxidant system of *C. oleifera* plant due to reduction of PPO, POD and CAT activity while MDA as well as relative conductivity were increased gradually (Cao et al. 2011b). In a preliminary study, effect of low water stress on *C. oleifera* was studied in detail, and it had been found that leaf was more venerable to dehydration than root and stem (Cao et al. 2011c). Simulated acid rains inhibited seed germination, increased MDA, reduced POD, SOD activity, chl a, chl b content as well as their ratio in young seedlings. Spraying paclobutrazol on the leaves of *C. oleifera* seedling increased shoot as well as root growth with increase of chl content, *Pn* and protein content (Hu et al. 1993). The growth of *C. sasanqua* pollen grains was found to be slow but when soaked in acetone or diethyl ether the same grew faster (Iwanami 1973). Under light stress, *C. nitidissima* changed both qualitative and quantitative characters. Low light intensity induced development of short-narrow leaves, while high light intensity stimulated longer and wider leaves of *C. nitidissima* (Hu et al. 2010).

Effect of cold on growth and development was compared between *C. japonica* and *C. rusticana.* Under prolonged snow cover, the leaves of *C. rusticana* showed no visible changes in comparison with *C. japonica*, while *C. japonica* showed a slow growth after 200 days. The decline in the rate of soluble carbohydrate content in *C. rusticana* remained about one-sixths of that in *C. japonica*. The *Pn* of *C. japonica* decreased to about half of its initial value after 140 days, while that of *C. rusticana* remained unchanged even after almost 1 year. Under dark treatment, the stomata of *C. rusticana* were closed, but in *japonica,* even after 90 days, remained open (Kume et al. 1998). Collectively, these physiological studies in different species are sporadic although wild species harbour many important characters of ABstresses, but systematic study is lacking.

7.7 Conclusion

Tea plants, being woody perennial, experience several ABstresses. Although conventional breeding is successful for varietal improvement of tea, they are focused onto limited traits such as improvement of yield, quality, drought and few diseases. While significant amount of achievements have been made in improving yield and quality parameters of tea, other attempts have marginally improved the target. This is primarily due to the fact that most of the ABstresses are complex in nature, controlled by several genetic elements except few cases of biotic stress. Thus, prior to any attempt to progress further; it is essential to understand the physiological as well as biochemical basis of such stresses, which later formulate the basis for studying the molecular mechanism. In tea, several studies have been accomplished which identified major physiochemical parameters. Detection of the key biochemical molecules associated with an assortment of traits and identification of the physiological parameters would aid to nursery selection for agronomically important traits and will assist molecular breeding ultimately. Although in several aspects of tea physiochemical study needs to be investigated, few specific areas are: (1) tea is a hyper-accumulator of Al and accumulates Al to such an extent that it is toxic to other plants. Thus, identification of the mechanism of tea plants will facilitate to understand the mechanism, which in turn will assist to develop the Al-tolerant plants in other species, (2) mild infestation of green leaf-hopper (*Empoasca vitis* Göthe) in tea leaves increases the aroma of made tea. Until now the physiochemical basis is not documented. However, most importantly, in tea, identification of extreme genotypes tolerant to ABstreses or at least development of trait specific 'core' panel through the approach of phenomics should be done in priority. This will be immensely useful for several applications including identification of major QTLs/genes. Besides, association mapping can only be successful if morphological, physiological as well as biochemical scorings are successful.

References

A-H-Mackerness S, Jordan BR, Thomas B (1997) UV-B effects on the expression of genes encoding proteins involved in photosynthesis. In: Lumsden PJ (ed) Plant and UV-B: responses to environmental change. Cambridge University Press, Cambridge, pp 113–114

Anandacoomaraswamy A, De Costa WAJM, Shyamalie HW, Campbell GS (2000) Factors controlling transpiration of mature field-grown tea and its relationship with yield. Agri Forest Meteor 103:375–386

Apel K, Hirt H (2004) Reactive oxygen species: metabolism, oxidative stress, and signal transduction. Ann Rev Plant Biol 55:373–399

Bagyalakshmi B, Ponmurugan P, Marimuthu S (2012) Influence of potassium solubilizing bacteria on crop productivity and quality of tea (*C. sinensis*). African J Agric Res 7:4250–4259

Barman TS, Baruah U, Saikia JK (2008) Irradiation influences tea leaf (*C. sinensis* L.) photosynthesis and transpiration. Photosynth 46:618–621

Barua DN (1969) Light as a factor in metabolism of tea plant pp 306-322. In: Luckwill LC and C V Cuttings (Eds) Physiology of tree Crops. Academic Press, New York

Barua DN (1989) Science and practice in tea culture. Tea Research Association, Kolkata, p P

Basak M, Sharma M, Chakraborty U (2001) Biochemical responses of *C. sinensis* (L.) O. Kuntze to heavy metal stress. J Environ Biol 22:37–41

Bhagat I, Chakraborty BN (2010) Defense response triggered by *Sclerotium rolfsii* in tea plants. Ecoprint 17:69–76

Bore, JK (2008) Weather and tea yields. *Tea* 29, 57-58

Burgess PJ, Carr MVK (1996) Responses of young tea (*C. sinensis*) clones to drought and temperature. II. Dry matter production and partitioning. Exper Agric 32:377–394

Cao P, Liu C, Li D (2011a) Effects of different autotoxins on antioxidant enzymes and chemical compounds in tea (*C. sinensis* L.) Kuntze. Afr J Biotech 10:7480–7486

Cao Z, Zhang S, Liu C, Hu J, Zhang X, Shu Q (2011b) Effects of dehydration degree on the physiological and chemical characteristics of *C. oleifera* seedlings. J Anhui Agric Univ 3:2011–2003

Cao Z, Hu J, Shu Q, Wen J, Zhan W, Xu G, Bu J (2011c) Effects of water stress on physical characteristics and survival rate of container seedling in *C. oleifera*. Nonwood Forest Res 2011–2004

Carr HP, Lombi E, Küpper H, McGrath SP, Wong MH (2003) Accumulation and distribution of aluminium and other elements in tea (*C. sinensis*) leaves. Agronomie 23:705–710

Chakraborty U, Dutta S, Chakraborty BN (2002) Response of tea plants to water stress. Biol Plant 45:557–562

Chakraborty U, Chakraborty BN (2005) Impact of environmental factors on infestation of tea leaves by *Helopeltis theivora*, and associated changes in flavonoid flavour components and enzyme activities. Phytopara 33:88–96

Chakraborty U, Chakraborty BN, Basnet M (2006) Plant growth promotion and induction of resistance in *C. sinensis* by *Bacillus megaterium*. J Basic Microb 46:186–195

Chakraborty U, Chakraborty BN, Chakraborty AP (2012) Induction of plant growth promotion in *C. sinensis* by *Bacillus megaterium* and its bioformulations. World J Agric Sci 8:104–112

Chamuah GS (1988) The effect of nitrogen in root growth and nutrient uptake of young tea plant (*C. sinensis*) grown in sand culture. Fert Res 16:59–65

Cheruiyot EK, Mumera LM, Ngetich WK, Hassanali A, Wachira FN (2007) Polyphenol as potention indicators for drought tolerance in tea (*C. sinensis L.*). Biosci Biotech Biochem 71:2190–2197

Cheruiyot EK, Mumera LM, Ng'etich WK, Hassanali A, Wachira FN, Wanyoko JK (2008) Shoot epicatechin and epigallocatechin contents respond to water stress in tea (*C. sinensis* (L.) O. Kuntze). Biosci Biotechnol Biochem 72:1–8

Chen YM, Tsao TM, Liu CC, Lin KC, Wang MK (2011a) Aluminium and nutrients induce changes in the profiles of phenolic substances in tea plants (*C. sinensis* CV TTES, No. 12 (TTE)). J Sci Food Agric 91:1111–1117

Chen Q, Yang L, Ahmad P, Wan X, Hu X (2011b) Proteomic profiling and redox status alteration of recalcitrant tea (*C. sinensis*) seed in response to desiccation. Planta 233:583–592

Chen L, Chen Y, Wang R, Ma L, Peng S, Wang X, Tang W (2011c) The effects of phosphorus deficiency stress on activities of acid phosphatase in different clones of *C. oleifera* Abel. Chinese Agric Sci Bull 6:2011–2031

Das GM (1965) Pests of tea in North East India and their control. Memorandom No. 27, Tocklai Experimental Station, Tea Research Association, Jorhat, Assam, India. pp 169–173

De Costa WAJM, Mohotti AJ, Wijeratne MA (2007) Ecophysiology of tea. Braz J Plant Physiol 19:299–332

Deng WW, Wang S, Chen Q, Zhang ZZ, Hu XY (2012) Effect of salt treatment on theanine biosynthesis in *C. sinensis* seedlings. Plant Physiol Biochem 56:35–40

Du YY, Chen H, Zhong WL, Wu LY, Ye JH, Lin C, Zheng XQ, Lu JL, Liang YR (2008) Effect of temperature on accumulation of chlorophylls and leaf ultrastructure of low temperature induced albino tea plant. Afr J Biotechnol 7:1881–1885

Forrest GI (1969) Effects of light and darkness on polyphenol distribution in the tea plant (*C. sinensis* L.). Biochem J 113:773–781

Frederick JE, Snell HE, Haywood EK (1989) Solar ultraviolet radiation at the Earth's surface. Photochem Photobiol 50:443–450

Fung KF, Zhang ZQ, Wong JWC, Wong MH (2003) Aluminum and fluoride concentrations of three tea varieties growing at Lantau island, Hong Kong. Environ Geochem Health 25:219–232

Ghanati F, Morita A, Yokota H (2005) Effects of aluminum on the growth of tea plant and activation of antioxidant system. Plant Soil 276:133–141

Gnanamangai BM, Ponmurugan P, Yazhini R, Pragadeesh SK (2011) PR enzyme activities of *Cercospora theae* bird's eye spot disease in tea plants (*C. sinensis* (L.) O. kuntze). Plant Patho J 10:13–21

Gohain T, Barbora AC, Deka A (2000) Effect of manganese on growth, yield and quality of tea (*C. sinensis L.* (O) Kuntez). Res Crops 1:91–97

Gulati A, Gulati A, Rabindranat SD, Gupta AK (1999) Variation in chemical composition and quality of tea (*C. sinensis*) with increasing blister blight (*Exobasidium vexans*) severity. Mycol Res 103:1380–1384

Guo CF, Sun Y, Tang YH, Zhang MQ (2009) Effect of water stresson chlorophyll fluorescence in leaves of tea plant. Chinese J Eco-Agri 17:560–564

Gupta D, Bhardwaj R, Nagar PK, Kaur S (2004) Isolation and characterization of brassinosteroides from leaves of *C. sinensis* (L.) O. Kuntze. Plant Growth Regu 42:97–100

Gurusubramanian G, Borthakur M (2005) Integrated management of tea pests. In: Dutta AK, Baruah SK, Ahmed N, Sarma AK, Burugohain D (ed) Field management in tea. Tocklai Experimental Station, TRA, Jorhat, Assam Printing Works Pvt Ltd, Jorhat, pp 159–172

Hajra NG, Kumar R (2002) Responses of young tea clones to subtropical climate: effect on photosynthesis and biochemical characteristic. J Plant Biol 29:257–264

Hajiboland R, Bastani S, Rad SB (2011a) Photosynthesis, nitrogen metabolism and antioxidant defense system in B-deficient tea (*C. sinensis* (L.) O. Kuntze) plants. J Sci 22:311–320

Hajiboland R, Bastani S, Rad SB (2011b) Effect of light intensity on photosynthesis and antioxidant defense in boron deficient tea plants. Acta Biol Szeged 55:265–272

Hajiboland R, Bastani S (2012) Tolerance to water stress in boron deficient tea (*C. sinensis*) plants. Folia Horti 24:41–45

Hajiboland R, Bahrami-Rad S, Bastani S (2013) Phenolics metabolism in boron-deficient tea (*C. sinensis* (L.) O. Kuntze) plants. Acta Biol Hung 64:196–206

Han WY, Shi YZ, Ma LF, Ruan JY, Zhao FJ (2007) Effect of liming and seasonal variation on lead concentration of tea plant (*C. sinensis* (L.) O. Kuntze). Chemosph 66:84–90

Handique AC, Manivel L (1990) Effect of certain antitranspirants in tea. Two Bud 37:20–23

He Y, Lu F (2008) Study on the photosynthesis of the different density of *C. oleifera*. Modern Agric Sci 5:2003–2008

He G, Liu Q, Peng S (2010) Effect of Aluminum toxicity on photosynthetic characters of wild *C. oleifera* under acidic conditions. Hubei Agric Sci 2010–2017

He XY, Ye H, Ma JL, Zhang RQ, Chen GC, Xia YY (2012) Semi-lethal high temperature and heat tolerance of eight *C.* species. Int J Exp Bot 81:177–180

Hemalatha K, Venkatesan S (2011) Impact of iron toxicity on certain enzymes and biochemical parameters of tea. Asian J Biotech 6:384–394

Hernandez I, Alegre L, Munné-Bosch S (2006) Enhanced oxidation of flavan-3-ols and proanthocyanidin accumulation in water-stressed tea plants. Phytochem 67:1120–1126

Hu Z (1998) Delayed action of 6-BA on detached-leaves senescence in *C. oleifera*. J Fujinan College For 3:1998–1901

Hu Z, Shen W, Zhang Y (1993) Physiological effect of paclobutrazol on the growth of *C. oleifera* seedling. J Fuji College For 1993–1902

Hu Q, Xu J, Pang G (2003) Effect of selenium on the yield and quality of green tea leaves harvested in early spring. J Agric Food Chem 51:3379–3381

Hu XH, Li JW, Jiang QS, Zhao RF (2010) Responses of leaf characters of *C. nitidissima* to different light environments. Guihaia 5:201–203

Huang Y, Yuan Z, Zhong W, Zhang H, Dong B (2009) Effects of ~ (60)Co-γ radiation on isozyme activity of *C. oleifera* seedling. Nonwood For Res 4:200–202

Huiqun M, Yuchen R (1987) Relationship between zinc and the metabolism of tea plant. J Tea Sci (China) 7:35–40

Iwanami Y (1973) Acceleration of the growth of *C. sasanqua* pollen by soaking in organic solvent. Plant Physiol 52:508–509

Jayaganesh S, Venkatesan S, Senthurpandian VK (2011) Impact of different sources and doses of magnesium fertilizer on biochemical constituents and quality parameters of black tea. Asian J Biochem 6:273–281

Jeyaramraja PR, Jayakumar D, Pius PK, Kumar RR (2002a) Screening of certain tea cultivars for productivity and drought hariness using biochemical markers. J Plant Crop 30:23–26

Jeyaramraja PR, Jayakumar D, Pius PK, Kumar RR (2002b) Application of rubisco, peroxidase and polyphenol oxidase as markers for productivity and drought tolerance in tea. J Plant Biol 29:315–320

Jeyaramraja PR, Pius PK, Kumar RR, Jayakumar D (2003a) Soil moisture stress-induced alterations in bioconstituents determining tea quality. J Sci Food Agric 83:1187–1191

Jeyaramraja PR, Raj Kumar R, Pius PK, Thomas J (2003b) Photoassimilatory and photorespiratory behavior of certain drought tolerant and susceptible tea clones. Photosynthe 41:579–582

Jinghua Y, Yongping C, Hesong L (1997) Effect of Zn^{++} on quality and lipid peroxidation in leaves of tea. J Anhui Agric Sci 25:30–32

Kakkar RK, Nagar PK (1997) Distribution and changes in endogenous polyamines during winter dormancy in tea (*C. sinensis* (L.) O. Kuntze). J Plant Physiol 151:63–67

Kaveh SH, Bernard F, Samiee K (2004) Growth stimulation and enhanced invertase activity induced by salicylic acid in tea cuttings (*C. sinensis* L.). Proceedings of the fourth international Iran and Russia conference, Shahr-e-Kord, Iran, pp 113–116

Kefei Z, Ma Q, Zang H (1997) Effect of water deficit on physiological activities of paddy rice and upland rice seedlings. J Shandong Agr Univ 28:53–57

Konishi S, Miyamoto S (1983) Alleviation of aluminum stress and stimulation of tea pollen tube growth by fluorine. Plant Cell Physiol 24:857–862

Konishi S, Ferguson IB, Putterill J (1988) Effect of acidic polypeptides on aluminium toxicity in tube growth of pollen from tea (*C. sinensis* L.). J Plant Sci 56:55–59

Kumar V, Yadav SK (2009) Proline and betaine provide protection to antioxidant and methylglyoxal detoxification systems during cold stress in *C. sinensis* (L.) O. Kuntze. Acta Physiol Plant 31:261–269

Kume A, Tanaka C, Matsumoto S, Ino Y (1998) Physiological tolerance of *C. rusticana* leaves to heavy snowfall environments: The effects of prolonged snow cover on evergreen leaves. Ecological Res 13:117–124

Lian CU, Wake YO, Yokota H, Wang G, Konishi S (1998) Effect of aluminum on callose synthesis in root tips of tea (*C. sinensis L.*) plants. Soil Sci Plant Nut 44:695–700

Li C, Xu H, Xu J, Chun X, Ni D (2011a) Effects of aluminium on ultrastructure and antioxidant activity in leaves of tea plant. Acta Physiol Plant 33:973–978

Li C, Zheng Y, Zhou J, Xu J, Ni D (2011b) Changes of leaf antioxidant system, photosynthesis and ultrastructure in tea plant under the stress of fluorine. Biol Plant 55:563–566

Li H, Zhou G, Zhang H, Song G, Liu J (2011c) Study on isolated pathogen of leaf blight and screening antagonistic bacteria from healthy leaves of *C. oleifera*. Afr J Agric Res 6:4560–4566

Lin Z-H, Chen L-S, Chen R-B, Peng A (2009a) Effects of phosphorus deficiency on nutrient absorption of young tea bushes. J Tea Sci 29:295–200

Lin Z-H, Chen L-S, Chen R-B, Zhang F-Z, Jiang H-X, Tang N (2009b) CO_2 assimilation, ribulose-1,5-bisphosphate carboxylase/oxygenase, carbohydrates and photosynthetic electron transport probed by the JIP-test, of tea leaves in response to phosphorus supply. BMC Plant Biol 9:43–55

Lin Z-H, Chen L-S, Chen R-B, Zhang F-Z, Jiang H-X, Tang N, Smith BR (2011) Root release and metabolism of organic acids in tea plants in response to phosphorus supply. J Plant Physiol 168:644–652

Lin Z-H, Chen L-S, Chen R-B, Zhang F-Z (2012) Antioxidant system of tea (*C. sinensis*) leaves in response to phosphorus supply. Acta Physiol Plant 34:2443–2448

Luo YP, Wu S, Kang ML (2002) Photosynthetic characteristic and cold resistance of two-year old grafted tea plant in winter. J Zhejiang Univ Agric Life Sci 28:397–400

Magambo MJS, Cannell MGR (1981) Dry matter production and partition in relation to yield of tea. Exp Agric 17:33–38

Mamati GE, Liang YR, Lu JL (2006) Expression of basic genes involved in tea polyphenol synthesis in relation

to accumulation of catechins and total tea polyphenols. J Sci Food Agric 86:459–464
Matsumoto H, Hirasawa E, Morimura S, Takahashi E (1976) Localization of aluminum in tea leaves. Plant Cell Physiol 17:627–631
Mokgalaka NS, McCrindle RI, Botha BM (2004) Multi-element analysis of tea leaves by inductively coupled plasma optical emission spectrometry using slurry nebulization. J Anal Atom Spectrom 19:1375–1378
Mohanpuria P, Rana NK, Yadav SK (2007) Cadmium induced oxidative stress influence on glutathione metabolic genes of *C. sinensis* (L.) O. Kuntze. Environ Toxicol 22:368–374
Mohotti AJ, Dennett MD, Lawlor DW (2000) Electron transport as a limitation to photosynthesis of tea (*C. sinensis* (L.) O. Kuntze): a comparison with sunflower (*Helianthus annus* L.) with special reference to irradiance. Trop Agric Res 12:1–10
Mohotti AJ, Lawlor DW (2002) Diurnal variation of photosynthesis and photoinhibition in tea: effects of irradiance and nitrogen supply during growth in the field. J Exp Bot 53:313–322
Mondal TK (2003) Frost management of Tea. Assam Rev Tea New 92:8–12
Mondal TK (2009) Tea In: Prydarsini M, Jain SM (Eds) Breeding plantation tree crops tropical species, pp 545–587
Mondal TK, Bhattacharya A, Sood A, Ahuja PS (1998) Micropropagation of tea using thidiazuran. Plant Growth Regu 26:57–61
Morita A, Fujii Y, Yokota H (2001) Effect of aluminium on exudation of organic acid anions in tea plants. In: Horst WJ et al. (ed) Plant nutrition-Food security and sustainability of agro-ecosystems. pp. 508–509
Morita A, Horie H, Fujii Y, Takatsu S, Watanabe N, Yagi A, Yokota H (2004) Chemical forms of aluminum in xylem sap of tea plants (*C. sinensis* L.). Phytochem 65:2775–2780
Morita A, Yanagisawa O, Takatsu S, Maeda S, Hiradate S (2008) Mechanism for the detoxification of aluminum in roots of tea plant (*C. sinensis* (L.) Kuntze). Phytochem 69:147–153
Mukhopadhyay M, Bantawa P, Das A, Sarkar B, Bera B, Ghosh PD, Mondal TK (2012) Changes of growth, photosynthesis and alteration of leaf antioxidative defence system of tea (*C. sinensis* (L.) O. Kuntze) seedling under aluminum stress. Biometals 25:1141–1154
Mukhopadhyay M, Ghosh PD, Mondal TK (2013a) Effect of boron deficiency on photosynthesis and antioxidant responses of young tea (*C. sinensis* (L.) O. Kuntze) plantlets. Russ J Plant Physiol 60:633–639
Mukhopadhyay M, Das A, Subba P, Bantawa P, Sarkar B, Ghosh PD, Mondal TK (2013b) Structural, physiological and biochemical profiling of tea plantlets (*C. sinensis* (L.) O. Kuntze) under zinc stress. Biol Plant. doi:10.1007/s10535-012-0300–2
Murugan AC, Thomas J, Rajagopal RK, Mandal AKA (2012) Metabolic responses of tea (*Camellia* sp.) to exogenous application of ascorbic acid. J Crop Sci Biotech 15:53–57
Nagar PK (1996) Changes in en dog e nous abscisic acid and phenols during winter dormancy in tea (*C. sinensis* (L.) O. Kuntze). Acta Physiol Plant 18:33–38
Nagar PK, Kumar A (2000) Changes in endogenous gibberellin activity during winter dormancy in tea (*C. sinensis* (L.) O. Kuntze). Acta Physiol Plant 22:439–443
Nagar PK, Sood S (2006) Changes in endogenous auxins during winter dormancy in tea (*C. sinensis* L.) O. Kuntze. Acta Physiol Plant 28:165–169
Nagarajah S (1979) The effect of potassium deficiency on stomatal and cuticular resistance in tea (*C. sinensis*). Physiol Plant 47:91–94
Nagarajah S (1981) The effect of nitrogen on plant water relations in tea (*C. sinensis*). Physiol Plant 51:304–308
Nagarajah S, Ratnasooriya GB (1977) Studies with antitranspirants on tea (*C. sinensis* L.). Plant Soil 48:185–197
Nagata T, Hayatsu M, Kosuge N (1992) Identification of aluminum forms in tea leaves by ^{27}Al NMR. Phytochem 31:1215–1218
Nandi SK, Palni LMS (1993) Shoot growth and winter dormancy in tea. In: Nair MK (ed) PLACROSYM X 1992. J Plant Crop 21:328–333
Napaporn SL, Orapin K, Natta L (2012) Chemical qualities and phenolic compounds of Assam tea after soil drench application of selenium and aluminium. Plant Soil 356:381–393
Nelson S (2006) Zinc deficiency in tea (*C. sinensis*). Plant Dis PD 34
Netto LA, Jayaram KM, Haridas P, Puthur JT (2005) Characterization of photosynthetic events and associated changes in various clones of tea (*C. sinensis* L.) under low temperature conditions. J Plant Biol 48:326–331
Ng'etich WK, Bore JK (2001) The effect of brassinolides (plant growth regulators) on yield and yield attributes of clonal tea (*C. sinensis*). Tea 22:8–12
Ng'etich WK, Wachira FN (2003) Variations in leaf anatomy and gas exchange in tea clones with different ploidy. J Hort Sci Biotech 78:173–176
Nikolaeva TN, Zagoskina NV, Zaprometov MN (2009) Production of phenolic compounds in callus cultures of tea plant under the effect of 2,4-D and NAA. Russ J Plant Physiol 56:45–49
Owuor PO (2001) Effects of fertilizers on tea yields and quality: a review with special reference to Africa and Sri Lanka. Int J Tea Sci 1:1–11
Premkumar R, Ponmurugan P, Manian S (2008) Growth and photosynthetic and biochemical responses of tea cultivars to blister blight infection. Photosynth 46:135–138
Qian D, Chang-quan W, Bing L, Huan-xiu L, Yang L (2009) Effects of Se, Zn and their interaction on polyphenol oxidase activity of tea leaves in summer season. Acta Metall Sin 15:930–935
Ramarethinam S, Rajalakshmi N (2004) Caffeine in tea plants (*C. sinensis* (L) O. Kuntze): in situ lowering by *Bacillus licheniformis* (Weigmann) Chester. Indian J Exp Biol 42:575–580

Rana NK, Mohanpuria P, Yadav SK (2008) Expression of tea cytosolic glutamine synthetase is tissue specific and induced by cadmium and salt stress. Biol Plant 52:361–364

Ruan J, Wong MH (2001) Accumulation of fluoride and aluminum related to different varieties of tea plant. Environ Geochem Health 23:53–63

Ruan JY, Wu X, Hardter R (1999) Effect of potassium and magnesium nutrition on the quality components of different types of tea. J Sci Food Agric 79:47–52

Ruan J, Ma L, Shi Y, Han W (2003) Uptake of fluoride by tea plant (*C. sinensis* L) and the impact of aluminium. J Sci Food Agric 83:1342–1348

Ruan J, Ma L, Shi Y, Han W (2004) The impact of pH and calcium on the uptake of fluoride by tea plants (*C. sinensis* L.). Ann Bot 93:97–105

Ruan J, Ma L, Shi Y (2006) Aluminium in tea plantations: mobility in soils and plants, and the influence of nitrogen fertilization. Environ Geochem Health 28:519–528

Ruan J, Gerendás J, Härdter R, Sattlemacher B (2007) Effect of nitrogen form and root-zone pH on growth and nitrogen uptake of tea (*C. sinensis*) plants. Ann Bot 99:301–310

Ruan J, Ma L, Yang Y (2012) Magnesium nutrition on accumulation and transport of amino acids in tea plants. J Sci Food Agric 92:1375–1383

Saha D, Mandal S, Saha A (2012) Copper induced oxidative stress in tea (*C. sinensis*) leaves. J Environ Biol 33:861–866

Saijo R (1980) Effect of shade treatment on biosynthesis of catechins in tea plants. Plant Cell Physiol 21:989–998

Saikia JK, Baruah U, Barman TS, Saikia H, Bandyopadhyay T (2011) Impact of tea mosquito infestation on endogenous hormones of tea (*C. sinensis* L.). Sci Cult 77:412–415

Salehi SY, Hajiboland RA (2008) A high internal phosphorus use efficiency in tea (*C. sinensis* L.) plants. Asian J Plant Sci 7:30–36

Sandanam S, Gee GW, Mapa RB (1981) Leaf water diffusion resistance in clonal tea (*C. sinensis* L.): Effects of water stress, leaf age and clones. Ann Bot 47:339–349

Sanderson GW, Sivapalan K (1966) Translocation of photosynthetically assimilated carbon in tea plants. Tea Quart 37:140–153

Sanjay R, Baby UI (2007) Physiological and biochemical changes in tea leaves due to *Pestalotiopsis* infection. J Plant Crop 35:15–18

Sharma P, Pandey S, Bhattacharya A, Nagar PK, Ahuja PS (2004) ABA associated biochemical changes during somatic embryo development in *C. sinensis* (L.) O. Kuntze. J Plant Physiol 161:1269–1276

Sharma V, Joshi R, Gulati A (2011) Seasonal clonal variations and effects of stresses on quality chemicals and prephenate dehydratase enzyme activity in tea (*C. sinensis*). Eur Food Res Technol 232:307–317

Shi Y, Ruan J, Ma L, Han W, Wang F (2008) Accumulation and distribution of arsenic and cadmium by tea plants. J Zhejiang Univ Sci B 9:265–270

Shu W, Zhang ZQ, Lan CY, Wong MH (2003) Fluoride and aluminium concentrations of tea plants and tea products from Sichuan Province, PR China. Chemosph 52:1475–1482

Singh S, Pandey A, Kumar B, Palni LMS (2010) Enhancement in growth and quality parameters of tea [*C. sinensis* (L.) O. Kuntze] through inoculation with arbuscular mycorrhizal fungi in an acid soil. Biol Fert Soils 46:427–433

Smirnoff N (1993) The role of active oxygen in the response of plants to water deficit and desiccation. New Phytol 125:27–58

Smith BG, Stephens W, Burgess PJ, Carr MKV (1993) Effects of light, temperature, irrigation and fertilizer on photosynthetic rate in tea (*C. sinensis*). Exp Agric 29:291–306

Smith BG, Burgess PJ, Carr MKV (1994) Effects of clone and irrigation on the stomatal conductance and photosynthetic rate of tea (*C. sinensis*). Exp Agric 30:1–16

Stapleton AE (1992) Ultraviolet radiation and plants: Burning questions. Plant Cell 4:1352–1358

Tan X, Yuan J, Li Z, Ye S, Jiang Z (2011) Effects of aluminum and phosphate on material distribution and roots characteristics and activity of *C. oleifera* seedlings. J Central South Univ Forestry Tech 20:11–12

Tang J, Xu J, Wu Y, Li Y, Tang Q (2012) Effects of high concentration of chromium stress on physiological and biochemical characters and accumulation of chromium in tea plant (*C. sinensis* L.). Afr J Biotechnol 11:2248–2255

Tang Q, Li X, Zhu X, Huang Y, Yang H (2008a) Effects of plumbum and chromium stress on the growth of tea plants. Southwest J Agri Sci 21:156–162

Tang Q, Zhu X, Li X, Tan H (2008b) Effects of As and Cd stress on the growth of tea plants. J Henan Agric Sci 208–211

Tanton TW (1979) Some factors limiting yields of tea (*C. sinensis*). Exp Agric 15:187–192

Thomas J, Ajay D, Raj Kumar R, Mandal AKA (2010) Influence of beneficial microorganisms during *in vivo* acclimatization of *in vitro*-derived tea (*C. sinensis*) plants. Plant Cell Tiss Organ Cult 101:365–370

Tolrà R, Vogel-Mikuš, Hajiboland R, Kump P, Pongrac P, Kaulich B, Gianoncelli A, Babin V, Barceló J, Regvar M, Poschenrieder C (2011) Localization of aluminium in tea (*C. sinensis*) leaves using low energy X-ray fluorescence spectro-microscopy. J Plant Res 124:165–172

Upadhyaya H, Panda SK (2004) Responses of *C. sinensis* to drought and rehydration. Biol Plant 48:597–600

Upadhyaya H, Panda SK, Dutta BK (2008) Variation of physiological and antioxidative responses in tea cultivars subjected to elevated water stress followed by rehydration recovery. Acta Physiol Plant 30:457–468

Upadhyaya H, Panda SK, Dutta BK (2011) $CaCl_2$ improves post-drought recovery potential in *C. sinensis* (L) O. Kuntze. Plant Cell Rep 30:495–503

Upadhyaya H, Dutta BK, Sahoo L, Panda SK (2012) Comparative effect of Ca, K, Mn and B on post-drought stress recovery in tea [*C. sinensis* (L.) O Kuntze]. Am J Plant Sci 3:443–460

Upadhyaya H, Dutta BK, Panda SK (2013) Zinc modulates drought induced biochemical damages in tea (*C. sinensis* (L) O Kuntze). J Agric Food Chem 61:6660–6670

Venkatesan S, Hemalatha KV, Jayaganesh S (2006) Zn toxicity and its influence on nutrient uptake in tea. Amer J Plant Physiol 1:185–192

Venkatesan S, Jayaganesh S (2010) Characterization of magnesium toxicity, its influence on amino acid synthesis pathway and biochemical parameters of tea. Res J Phytochem 4:66–67

Vyas D, Kumar S (2005a) Purification and partial characterization of a low temperature responsive Mn-SOD from tea (*C. sinensis* (L.) O. Kuntze). Biochem Biophys Res Comm 329(:):831–838

Vyas D, Kumar S (2005b) Tea (*C. sinensis* (L.) O. Kuntze) clone with lower period of winter dormancy exhibits lesser cellular damage in response to low temperature. Plant Physiol Biochem 43:383–388

Vyas D, Kumar S, Ahuja PS (2007) Tea (*C. sinensis*) clones with shorter periods of winter dormancy exhibit lower accumulation of reactive oxygen species. Tree Physiol 27:1253–1259

Waheed A, Hamid FS, Shah AH, Ahmad H, Khalid A, Abbasi FM, Ahmad N, Aslam S, Sarwar S (2012) Response of different tea (*C. sinensis* L.) clones against drought stress. J Master Environ Sci 3:395–410

Wan Q, Xu RK, Li XH (2012) Proton release by tea plant (*C. sinensis* L.) rootsas affected by nutrient solution concentration and pH. Plant Soil Environ 58:429–434

Wang X, Liu P, Luo H, Xie Z, Xu G, Yao J, Chen K (2009a) Effect of Al and F interaction on physiological characteristics of tea plants. Acta Hort Sin 7:207–209

Wang R, Chen Y, Wang X, Peng S, Yang X, Wang Y, Yang Y (2009b) Influencing factors on photosynthetic characteristic of superior clones of *C. oleifera*-leaf age and leaf position. Chinese Agric Sci Bull 2:209–217

Wang LY, Wei k, Jiang YW, Cheng H, Zhou J, He W, Zhang CC (2011) Seasonal climate effects on flavanols and purine alkaloids of tea (*C. sinensis* L.). European Food Res Technol 233:1049–1055

Wang G, Chen L, Kou L, Yang Y, Feng F, Cao F, Fang Y (2012) Effects of high temperature stress on osmotic adjustment substances of 25 varieties of *C. oleifera*. J Henan Agric Sci 2012–04

Wijeratne MA, Formham R, Anandacoomaraswamy A (1998) Water relations of clonal tea (*C. sinensis* L.) with reference to drought resistance: II. Effect of water stress. Tropical Agric Res Ext 1:74–80

Wong MH, Zhang ZQ, Wong JWC, Lan CY (1998) Trace metal contents (Al, Cu and Zn) of tea: Tea and soil from two tea plantations, and tea products from different provinces of China. Environ Geochem Health 20:87–94

Wu C, Fang X (1994) Effect of zinc on carbon and nitrogen metabolism in tea plant (*C. sinensis* L.). Sci Agric Sinica 27:72–77

Wu B, Pan G (1995) Studies on physiological and biochemical response to water stress in tea plant. J Zhejiang Agric Univ 4:1995–2005

Wu Y, Liang Q, Tang Q (2011) Effect of Pb on growth, accumulation and quality component of tea plant. Procedia Eng 18:214–219

Xia J, Lan H (2008) Effects of cadmium stress on growth of tea plant and physiological index in leaves of mengshan tea. J Tea Sci 3:205–207

Xu J, Yang F, Chen L, Hu Y, Hu Q (2003) Effect of selenium on increasing the antioxidant activity of tea leaves harvested during the early spring tea producing season. J Agric Food Chem 51:1081–1094

Yadav SK, Mohanpuria P (2009) Responses of *C. sinensis* cultivars to Cu and Al stress. Biol Plant 53:737–740

Yang Y, Zhuang X, Hu H (1987) Effect of the soil water content on the physiological process of tea plant. J Tea Sci. doi:cnki:ISSN:1000-369X.0.1987-01-006

Yang Z, Kobayashi E, Katsuno T, Asanuma T, Fujimori T, Ishikawa T, Tomomura M, Mochizuki K, Watase T, Nakamura Y, Watanabe N (2012) Characterisation of volatile and non-volatile metabolites in etiolated leaves of tea (*C. sinensis*) plants in the dark. Food Chem 135:2268–2276

Yu C, Pan Z, Chen J, Fan D, Wang X (2012) Effects of Al^{3+} on growth and physiological characteristics of tea plant (*C. sinensis*). Plant Nutr Fert Sci 2012–2021

Zagoskina NV, Dubravina GA, Alyavina AK, Goncharuk EA (2003) Effect of ultraviolet (UV-B) radiation on the formation and localization of phenolic compounds in tea plants callus cultures. Russ J Plant Physiol 50:270–275

Zagoskina NV, Alyavina AK, Gladyshko TO, Lapshin PV, Egorova EA, Bukhov NG (2005) Ultraviolet rays promote development of photosystem II photochemical activity and accumulation of phenolic compounds in the tea callus culture (*C. sinensis*). Russ J Plant Physiol 52:731–739

Zagoskina NV, Goncharuk EA, Alyavina AK (2007) Effect of cadmium on the phenolic compounds formation in the callus cultures derived from various organs of the tea plant. Russ J Plant Physiol 54:237–243

Zheng XQ, Jin J, Chen H, Du YY, Ye JH, Lu JL, Lin C, Dong JJ, Sun QL, Wu LY, Liang YR (2008) Effect of ultraviolet B irradiation on accumulation of catechins in tea (*C. sinensis* (L) O. Kuntze). Afr J Biotech 7:3283–3287

Zeng QL, Chen RF, Zhao XQ, Wang HY, Shen RF (2011) Aluminium uptake and accumulation in the hyperaccumulator *C. Oleifera* Abel. Pedosphere 21:358–364

Functional Genomics

8

8.1 Introduction

Functional genomics play important role in plant improvement (Mondal 2013; Mukhopadhaya et al. 2013). In tea, functional genomics work was initiated with the isolation of chalcone synthase gene from a Japanese green tea cultivar 'Yabukita' (Takeuchi et al. 1994a). Since then a significant amount of genomic work has been done in tea and its wild relatives targeting isolation and characterization of trait-specific genes and their expression under varying conditions (Table 8.1). It has been evident that two types of efforts have been made: (i) cloning of individual gene associated with particular trait and (ii) differential gene expression which leads to identification of genes that are associated to a trait.

8.2 Cloning and Characterization of Individual Genes

Interestingly, majority of the genes cloned so far belongs to quality followed by yield as these two parameters that have major demand in the commercial tea cultivations.

8.2.1 Quality Related Genes

Tea is valued for its cup quality. Therefore, research efforts of tea are biased towards the improvement of quality. Similarly, in the area of genomics, several individual genes of various biochemical pathways were targeted, cloned and characterized which are discussed below (Fig. 8.1).

8.2.1.1 Theanine Biosynthesis Related Genes

Theanine, a glutamate derivative is the most dominant amino acid in tea plant, which although is found in all the organs and tissues, yet with maximum amount in tender leaves (Sugiyama and Sadzuka 2004). Theanine is one of the constituents which provide the briskness of tea, a parameter which is used to determine the quality of black tea. It had been documented that theanine, makes up approximately 50 % of total the amino acids of tea. It is synthesized in the roots and is transported to the tender parts of the plant for storage. In the presence of sunlight, it is converted to polyphenols. Apart from its contribution in the taste of tea, it has some medicinal values as well (Sugiyama and Sadzuka 2003). Glutamine synthase, an important enzyme of theanine biosynthesis pathway is involved in assimilation of ammonia generated by various biochemical processes, had been cloned from tea (Rana et al. 2008a, b). Based on their cellular locations of isozymes, they were classified into two groups-GS3A (cytosolic GS) and GS2 (chloroplast GS). The distinct, cell-specific patterns of expression conferred by the promoters for GS2 and GS3A suggesting the different roles in metabolism (Edwards et al. 1990).

T. K. Mondal, *Breeding and Biotechnology of Tea and its Wild Species*,
DOI 10.1007/978-81-322-1704-6_8,

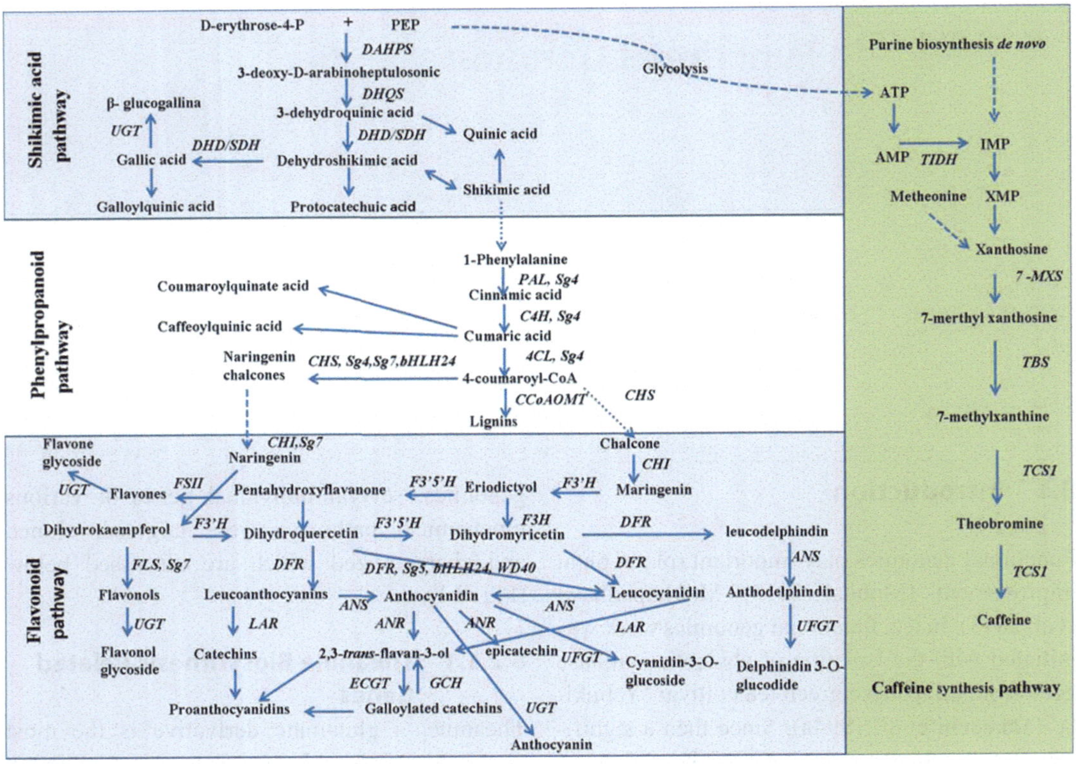

Fig. 8.1 Summary of major biochemical pathways involved in quality of made tea and genes of enzyme that are cloned (partially or fully) from tea, denoted by italics. : anthocyanidin reductase (*ANR*), anthocyanidin synthase (*ANS*), cinnamate 4-hydroxylase (*C4H*), 4-coumaroyl-CoA ligase (*4CL*), chalcone synthase (*CHS*), chalcone isomerase (*CHI*), Caffeoy-CoA 3-O-methyl transferase (*CCoAOMT*), 3-deoxy-D-arabinoheptulosonate 7-phosphate synthase (*DAHPS*), dihydroflavonol 4-reductase (*DFR*), 3-dehydroquinate synthase (*DHQS*), 3-dehydroquinate dehydratase (*DHD*), epicatechins 1-O-galloyl-β-Dglucose O-galloyltransferase (*ECGT*), flavanone 3-hydroxylase (*F3H*), flavonoid 39-hydroxylase (*F3'H*), flavonoid 39, 59-hydroxylase (*F3'5'H*), flavonol synthase (*FLS*), flavone synthase II (*FSII*), galloylated catechins hydrolase (*GCH*), inosine-5-monophosphate (*IMP*), leucoanthocyanidin reductase (*LAR*), 7 methyl xanthosine synthase (*7-MXS*), phenylalanine ammonia lyase (*PAL*), shikimate dehydrogenase (*SDH*), Theobromine synthase (*TBS*), caffeine synthase (*TCS1*), inosine-5-monophosphate dehydrogenase (*TIDH*), UDP-glucose:flavonoid 3-O-glucosyltransferase (*UFGT*), UDP glucose galloyl-1-O-β-D-glucosyltransferase (*UGGT*), uridine diphosphate Glycosyltransferase (*UGT*), xanthosine-5-monophosphate (*XMP*), Sg4, Sg7, WD40, bHLH24 are the transcription factors as reported by Zhao et al. (2012)

8.2.1.2 Flavonoid Biosynthesis Related Genes

Polyphenols content in tea leaf account for almost 15–35 % of the total dry weight of the plants. Polyphenols are chemical compounds such as flavonoids and tannins found naturally in tea. Among the polyphenols, catechins are the most abundant, account for 80 % of the total polyphenols. Owing to its importance, several genes of this pathway such as *PAL*, chalcone synthase (*CHS*), anthocyanidin synthase (*ANS*), anthocyanidin reductase (*ANR*), chalcone isomerase (*CHI*), flavanone-3-hydroxylase (*F3'H*), flavanone-3-hydroxylase (*F3'5'H*), leucoanthocyantin reductase (*LAR*), dihydroflavonol 4-reductase (*DFR*) had been isolated, cloned and characterized from tea plants (Table 8.1). Among them, PAL catalyses the initial step for catechins biosynthesis, via phenyl-propanoid pathway, by deamination of phenylalanine evolving ammonium ion and trans-cinnamic acid. On the other hand, CHS and CHI (chalcone isomerase) catalyse the formation of flavonoids from flavanone (2S)-naringenin by sequential reactions of 4-coumaroyl-CoA and three molecules of malonyl-CoA (Matsumoto et al. 1994a).

Table 8.1 Genes cloned from *Camellia* species

Stress	Gene	Encoding product	Expressed tissue	Reference
Growth and development	*CsTUA 1*	Tubulin	Pollen tube	Fang et al. (2006)
	CsLOX1	Lipoxygenase	Flower	Liu and Han (2010)
	Cs14-3-3	14-3-3 protein	Late stage flower bud	Liu et al. (2010)
	CsELIP	Early light-induced proteins	Leaf	Li et al. (2013)
	CsTUA	α-tubulin	Leaf	Paul et al. (2012a)
	Cs26srRNA	26S rRNA gene	Different plant parts	Singh et al. (2004)
	CsH3	Histone H3.1	Leaf	Singh et al. (2009d)
	CsQM	Ribosomal protein L10	Leaf	Singh et al. (2009e)
	CsTUB	β-tubulin	Leaf	Takeuchi et al. (2004b)
	CsCYS	Cystatin	Leaf	Wang et al. (2005)
	CsBDP	BURP domain-containing protein	Leaf	Wang et al. (2011a)
	CsARP1	Auxin repressed protein	Dormant bud	Wang et al. (2011b)
	CsCYC1	Cyclin	Sprouting bud	Wang et al. (2011c)
	CsCDK1	Cyclin dependent kinase	Bud	Wang et al. (2012a)
	CsACT	Actin	Leaf	Yang et al. 2012b
	CsRbcS	ribulose-1,5-bisphosphate carboxylase/oxygenase	Leaf	Ye et al. (2009)
	CsCYP	Cyclophilin	Meristem tissue	Zhang et al. (2007)
Abiotic stress	*CsCBF*	C-repeat binding factor	Leaf	Chen et al. (2009)
	CsCSH1	H1-histone	Leaf	Fang et al. (2009); Fang et al. (2013)
	CsCOR1	CsCOR1	Leaf	Li et al. (2010)
	CsDHY	Dehydrin1, Dehydrin2	Leaf	Paul and Kumar (2013)
	CsNAM	Nuclear localized protein	Stem, flower, bud	Paul et al. (2012b)
	CsGS	Glutamine synthetase	Leaf, bud	Rana et al. (2008a)
	CsICE1, CsCBF1	CsICE1, CsCBF1	Leaf	Wang et al. (2012b)
	CsVDE	Violaxanthin de-epoxidase	Leaf	Wei et al. (2003)
	CsSMT	Selenocysteine methyltransferase	NA	Zhu et al. (2008b)
Biotic stress	*CsCHIi*	Chitinase	Shoot	Roy and Chakvaaborty (2009)
	CsPR	*PR-1 like protein*	Petals, anthers, Style and young leaf	Tomimoto et al. (1999)
	CsCAD	Cinnamyl alcohol dehydrogenase	Leaf	Deng et al. (2013b)
Quality	*CsLCYE*	Lycopene-epsilon -cyclase	Leaf	Borchetia et al. (2011)
	CsPSY	*Phytoene synthase*	Leaf	Borthakur et al. (2008)
	CsTS	Theanine synthetase	Root and shoot	Deng et al. (2012), (2013a)
	CsSAM	S-adenosylmethionine synthase	Leaf	Feng and Liang (2001)

Table 8.1 (continued)

Stress	Gene	Encoding product	Expressed tissue	Reference
	CsTPS1	Sesquiterpene synthase	Root, leaf	Fu (2012)
	CsTBS	Theobromine synthase	Leaf	Ishida et al. (2009)
	CsGUS	β-glucosidase	Cytoplasm	Jiang (2005)
	CsTCS1	Caffeine synthase	Leaf	Kato et al. (2000)
	CsO-MTF	O-methyltransferase	Shoot	Kirita et al. (2010)
	CsCS	Caffeine synthase	Leaf	Li et al. (2007)
	CsFLS	Flavonol synthase	Leaf	Lin et al. (2007); SuXia et al. (2009)
	CsLAR	Leucoanthocyantin reductase	Leaf	Ma et al. (2010)
	CsPAL	Phenylalanine ammonia lyase	Leaf	Matsumoto et al. (1994b)
	CsPMO	β-Primeverosidase	Leaf	Mizutani et al. (2002)
	CsLCR	Leucoanthocyanidin reductase	Leaf	Park et al. (2004)
	Cs4CL	p-coumarate:CoA ligase	Shoot	Rani et al. (2009)
	CsCHI	Chalcone isomerase	Leaf	Rani et al. (2012)
	CsF3'-5'H	Flavanone-3-hydroxylase	Leaf	Rani et al. (2012)
	CsANS	Anthocyanidin synthase	Leaf	Rani et al. (2012)
	CsF3'H	Flavanone-3-hydroxylase	Leaf	Singh et al. (2008)
	CsDFR	Dihydroflavonol 4-reductase	Leaf	Singh et al. (2009a)
Punyasiri et al. (2004)				
	CsANR	Anthocyanidin reductase	Leaf	Singh et al. (2009b)
Zhang et al. (2012); Pang et al. (2013)				
	CsC4H	Cinnamate-4-hydroxylase	Leaf	Singh et al. (2009c)
	CsPAL	Phenylalanine ammonia-lyase	Leaf	Singh et al. (2009c)
	CsCHS1, CsCHS2, CsCHS3	*Chalcone synthase*	Leaf	Takeuchi et al. (1994a)
	CsCHS	Chalcone synthase	Leaf	Takeuchi et al. (1994a)
Rani et al. (2012)				
	CsPPO	Polyphenol oxydase	Leaf	Wu et al. (2010)
	CsACCSy	ACC synthase	Leaf	Zhang et al. (2008a)
	CsACCO	ACC oxydase	Leaf	Zhang et al. (2008b)
	CsPPO	Polyphenol oxidase	Leaf	Zhao et al. (2001)
Ion homeostasis	*CsNDPK1*	NDPK	Leaf	Prabu et al. (2012)
	CsGS1	Cytosolic glutamine synthetase	Root	Taniguchi and Tanaka (2004)
	CsAMT	Ammonium transporter protein	Root	Taniguchi and Tanaka (2004)
	CsGDH2	Glutamate dehydrogenase α-subunit	Leaf	Wang et al. (2012d)
	CsAPS1, CsAPS2	Adenosine phosphosulfate	Leaf and root	Zhu et al. (2008a)

In a pioneer work, young leaves of the widely cultivated Japanese cultivar 'Yabukita' was used by Matsumoto et. al. (1994b) to construct the cDNA library and hybridized with rice *PAL* cDNA probe to clone the tea PAL gene. Alignments of amino acid sequence of tea PAL protein revealed higher homology with dicots than monocots confirming its evolutionary proximity in relation with dicot *PAL* genes. Later, a 280 bp from 3′ UTR of this gene was used as probe to test the segregation pattern among the mapping population. It showed that *PAL* in tea was multiallelic and inherited as per the Mendelian monogenic ratio. Similarly, using heterologus probe, *CHS* gene of tea was cloned by screening the cDNA library which showed 85 % similarity between parsley *CHS* and *CsCHS* (Takeuchi et al. 1994a). Later several genes related to biosynthesis of flavonoids including LAR (Ma et al. 2010) had been cloned in tea (Table 8.1).

Further DFR, from tea (*CsDFR*) was also cloned. *In vivo* expression analysis through Q-PCR indicated younger leaves had higher level of expression than older leaves under controlled condition. However, it was down-regulated under dehydration stress and ABA treatment, but up-regulated upon wounding. Interestingly, all the regulation of this gene was modulated with catechins content in the respective tissue or stress (Singh et al. 2009a). Anthocyanidin reductase, a key enzyme of epicatechins pathway of tea (*CsANR*) had been cloned (Singh et al. 2009b). The functional protein, upon its expression in *E. coli*, catalysed the conversion of cyanidin to epicatechins in the presence of NADPH. PAL and C4H are two important enzymes in catechin biosynthesis. Fragment of *C4H* gene was isolated from tea by differential display and later full length gene was cloned using RACE technology. It had been found that catechin contents decreased in response to drought, ABA and GA_3 treatments but increased in response to wounding. Interestingly, the expression of *CsPAL* and *CsC4H* transcripts were in similar trend along with catechin contents under the above treatments. Therefore, a positive correlation between catechin contents and gene expression was established during catechin biosynthesis (Singh et al. 2009c). Flavanone 3-hydroxylase (F3′H), an important enzyme of catechin biosynthesis pathway catalyses the stereospecific hydroxylation of (2S)-naringenin to form (2R, 3R)-dihydrokaempferol. The *CsF3′H* was cloned from tea and expressed in *E. coli* which yielded a functional protein. Furthermore, a positive correlation was detected between the concentration of catechins and *CsF3′H* expression at different developmental stages of tea plants (Singh et al. 2008).

Polyphenol oxidase is an important enzyme which plays a vital role during fermentation of black tea processing. Using degenerate primers, Zhao et al. 2001 were first to clone this gene from young tea leaf. Phylogenetic analysis indicated a close proximity of *CsPPO* to other woody plants. Later, in order to understand the allelic variation, *CsPPO* was cloned from five different genotypes of tea and potential genic-SNPs were detected. Wu et al. 2010 also cloned PPO from tea. It had a signal peptide targeting the chloroplast and three Cu binding domains. The mature *PPO* showed high expression and enzyme activity in *E. coli* strain BL21 under pET30c expression vector. 4-Coumaroyl-CoA (4CL) is another gene of catechin biosynthesis pathway. The full-length cDNA cloning of *Cs4CL* and its association with catechin accumulation in tissue was established. Altering catechin content through drought stress, ABA, GA_3 treatments and wounding established a strong positive correlation coefficient between catechins content and the expression of *Cs4CL* (Rani et al. 2009).

To understand the molecular mechanism, Pang et al. (2013) characterized the gene of key enzymes involved in polyphenolic proanthocyanidins pathway (PA) of tea. Recombinant protein of tea leucoanthocyanidin reductase (*CsLAR*) expressed in *E. coli* was found to be functional and with leucocyanidin as substrate, could produce the product 2R, 3S-*trans*-flavan-ol (+)-catechin under *in vitro* conditions. Two genes encoding anthocyanidin reductase, *CsANR1* and *CsANR2*, were also expressed in *E. coli*, and the recombinant proteins exhibited similar kinetic properties. Both converted cyanidin to a mixture of (+)-epicatechin and (–)-catechin, although in different proportions, indicating that both enzymes possess epimerase activity and hence functional in nature (Pang et al. 2013).

8.2.1.3 Purine Biosynthesis Related Genes

Caffeine, theobromine and theophylline are considered as purine alkaloids and flavour forming chemicals present in some species of *Camellia*. Caffeine and theobromine are purine alkaloids that are present in high concentrations in tea. However, most members of the genus *Camellia* contain no purine alkaloids. The major steps of caffeine biosynthetic pathway are: xanthosine → 7-methylxanthosine → 7-methylxanthine → theobromine → caffeine (Suzuki 1972; Ashihara and Kubota 1987; Ashihara et al. 1996, 1997; Kato et al. 1996; Deng et al. 2008)).

Among the pivotal enzymes, S-adenosylmethionine synthase, convert methionine to S-adenosylmethionine. It serves as the methyl donor in transmethylation reactions and also acts as the precursor in ethylene and polyamine biosynthesis. So, isolation and cloning of *SAM* was might pave the way for further study and manipulation of caffeine biosynthesis, stress and senescence physiology. Accordingly a *CsSAM* gene was cloned and ORF was found to be 1303 bp in length with 349 amino acids. This is an essential step to manipulate the caffeine biosynthesis in tea plant (Feng and Liang 2001).

In tea, among the three genes namely 7-N-methyltransferase (*7-NMT*), 3-N-methyltransferase (*3-NMT*) and theobromine N-methyltransferase (*1-NMT*), two (3-NMT and 1-NMT) are highly homologous due to similarity of many properties and are considered them as single enzyme, caffeine synthase (*CS*). Initially using degenerate primer, partially *CsCS* gene and later full length was cloned from young leaf of tea using RACE-PCR (Kato et al. 1999). The substrate specificity of the recombinant enzyme, after expression of cDNA in *E. coli*, remained similar to that of the original tea enzyme. The cloning of *CS* gene led to the development of transgenic caffeine deficient tea plants by antisense technology (Kato et al. 2000).

Tracer experiments using [8-^{14}C] adenine and [8-^{14}C] theobromine showed that the purine alkaloid pathway is not fully functional in leaves of purine alkaloid-free species. Interestingly, in purine alkaloid-free *Camellia* species, sufficient evidence was obtained to show the occurrence of genes that are homologous to caffeine synthase. Recombinant enzymes derived from purine alkaloid-free species showed only theobromine synthase activity. Unlike the caffeine synthase gene, these genes were expressed more strongly in mature tissue than in young tissue (Isida et al. 2009).

Mohanpuria et al. (2009) suggested that caffeine metabolism in tea appears to be dependent on developmental and seasonal factors. Expression of caffeine synthase as well as caffeine content were found to be higher in harvestable plant parts such as apical bud, 1st leaf, 2nd leaf, young stem, and was lower in older leaf during non-dormant growth periods compared to dormant growth phase.

8.2.1.4 Floral Aroma Formation Related Genes

Aroma is an important yardstick to determine quality of black tea and primarily the aroma forming compounds are produced by endogenous enzymes such as β-glucosidases and β-primeverosides. In order to increase quality of made tea more, it is obvious to pursue ways that ensure an increment in tea aroma forming compounds. The β-glucosidases a hydrolases was found to be involved in the formation of floral aroma (Takeo 1981). Later on many aroma forming compounds had been isolated and identified from green, as well as black tea leaves. Off late, the concept of hydrolyzation of polyglycoside by β-glucosidase vis-a-vis β-primeverosides and subsequent removal of floral aroma were established. It is a disaccharide specific glycosidase which hydrolyzes aroma precursors of β-primeverosides to liberate aroma compounds. Using degenerate primer, full length cDNA sequence of β-primeverosidase genes were cloned and characterized (Mizutani et al. 2002). It had 50–60% identical protein sequence with β-glucosidases of other plants and classified this as family 1 glycosyl hydrolase. The enzyme expressed in *E. coli* hydrolyzed β-primeverosides to liberate a primeverose unit and aglycons under *in vitro* condition. It was confirmed that the β-primeverosidase selectively hydrolyzed the β-glycosidic bond of β-primeverosides between the disaccharide and the aglycons (Mizutani et al. 2002). Later, the β-glucosidase gene of tea was also cloned (Li et al. 2004b). It shared 40–60 % similarity in nucleotide sequence with other plants. Upon expressing in heterologus system in *E. coli*, it possessed normal bioactivity by breaking down the glycosidic bonds. Subsequently, the spatio-temporal expression pattern of the aforesaid genes in different leaves of young shoot was determined in tea (Zhao et al. 2006a). Also, it revealed the variation of expression pattern according to leaf position.

8.2.2 Abiotic Stress Related Genes

Differential expression of a cytosolic glutamine synthetase (*CsGS*) from tea during developmental stages and light or dark conditions on the utilization of nitrate as well as ammonia had been reported (Rana et al. 2010). The *CsGS* transcript expression level was highest in apical bud, gradually reduced towards the older leaf and was found to be lowest in the fourth leaf. Exposure to both nitrate and ammonia N-sources showed enhancing effect on *CsGS* enzyme activity during light but under dark, ammonium increased *CsGS* activity and nitrate had inhibitory effect (Rana et al. 2010).

Ammonium transporters (*AMT*) are involved in nitrogen absorption of plant roots, therefore important for utilization in nitrogenous fertilizer. This gene had also been cloned from tea root (Taniguchi and Tanaka 2004). Cold is an important abiotic stress for tea cultivation as it hampers the productivity of tea significantly. However not much cold responsive genes had been cloned from tea, except, *CsCOR1*, which had been cloned recently from tea leaves (Li et al. 2010). The deduced amino acid sequence contained hydrophobic N-terminus as a signal peptide and a chareteristic glycine, arginine and proline rich C-terminal domain. The gene was found to be localized in the cell-wall of transgenic tobacco by *CsCOR1::GPF* fusion approach. Its (*CsCOR1*) expression was up-regulated under cold and dehydration stress.

Sulphate assimilation is an important metabolic pathway for plants and ATP sulfurylase, the first enzyme of this pathway, converts ATP and sulphate to adenosine phosphosulphate. Two cDNAs encoding ATP sulfurylase (*APS1* and *APS2*) were cloned from tea (Zhu et al. 2008a). While, ORF of *APS1* was 1,415 bp in length with 360 amino acids, yet *APS2* ORF was found to be 1706 bp in length which encoded 465 amino acids. They shared 59.6 % identity with each other. The predicted amino acid sequences exhibited 86 % and 84 % similarity to ATP sulfurylases of *Medicago truncatula* and *Solanum tuberosum* respectively.

A higher level of selenium is toxic to plants. Enzyme selenocysteine methyltransferase (SMT) methylates selenocysteine to Se-methyl selenocysteine and thus play vital role in toxicity removal. Full length cDNA of *CsSMT* had been cloned from tea (Zhu et al. 2008b). This sequence of amino acids shared 74 % and 69 % identity with *Astragalus bisulcatus* SMT and *Brassica. oleracea* SMT respectively. Expression of *CsSMT* correlated with the presence of SMT enzyme activity in cell extracts and bacterial recombinant *CsSMT* had higher tolerance to selenate and selenite.

8.2.3 Biotic Stress Related Genes

Pests and pathogens are harmful agents that act upon the bush health and productivity. Although both pest and diseases are important for tea cultivation, yet not much works have been done, primarily due to lack of tolerant genotype in tea. Wang et al. (2005) cloned the cystain gene with the help of degenerate primers from tea. The deduced amino acid sequence contained the motif QXVXG conserved among most members of cystatin superfamily. In an another study, eight wound/pathogen inducible cDNAs were cloned and the structure was analysised through *in silico* (Yoshida and Homma 2005). It had been found with high sequence homology with other wound/pathogen inducible genes of herbaceous plants.

8.2.4 Energy Metabolism Related Genes

Green plants, by virtue of chlorophyll, absorb solar energy to produce organic substance and release oxygen through the process of photosynthesis. Though solar energy is the basis for photosynthesis, excessive energy may result in photo-oxidative damage to photosynthetic apparatus and other cell components. There exists some mechanism to protect the plants from such adverse effects. In tea plants two photosynthesis related genes had been cloned, which were photosystem II protein D1 and violaxanthin de-epoxidase genes (*VDE*). Among them, *VDE* had been studied extensively. Under excess light intensity, violaxanthin converted to antheraxanthin and ultimately to zeaxanthin by *VDE*, whereas under low light intensity zeaxanthin epoxidase (ZE) catalyse the reverse reaction. Zeaxanthin can protect photosynthetic apparatus from excessive irradiation. Thus VDE can be considered as a vital enzyme in the xanthophyll cycle-dependent photo protective mechanism. Using degenerate primers, Wei et al. (2003) cloned full length cDNA of *VDE* gene. It was demonstrated that the gene contained lipocalin signature which considered to be most active domains in VDE (Wei et al. 2004).

RUBISCO (Ribulose 1, 5-bisphosphate carboxylase) is a key enzyme in energy metabolism of the Calvin cycle pathway of photosynthesis. The full length RUBISCO small subunit (*RbcS*) was cloned from tea. The amino acids sequence had a high identity to those of other plant *RbcS* genes. It had three conserved domains as well as a protein kinase C phosphorylation site, one tyrosine kinase phosphorylation site and two N-myristoylation sites. The Q-PCR analysis showed that the expression of *RbcS* was the highest in youngest tea leaf (Ye et al. 2009).

8.2.5 Developmentally Regulated Genes

Microtubules composed of α-tubulin and β-tubulin dimers are one of the most important constituents of eukaryotic cytoskeleton. It is an important structural element involved in mitosis, cytokinesis and vesicular transport. Sediment directions of microfilaments on cell-walls, which control the growth of cells, are also regulated by microtubules. The β-tubulin gene had been cloned from tea (Takeuchi et al. 1994b). The putative amino acid sequence had a high similarity to β-tubulin genes from other plants.

The eukaryotic nucleosome octamer core consists of the four histones, H3, H2A, H2B and H4. Using differential display and RACE-PCR, a full-length histone H3.1 cDNA (*CsH3*) was cloned from tea leaves. During active growth, higher expression was observed in apical buds and decreased gradually with maturity of the leaf. During dormancy, drought stress and ABA treatments, the expression of *CsH3* was severely down-regulated but up-regulated by GA_3 treatment. Positive co-relations between *CsH3* and active cellular growth suggested its role in plant growth and development (Singh et al. 2009d).

On the other hand, lipoxygenase (*LOXs*) is a large gene family which catalyzes the hydroperoxidation of free polyunsaturated fatty acids into different biologically active compounds and had been isolated from tea plant (*CsLOX1*). Heterologous expression in yeast showed that *CsLOX1* protein conferred a dual positional specificity and hence it was named 9/13-*CsLOX1*. Analysis of the isolation and expression of the *LOX* gene in tea plant indicated that the acidic *CsLOX1* together with its primary and final products played an important role in regulating cell death related to flower senescence and the JA-related defensive reaction of the plant to phloem-feeders (Liu and Han 2010). Pollen tube growth in tea was promoted by *Tua 1* gene. The cDNA-AFLP technique was used by Fang et al. (2006) to isolate transcript-derived fragments corresponding to these genes. The complete cDNA sequence of this gene had been cloned which had two N-glycosylation sites and four protein kinase C phosphorylation sites.

8.2.6 Other Genes

Ribosomes play an important role in plant evolution with 5.8S, 16S and 26S rRNA being the prime components of ribosome. The QM like protein gene encodes for ribosomal protein which has a function of ribosome stability. Full length QM cDNA from tea (*CsQM*) had been cloned (Singh et al. 2009d) which shared 71–87 % and 85–91 % identity at nucleotide and amino acid sequences, respectively with QM genes isolated from other woody plants.

8.3 Differentially Expressed Genes

Understanding the global transcript changes under a stress, cell types, tissue or developmental stage is a robust approach. Although several techniques are available (Mondal and Sutoh 2013; Das et al. 2013), suppression subtractive hybridization (SSH), cDNA AFLP have been widely used to study the differential gene expression in tea. For an example, SSH had been used to understand the molecular regulation of secondary metabolism pathways, in the young leaves (Park et al. 2004; Chen et al. 2005), to identify the cold responsive genes (Wang et al. 2009), to identify dormancy-associated genes (Wang et al. 2010; Krishnaraj et al. 2011; Paul and Kumar 2011; Yang et al. 2012a; Phukon et al. 2012; Thirugnanasambantham et al. 2013) and to identify the drought responsive genes (Sharma and Kumar 2005; Das et al. 2012; Gupta et al. 2012; Muoki et al. 2012; Gupta et al. 2013). A forward SSH library under drought stress was constructed from TV-23 (a drought tolerant cultivar) roots, to ascertain 123 drought responsive genes (Das et al. 2012). Recently 7 cDNA libraries from various organs of tea plants were used to generate 17,458 ESTs of tea (Taniguchi et al. 2012). Interestingly, to investigate the molecular mechanism by which light regulates phenylpropanoid metabolism, a light-induced SSH library of tea calli was constructed. Several genes in lignin biosynthesis pathways were identified in the light-induced library (Wang et al. 2012). To understand molecular mechanism of enhanced aroma under mild infestation of green leaf hopper (*Empoasca vitis* Göthe) in tea leaf, a SSH library was constructed and several genes were identified (Yang et al. 2011) and interestingly it was found that insect infection triggered the several genes that were responsible for quality. Furthermore, through cDNA-AFLP approach, they identified 108 transcript-derived fragments that were expressed differentially in the leaf of drought tolerant tea cultivars (Gupta et al. 2013). Eungwanichayapant and Popluechai (2009) reported that the expression of phenylalanine ammonia-lyase 1, chalcone synthase, dihydroflavonol 4-reductase, leucoanthocyanidin reductase, and flavanone 3-hydroxylase genes were higher in the young shoots than in the mature leaves which were obvious as most of the quality compounds were produced by these genes. These genes are normally found to be more expressive in young leaves. Nineteen genes were found to be up-regulated with *Ectropic obliqua* infection as detected by SSH, differential screening, and Q-PCR analysis. These genes were involved in JA synthesis, cell-wall modification, metabolism of secondary metabolites as well as carbohydrate, pathogen defensive response and oxidative stress protection. The results showed that *Ectropic obliqua* infections activated defense related genes and induced tea defense responses (Qiao et al. 2011).

In Taiwan, 'oriental beauty', a high flavoured made tea from leaf hopper infestation is very popular. In order to know the genes responsible for the same, gene expression profile and biochemical profile were done by DNA microbead array, a fluorescence labeled based technique where different fluorescence DNAs are sorted out by cell sorter machine. Interestingly several stress responsive genes were found to be up-regulated in infested tea leaves (Choi et al. 2007). Darjeeling teas of India are the highest grown teas in the world and best -known for its flavour, aroma and quality. Apart from the genetic makeup of the plant, reports suggested that insect infestation, particularly jassids and thrips triggered the aroma and flavour formation in Darjeeling tea. Several genes and transcription factors were identified through the SSH library of leaf that were highly infested by these insects (Gohain et al. 2012).

Expression of genes encoding enzymes involved in flavan-3-ol biosynthesis pathway such as, *CHS*, *CHI*, *F3H*, *F3'-5'H*, *DFR*, *ANS*, *ANR* and *LAR* was investigated. Transcripts of all genes, except *LAR*, were more abundant in leaves and stems than in roots and cotyledons. No significant difference of expression was found for the transcript of *LAR*. In tea, flavan-3-ols are produced by a naringenin-chalcone to naringenin to dihydrokaempferol pathway. Dihydrokaempferol is a branch point in the synthesis of (-)-epigallocatechin-3-O-gallate and other flavan-3-ols which can be formed by routes beginning with either a flavonoid 30-hydroxylase mediated conversion of the flavonol to dihydroquercetin or a flavonoid 30,50-hydroxylase-catalysed conversion to dihydromyricetin with subsequent steps involving sequential reactions catalysed by dihydroflavanol 4-reductase, anthocyanidin synthase, anthocyanidin reductase and flavan-3-ol gallate synthase (Ashihara et al. 2010). Expression of *PAL*, *CHS*, *DFR* were differentially expressed under red and blue light, however, their expression were lower in mature leaf than the younger leaf (Takechi and Matsumoto 2003).

R2R3-MYB, *bHLH*, and *WD40* transcription factors are well-known to control multiple enzymatic steps in the biosynthetic pathway responsible for the production of flavonoids, important secondary metabolites of tea. The presence of 73 *R2R3-MYB*, 49 *bHLH*, and 134 *WD40* transcripts were identified through *in silico* analysis of 127,094 unigenes of tea. *In silico* analysis of protein sequences of CsMYB4-1, CsMYB4-2, CsMYB4-3, CsMYB4-4, CsMYB5-1, and CsMYB5-2 indicated presence of [DE]Lx2[RK] x3 Lx6 Lx3R motif, potentially contributing to the specificity of the bHLH partner in the stable MYB-bHLH complex. Q-PCR analysis validated selected genes and their expression profiles under various developmental stages and treatment conditions, including hormones (Zhao et al. 2012).

To investigate the molecular mechanisms by which light regulates phenylpropanoid metabolism, light-induced SSH cDNA libraries from tea calli was prepared. Nine diverse ESTs involved in phenylpropanoid biosynthesis were detected in the library. Among them, CsDFR gene (*CsDFR2*) was found to be up-regulated by light but had sequence difference with previously reported *CsDFR* gene (*CsDFR1*) of tea. The key phenylpropanoid compounds and representative genes expression analysis implied that light could be effective for activation of the biosynthesis of phenylpropanoids. Compared to the dark treatment as control, levels of lignins, catechins, and PAs were increased 3.46, 3.00, and 1.21-fold, in light-induced calli, respectively. Lignin biosynthesis genes, involved in *CCoAOMT*, *HCT* and *CCR*, were identified in the light-induced SSH library (Wang et al. 2012a). The plant hormone, auxin plays a key role in adventitious rooting. To increase our understanding of genes involved in adventitious root formation, Wei et al. (2013) identified differentially expressed transcripts in single nodal cuttings of tea treated with or without IBA. A total of 77 differentially expressed transcripts, including 70 up-regulated and 7 down-regulated sequences, were identified that were expressed under IBA treatment. Influences of shade on flavonoid biosynthesis in relation to expression of the flavonoid pathway genes in tea leaves were analyzed. It was found that shade had notable effects on both flavonoids (including catechins, O-glycosylated flavonols) and proanthocyanins (PAs) and lignin biosynthesis, but had no significant effect on anthocyanin accumulation. Among all the compounds, the concentration of PAs and O-glycosylated flavonols under shaded leaves reduced more than other compounds, compared to the sunlight-exposed leaves. Expressions of *PAL*, flavanone 3-hydroxylase, flavonoid 3-hydroxylase, *DFR* and *ANR1* were notably correlated with the concentration of PAs in leaves, and expression of *CHS* and flavonoid 3,5-hydroxylase (*F3'-'5H*) were correlated with the concentration of O-glycosylated flavonols. It was suggested that polymerization of catechins and glycosylation of flavonols might be key pathways of flavonoid metabolism in tea leaves affected by shade treatment. On the other hand, phenolic acids were increased under shade and a negative correlation with lignin accumulation suggested that phenolic acids might compete for the same substrate with lignins and flavonoids in tea leaves under different illumination conditions (Wang et al. 2012c).

In order to identify the differentially expressed genes related to morphological variation of callus, Yang et al. (2012a) analyzed two contrasting callus lines (Yunjing 63Y and Yunjing 63X) that showed different morphological characteristics and catechin contents. Yunjing 63Y callus was yellow and compact, while yunjing 63X callus was white and loose. Using cDNA AFLP, they identified 68 genes that were differentially expressed between the two lines. Of the 68 differentially expressed ESTs, 40 showed higher expressions in Yunjing 63Y and remaining 28 showed higher expressions in Yunjing 63X. Gray blight disease (*Pestalotiopsis theae*) is an economically important diseases of tea which account enormous loss to the tea industry. Senthilkumar et al. (2012) used SSH technique to identify the differential gene expression pattern during gray blight disease infection in tolerant cultivar UPASI-10. Majority of the up-regulated genes were related to hypersensitive response and reactive oxygen species production.

Three tea clones with varying degree of polyphenol content were selected to identify genes that were related to polyphenol biosynthetic pathway. Totally, 1,680 gene tags were obtained from the cDNA library to develop the first cDNA microarray of tea plant. Each gene probe in the microarray was duplicated in the slides. Finally, two genes that were closely related to tea aroma were selected for Q-PCR analysis to validate the cDNA microarray. The newly developed cDNA microarray could be applied in various tea research fields to make the high-throughput detection in the gene expression profiling (Zhao et al. 2006b). Cross species hybridization were also attempted to understand the gene expression of tea. Using *Arabidopsis* cDNA array, Venkatesh et al. (2006) found that highest level of induction of CHS gene in the tea leaf during the second flush (a period where flavonoids content of tea was found to be highest in tea planataion of Assam, India). On the other hand, expressions of cinnamyl alcohol dehydrogenase (CAD), cinnamyl CoA reductase (CCR) and POD involved in the synthesis of lignin were reduced during the second flush. These observations suggested that CHS may be a key enzyme involved in catechin biogenesis in tea leaf. Similarly, to identify, the Al uptake responsive genes in tea, Murayama et al. (2007) used cross-species microarray analysis. For the target gene of microarray analysis, approximately 8,500 cDNA fragments were identified from *Arabidopsis thaliana*. According to the signal intensity of cDNA detected by microarray analysis, putative genes including auxin-induced protein, zinc finger family protein, blue copper binding protein, were found to be expressed differentially under Al treatment in tea. Therefore, it was suggested that antioxidant enzymes genes and synthesis of auxin as well as cell-wall genes might be induced by Al treatment in tea root tips.

To understand the molecular mechanism of albinism, tea cultivar 'Anji Baicha', was selected which was a special green-revertible albino mutant, widely cultivated in China for producing high quality green tea. A total of 671 differentially expressed genes at different albescent stages were identified using cDNA microarray analysis. The corresponding genes were involved in energy metabolism, carbon fixation, cell expansion, secondary metabolism, plant growth and defence, and other physiological processes including protein, nucleotide synthesis, etc. Particularly, some differentially expressed genes encoding important catalyzing enzymes or regulatory proteins which took part in chlorophyll biosynthesis or chloroplast development were identified. Some candidate genes possibly related to the albino process were further analyzed by Q-PCR. The present study gave some useful clues for genes worthy of further understanding the albino phenotype of 'Anji Baicha' and also provided a model for utilization of the microarray technology in the tea plant (Ma et al. 2012).

8.4 Proteomics and Metabolomics

Proteomics study correlates the potential protein modifications to particular phenotypes through the techniques such as HPLC, mass spectrometry, SDS-PAGE, two-dimensional gel electrophoresis, *in silico* protein modeling, matrix-assisted laser desorption/ionization (MALDI) etc.

Li et al. (2008) did the proteomics analysis to identify the differentially expressed proteins of pollen that were associated with cold stress. Further, changes in protein expression, in the embryo of tea in response to desiccation, had been investigated and differentially expressed proteins were identified. Around twenty-three different proteins related to defense response, metabolism and redox status were up-regulated under desiccation. This finding is particularly important as tea seeds being recalcitrant, maintain high rate of metabolism and water content during the seed maturation stage (Chen et al. 2011). In China, an albino mutant of tea were identified which had typical white leaf phenotype. In order to understand the protein involve in albino leaf, proteomics analysis were done which identified 61 proteins during the three developmental stages of an albino cultivar of tea. Important proteins that played crucial roles in the periodic albinism were identified (Li et al. 2011). Altogether these studies basically identified the differentially expressed proteins with compared to their control counterparts which at this stage mere indicate the various pathways that are involved in that particular situation.

Metabolomics is also powerful technique and in tea particularly very relevant as quality is a complex trait, mixture of several metabolites of various nature. Metabolic characteristics associated with climatic variables, were investigated by ^{1}H NMR spectroscopy on tea grown in different areas of South Korea, thereby allowing for the assessment of quality strategy in green tea production (Lee et al. 2010). Differential metabolites expressed under shade grown tea plants had been detected by LC-MS and GC-MS spectrometry coupled with a multivariate data set which helped to understand the effect of shade on tea growth (Ku et al. 2010). Further, the dependence of global green tea metabolome on plucking positions was investigated through NMR analysis coupled with multivariate statistical data set which correlated the age of plucking shoots with quality of made tea (Lee et al. 2011).

8.5 System Biology

Systems biology is a biology-based inter-disciplinary field of study that focuses on complex interactions within biological systems, using a more holistic perspective. Tea quality is a highly complex trait and tea manufacture induces a variety of stresses that affect quality. Using cross species Affymetrix *Arabidopsis* genome arrays, Marshall et al. (2007) attempted to track transcriptional changes occurring during wounding and withering of the leaves to identify metabolic pathways that could influence tea aroma and flavour. *Arabidopsis* metabolic SBML (Hucka et al. 2003) network data from AraCyc (Mueller et al. 2003), KEGG and Reactome were collated and merged, then subsequently overlaid with the tea expression data. Sub networks were constructed by connecting the shortest paths between the differentially expressed genes and the downstream aroma-related compounds, therefore identifying the pathways involved in aroma.

8.6 Bioinformatics

In silico analysis is an integral part of the functional genomics today. With the accumulation of large scale ESTs in the public domain, researchers started to use for various studies. The emerging computational approach provides a better alternative process of development of SSRs from the ESTs than the conventional methods. In the present study, 12,851 ESTs of tea, were mined for the development of microsatellites. Finally, 6,148 (4,779 singletons and 1,369 contigs) non-redundant ESTs were found using various computational tools. Out of total 3822.68 kb sequence examined, 1,636 (26.61 %) ESTs containing 2,371 SSRs, were detected with a density of 1 SSR/1.61 kb leading to development of 245 primer pairs. These mined EST-SSR markers will help further in the study of variability, mapping, evolutionary relationship in tea. In addition, these developed SSRs can also be applied for various studies across species (Sahu et al. 2012). Using similar approaches, miRNAs of tea were identified along with their stem-loop structure from the ESTs (Das and Mondal 2010; Prabu and Mandal et al. 2010).

Structural analysis of a cytosolic glutamine synthetase from tea (*CsGS*) had been conducted employing computational techniques. This was conducted to compare its structural aspects with other known structures of *CsGS*. The disordered residues and their distribution in *CsGS* were in close comparison to earlier reported *GS*. The 3-D structure of CsGS protein also showed high degree of similarity with the only known crystal structure of GS from maize (Yadav 2009). Similarly, coding sequence of the *ICE1* in tea was analyzed with CodonW, CHIPS (codon heterozygosity in a protein coding sequence) and CUSP (create a codon usage table) programs, while compared with the genome of tea plant and *ICE1* from seven plant species. The results showed that *ICE1* of tea plant was bias toward the synonymous codons with A and T. (Shi et al. 2012).

The stearoyl-acylcarrier protein desaturase (*SAD*) gene is widely present in all kinds of plants. Tea SAD gene (*CsSAD*) sequence was analyzed by various bioinformatics tools such as CodonW, CHIPS, and CUSP programs, and compared with publicly available tea sequences, other *SAD* genes from 11 plant species. It had been found that *CsSAD* gene had similar codon usage bias. The *CsSAD* gene had a bias toward the synonymous codons with A and T at the third codon position. Compared with monocotyledons such as *Triticum aestivum* and *Zea mays*, the differences in codon usage frequency between the *CsSAD* gene and dicotyledons such as *Arabidopsis thaliana* and *Nicotiana tobacum* were less. Therefore, *A. thaliana* and *N. tobacum* expression systems may be more suitable for the expression of the *CsSAD* gene. The analysis result of *SAD* genes from the 12 plant species also showed that most of the *SAD* genes were biased toward the synonymous codons with G and C at the third codon position (Pan et al. 2013).

8.7 High Throughput Sequencing

Recently, the next generation sequencing technology, specially RNA-seq is widely used for identification of differentially expressed genes (Mondal and Rana 2013; Mukhopadhayay et al. 2013). For the first time, an extensive transcriptome dataset had been generated by RNA-Seq from the young leaf of tea plants. The coverage of the transcriptome was comprehensive enough to discover all known genes of several major metabolic pathways. This transcriptome dataset can serve as an important public information platform for gene expression, genomics, and functional genomics studies in tea (Shi et al. 2011). In the same year, Jiang et al. (2011) did deep transcriptome sequencing of *C. oleifera*, *C. chekiangoleosa* and *C. brevistyla* using 454 GS FLX platform. After removal of the adaptors as well as low complexity sequences, finally 182,766, 190,545 and 132,147 reads of *C. oleifera*, *C. chekiangoleosa* and *C. brevistyla*, were generated respectively. These reads were assembled into 49,909 contigs which were annotated using BLAST searches. Based on the information, they could estimated the similarity of gene expression in different species, detected and validated SNPs.

In order to know the genes involved secondary metabolism, high-throughput sequencing of tea leaf transcriptoms was done. A total of 437,908 reads were generated using 454 GS FLX platform. *De novo* assembly of those reads yielded 25,637 unigenes, 22,872 of which were annotated by BLAST searches against public databases. Some abundant transcripts were found to be related to developmentally regulated in plants, including ubiquitin/26S proteasome, lipid transfer protein, PPR containing protein, small GTPase, expansin, transport inhibitor response 1 and thioredoxin. Additionally, 3,767 EST-SSRs were identified as potential molecular markers in our unigenes. A total of 100 PCR primer pairs used in initial screening test among the 20 tea genotypes successfully identified 36 polymorphic loci (Wu et al. 2013).

Changes that occur at the molecular level in response to low temperature are poorly understood in tea plants. To elucidate the molecular mechanisms of cold acclimation, Wang et al. (2013) employed RNA-Seq and digital gene expression (DGE) technologies to the study of genome-wide expression profiles during cold acclimation in tea plants. In total, 1,770 differentially expressed transcripts were identified, of which 1,168 were up-regulated and 602 down-regulated. These included a group of cold sensor or signal transduction genes, cold-responsive transcription factor genes, plasma membrane stabilization related genes, osmosensing-responsive genes, and detoxification enzyme genes. DGE and Q-PCR analysis further confirmed the results from RNA-SEq. Pathway analysis indicated that the 'carbohydrate metabolism pathway' and the 'calcium signaling pathway' might play a vital role in tea plants' responses to cold stress (Wang et al. 2013).

8.8 Conclusion

Functional and structural genomics both are new development of biotechnology. These 'omics' approaches along with other advance techniques of microscopy as well as DNA sequencing are very potential for deciphering several fundamental questions of gene expression. However, large-scale application of 'omics' has not been done perhaps primarily due to the reason of non-availability of complete genome sequence of tea. Although in several aspects of tea molecular biology work can be intended for, yet priority should be given: (i) to undertake a *de novo* whole genome sequencing for development of reference genome, (ii) to re-sequence popular trait-specific cultivars or extreme genotypes to generate large scale SNPs markers and their utilizations for association mapping, gene introgression, high density linakge map constructions etc., (iii) some other aspects which need an urgent priority in tea are genome-wide discovery of small RNAs, high throughput allele-mining, sequence information-based conservation and utilization of tea genetic resources etc.

References

Ashihara H, Kubota H (1987) Biosynthesis of purine alkaloids in *Camellia* plants. Plant Cell Physiol 28:535–539

Ashihara H, Gillies FM, Crozier A (1997) Metabolism of caffeine and related purine alkaloids in leaves of tea (*Camellia sinensis* L.). Plant Cell Physiol 38:413–419

Ashihara H, Monterio AM, Gillies FM, Crozier A (1996) Biosynthesis of caffeine in leaves of coffee. Plant Physiol 111:747–753

Ashihara H, Deng WW, Mullen W, Crozier A (2010) Distribution and biosynthesis of flavan-3-ols in *Camellia sinensis* seedlings and expression of genes encoding biosynthetic enzymes. Phytochem 71:559–566.

Anon (2005) L-Theanine. Alternat Med Rev 10:136–138.

Borchetia S, Bora C, Gohain B, Bhagawati P, Agarwala N, Bhattacharya N, Bharalee R, Bhorali P, Bandyopadhyay T, Gupta S, Das SK, Singh HR, Ahmed P, Gogoi M, Das S (2011) Cloning and heterologous expression of a gene encoding lycopene-epsilon-cyclase, a precursor of lutein in tea (*Camellia sinensis var assamica*). Afr J Biotech 10:5934–5939

Borthakur D, Du YY, Chen H, Lu JL, Lin C, Dong JJ, Ye JH, Zheng XQ, Liang YR (2008) Cloning and characterization of a cDNA encoding phytoene synthase (*PSY*) in tea. Afr J Biotech 7:3577–3581

Chen L, Zhao LP, Gao QK (2005) Generation and analysis of expressed sequence tags from the tender shoots cDNA library of tea plant (*Camellia sinensis*). Plant Sci 168:359–363

Chen Q, Yang L, Ahmad P, Wan X, Hu X (2011) Proteomic profiling and redox status alteration of recalcitrant tea (*Camellia sinensis*) seed in response to desiccation. Planta 233:583–592

Chen X, Fang WP, Zou Z, Wang YH, Cheng H, Li XH (2009) Cloning and expression analysis of CBF gene in cold induced tea plant [*Camellia sinensis* (L.) O. Kuntze]. J Tea Sci 29:53–59

Choi JY, Mizutani M, Shimizu BI, Kinoshita T, Ogura M, Tokoro K, Lin ML, Sakata K (2007) Chemical profiling and gene expression profiling during manufacturing process of Taiwan oolong tea, "Oriental Beauty". Biosci Biotech Biochem 71:1476–1486

Das A, Mondal TK (2010) *In silico* analysis of miRNA and their targets in tea. Amer J of. Plant Sci 1:77–86

Das A, Das S, Mondal TK (2012) Identification of differentially expressed gene profiles in young roots of tea [*Camellia sinensis* (L.) O. Kuntze] subjected to drought stress using suppression subtractive hybridization. Plant Mol Biol Repo 30:1088–1101

Das A, Saha D, Mondal TK (2013) An optimized method for extraction of RNA from tea roots for functional genomics analysis. Indian J Biotech 12:129–132

Deng WW, Ogita S, Ashihara H (2008) Biosynthesis of theanine (γ-ethylamine –L-Glutamine acid) in seedling of *Camellia sinensis*. Phytochem Lett 1:115–119

Deng W, Wang S, Chen Q, Zhang Z, Hu X (2012) Effect of salt treatment on theanine biosynthesis in *Camellia sinensis* seedlings. Plant Physiol Biotech 56:35–40

Deng W-W, Fei Y, Wang S, Wan X-C, Zhang Z-Z, Hu X-Y (2013a) Effect of shade treatment on theanine biosynthesis in *Camellia sinensis* seedlings. Plant Growth Regul. doi:10.1007/s10725-013-9828-1

Deng WW, Zhang M, Wu JQ, Jiang ZZ, Tang L, Li YY, Wei CL, Jiang CJ, Wan XC (2013b) Molecular cloning, functional analysis of three cinnamyl alcohol dehydrogenase (CAD) genes in the leaves of tea plant, *Camellia sinensis*. J Plant Physiol 170:272–282

Edwards JW, Walker EL, Coruzzi GM (1990) Cell-specific expression in transgenic plants reveals nonoverlapping roles for chloroplast and cytosolic glutamine synthase. Proc Nat Acad Sci 87:3459–3463

Eungwanichayapant PD, Popluechai S (2009) Accumulation of catechins in tea in relation to accumulation of mRNA from genes involved in catechin biosynthesis. Plant Physiol Biotech 47:94–97

Fang WP, Jiang CJ, Yu M, Ye AH, Wan ZX (2006) Differentially expression of Tua 1, a tubulin encoding gene, during flowering of tea plant *Camellia sinensis* (L.) O. Kuntze using cDNA amplified fragment length polymorphism technique. Acta Biochim Biophys Sin 38:653–662

Fang W, Zhang Y, Zhou L, Wang W, Li X (2013) Isolation and characterization of *Histone1* gene and its promoter from tea plant (*Camellia sinensis*). Mole Biol Repo 40:3641–3648

Fang W, Zou Z, Hou X, Zhang D, Duan Y, Yang Y, Li X (2009) Cloning and sequence analysis of cold-induced H1-histone gene from *Camellia sinensis*. Acta Bot Boreali-Occidentalia Sin 8:1514–1519

Feng YF, Liang YR (2001) Cloning and sequencing of S-adenosylmethionine synthetase gene in tea plant. J Tea Sci 21:21–25

Fu J (2012) Molecular cloning and expression analysis of a putative sesquiterpene synthase gene from tea plant (*Camellia sinensis*). Acta Physiol Plant. doi:10.1007/s11738

Gohain B, Borchetia S, Bhorali P, Agarwal N, Bhuyan LP, Rahman A, Sakata K, Mizutani M, Shimizu B, Gurusubramaniam G, Ravindranath R, Kalita MC, Hazarika M, Das S (2012) Understanding Darjeeling tea flavour on a molecular basis. Plant Mol Biol 78:577–597

Gupta S, Bharalee R, Bhorali P, Bandyopadhyay T, Gohain B, Agarwal N, Ahmed P, Saikia H, Borchetia S, Kalita MC, Handique K, Das S (2012) Identification of drought tolerant progenies in tea by gene expression analysis. Funct Integ Genom 12:543–563

Gupta S, Bharalee R, Bhorali P, Das SK, Bhagawati P, Bandyopadhyay T, Gohain B, Agarwal N, Ahmed P, Borchetia S, Kalita MC, Handique AK, Das S (2013) Molecular analysis of drought tolerance in tea by cDNA-AFLP based transcript profiling. Mol Biotech 53:237–248

Hucka M, Finney A, Sauro HM, Bolouri H, Doyle JC, Kitano H, Arkin AP, Bornstein BJ, Bray D, Cornish-Bowden A (2003) The Systems Biology Markup Language (SBML): A medium for the representation and exchange of biochemical network models. Bioinformat 19:524–531

Ishida M, Kitao N, Mizuno K, Tanikawa N, Kato M (2009) Occurrence of theobromine synthase genes in purine alkaloid-free species of *Camellia* plants. Planta 229:559–568

Jiang C, Li Y, Fang W (2005) cDNA cloning and prokaryotic expression of β-glucosidase in tea plant [*Camellia sinensis* (L.) O. Kutze]. Chinese J Agril Biotech 2:107–111

Jiang C, Wen Q, Chen Y, Xu LA, Huang MR (2011) Efficient extraction of RNA from various *Camellia* species rich in secondary metabolites for deep transcriptome sequencing and gene expression analysis. Afr J Biotech 10:16769–16773

Kato M, Kanehara T, Shimizu H, Suzuki Gillies FM, Crozier A, Ashihara H (1996) Caffeine biosynthesis in young leaves of *Camellia sinensis*: *in vitro* studies on *N*-methyltransferase activity involved in the conversion of xanthosine to caffeine. Physiol Plant 98:629–636

Kato M, Mizuno K, Crozier A, Fujimura T, Ashihara A (2000) Caffeine synthase gene from tea leaves. Nat 406:956–957

Kato M, Mizuno K, Fujimura T, Iwama M, Irie M, Crozier A, Ashihara A (1999) Purification and characterization of caffeine synthase from tea leaves. Plant Physiol 120:579–586

Kirita M, Honma D, Tanaka Y, Usui S, Shoji T, Sami M, Yokota T, Tagashira M, Muranaka A, Uchiyama M, Kanda T, Maeda-Yamamoto M (2010) Cloning of a novel O-methyltransferase from *Camellia sinensis* and synthesis of o-methylated EGCG and evaluation of their bioactivity. J Agric Food Chem 58:7196–7201

Krishnaraj T, Gajjeraman P, Palanisamy S, Chandrabose SRS, Mandal AKA (2011) Identification of differentially expressed genes in dormant (banjhi) bud of tea (*Camellia sinensis* (L.) O. Kuntze) using subtractive hybridization approach. Plant Physiol Biochem 49:565–571

Ku KM, Choi JN, Kim J, Kim JK, Yoo LG, Lee SJ, Hong YS, Lee CH (2010) Metabolomics analysis reveals the compositional differences of shade grown tea (*Camellia sinensis* L.). J Agric Food Chem 58:418–426

Lee JE, Lee BJ, Chung JO, Hwang JA, Lee SJ, Lee CH, Hong YS (2010) Geographical and climatic dependencies of green tea (*Camellia sinensis*) metabolites: A ^{1}H NMR-based metabolomics study. J Agri Food Chem 58:10582–10589

Lee JE, Lee BJ, Hwang JA, Ko KS, Chung JO, Kim EH, Lee SJ, Hong YS (2011) Metabolic dependence of green tea on plucking positions revisited: A metabolomic study. J Agri Food Chem 59:10579–10585

Li YH, Jiang CJ, Wan XC (2004a) Study on the expression of caffeine synthase gene mRNA in tea plant. J Tea Sci 24:23–28

Li YH, Jiang CJ, Yang SL, Yu YB (2004b) β-glucosidase cDNA cloning in the tea (*Camellia sinensis*) and its prokaryotic expression. J Agri Biotech 12:625–629

Li Y, Gu W, Ye S (2007) Expression and location of caffeine synthase in tea plants. Russ J Plant Physi 54:698–701

Li J, Chen J, Zhang Z, Pan Y (2008) Proteome analysis of tea pollen (*Camellia sinensis*) under different storage conditions. J Agric Food Chem 56:7535–7544

Li Q, Huang J, Liu S, Li J, Yang X, Liu Y, Liu Z (2011) Proteomic analysis of young leaves at three developmental stages in an albino tea cultivar. Proteome Sci 9:44–56

Liu Q, Tan X-F, Hu X-Y, Tian X-M (2010) Cloning and sequence analysis of a 14–3-3 protein gene from Camellia oleifera Acta Agri Univer 5:13–17

Li XW, Liu HJ, Xie SX, Yuan HY (2013) Isolation and characterization of two genes of the early light-induced proteins of *Camellia sinensis*. Photosynthe 51:305–311

Li XW, Feng ZG, Yang HM, Zhu XP, Liu J, Yuan HY (2010) A novel cold-regulated gene from *Camellia sinensis*, *CsCOR1*, enhances salt- and dehydration-tolerance in tobacco. Biochem Biophys Res Commun 394:354–359

Lin GZ, Lian YJ, Ryu JH, Sung MK, Park JS, Park HJ, Park BK, Shin JS, Lee MS, Cheon CL (2007) Expression and purification of His-tagged flavonol synthase of *Camellia sinensis* from *Escherichia coli*. Protein Exp Puri 55:287–292

Liu S, Han B (2010) Differential expression pattern of an acidic 9/13-lipoxygenase in flower opening and senescence and in leaf response to phloem feeders in the tea plant. BMC Plant Biol 10:228–238

Marshall A, Gollapudi S, de Silva JD, Hodgman C (2007) A systems biology approach to modelling tea (*Camellia sinensis*). BMC Sys Biol 1(Suppl. 1):13

Ma C, Qiao X, Chen L (2010) Cloning and expression analysis of leucoanthocyantin reducase gene of tea plant (*Camellia sinensis*). J Tea Sci. doi:CNKI:SUN :CYKK.0.2010–01-006

Ma C-L, Chen L, Wang X-C, Jin J-Q, Ma J-Q, Yao M-Z, Wang Z-L (2012) Differential expression analysis of different albescent stages of 'Anji Baicha' (*Camellia sinensis* (L.) O. Kuntze) using cDNA microarray. Scientia Hort 148:246–254

Matsumoto S, Takeuchi A, Hayatsu M (1994a) Chalcone synthase from *Camellia sinensis*: isolation of the cDNAs and the organ-specific and sugar-responsive expression of the genes. Plant Cell Physiol 35:1011–1018

Matsumoto S, Takeuchi A, Hayatsu M (1994b) Molecular cloning of phenylalanine ammonia-lyase cDNA and classification of varieties and cultivars of tea plants (*Camellia sinensis*) using the tea PAL cDNA probe. Theor Appl Genet 89:671–675

Mizutani M, Nakanishi H, Ema J, Ma SJ, Noguchi E, Inohara-Ochiai M, Fukuchi-Mizutani M, Nakao M, Sakata K (2002) Cloning of β-primeverosidase from tea leaves, a key enzyme in tea aroma formation. Plant Physiol 130:2164–2176

Mohanpuria P, Kumar V, Joshi R, Gulati A, Ahuja PS, Yadav SK (2009) Caffeine biosynthesis and degradation in tea (*Camellia sinensis* (L.) O. Kuntze) is under developmental and seasonal regulation. Mol Biotech 43(2):104–111

Mondal TK (2013) Application of highthroughput DNA sequencing for Plant Science research. In: Pullaiah T (ed) Abiotic Stress and Biotechnology. Abiotic stress and Biotechnology Regency Publications, New Delhi, pp 57–74

Mondal TK, Rana MK (2013) High throughput DNA sequencing and its implication in plant science research. Everyman Sci 1:15–25

Mondal TK, Sutoh K (2013) Application of Next-Generation Sequencing for Abiotic Stress Tolerance. In: Barh D, Zambare V, Azevedo V (eds) The world of omics and applications in plant and agricultural sciences. CRC Press, USA, pp. 383–402

Mukhopadhaya M M, Sarkar B B, Mondal TK (2013) Omics advances of tea (Calliame sinensis). In: Omics Applications in crop science (Ed Bhar D) CRC Pres New York pp 439–466

Murayama Y, Uematsu K, Koyama H, Morita A (2007) Detection of aluminium responsive genes in tea plant by hybridization with *Arabidopsis thaliana* cDNA. The 3rd Int Conf on O-Cha tea culture and science (ICOS) Shizuoka, Japan, 29 Feb 2008

Muoki RC, Paul A, Kumar S (2012) A shared response of *thaumatin like protein*, *chitinase*, and *late embryogenesis abundant protein3* to environmental stresses in tea [*Camellia sinensis (L.) O. Kuntze*]. Funct Integr Genom 12:565–571

Mueller LA, Zhang P, Rhee SY (2003) AraCyc: a biochemical pathway database for *Arabidopsis*. Plant Physiol 132:453–460

Pan L-L, Wang Y, Hu J-H, Ding Z-T, Li C (2013) Analysis of codon use features of stearoyl-acyl carrier protein desaturase gene in *Camellia sinensis*. J Theor Biol 334:80–86

Pang Y, Abeysinghe ISB, He J, He X, Huhman D, Mewan KM, Sumner LW, Yun J, Dixon RA (2013) Functional characterization of proanthocyanidin pathway enzymes from tea (*Camellia sinensis*) and their application for metabolic engineering. Plant Physiol 161:1103–1116

Park JS, Kim JB, Hahn BS, Kim KH, Ha SH, Kim JB, Kim YH (2004) EST analysis of genes involved in secondary metabolism in *Camellia sinensis* (tea), using suppression subtractive hybridization. Plant Sci 166:953–961

Paul A, Kumar S (2011) Responses to winter dormancy, temperature, and plant hormones share gene networks. Funct Integr Genom 11:659–664

Paul A, Kumar S (2013) Dehydrin2 is a stress-inducible, whereas Dehydrin1 is constitutively expressed but up-regulated gene under varied cues in tea [*Camellia sinensis (L.) O.* Kuntze]. Mol Biol Rep 40:3859–3863

Paul A, Lal L, Ahuja PS, Kumar S (2012a) Alpha-tubulin (CsTUA) up-regulated during winter dormancyis a low temperature inducible gene in tea [*Camellia sinensis* (L.) O. Kuntze]. Mol Biol Rep 39:3485–3490

Paul A, Muoki RC, Singh K, Kumar S (2012b) CsNAM-like protein encodes a nuclear localized protein and responds to varied cues in tea [*Camellia sinensis (L.) O. Kuntze*]. Gene 502:69–74

Phukon M, Namdev R, Deka D, Modi SK, Sen P (2012) Construction of cDNA library and preliminary analysis of expressed sequence tags from tea plant [*Camellia sinensis* (L) O. Kuntze]. Gene 506:202–206

Prabu GR, Mandal AKA (2010) Computational identification of miRNAs and their target genes from expressed sequence tags of tea (*Camellia sinensis*). Genom Proteom Bioinform 8:113–121

Prabu GR, Thirugnanasambantham K, Mandal AKA, Saravanan A (2012) Molecular cloning and characterization of nucleoside *diphosphate kinase* 1 cDNA in tea. Biol Planta 56(1):140–144

Punyasiri PA, Abeysinghe IS, Kumar V, Treutter D, Duy D, Gosch C, Martens S, Forkmann G, Fischer TC (2004) Flavonoid biosynthesis in the tea plant *Camellia sinensis*: properties of enzymes of the prominent epicatechin and catechin pathways. Arch Biochem Biophys 431:22–30

Qiao J, Zhang YT, Zhu XP (2011) Identification of genes up-regulated by *ectropic obliqua* feeding in tea seedlings. Acta Hort Sinica 38:783–789

Rana NK, Mohanpuria P, Yadav SK (2008a) Cloning and characterization of a cytosolic glutamine synthetase from *Camellia sinensis* (L.) O. Kuntze that is upregulated by ABA, SA, and H_2O_2. Mol Biotech 39:49–56

Rana NK, Mohanpuria P, Yadav SK (2008b) Expression of tea cytosolic glutamine synthetase is tissue specific and induced by cadmium and salt stress. Biol Planta 52:361–364

Rana NK, Mohanpuria P, Kumar V, Yadav SK (2010) A CsGS is regulated at transcriptional level during developmental stages and nitrogen utilization in *Camellia sinensis* (L.) O. Kuntze. Mol Biol Rep 37:703–710

Rani A, Singh K, Ahuja PS, Kumar S (2012) Molecular regulation of catechins biosynthesis in tea [*Camellia sinensis* (L.) O. Kuntze]. Gene 495:205–210

Rani A, Singh K, Sood P, Kumar S, Ahuja PS (2009) p-Coumarate: CoA ligase as a key gene in the yield of catechins in tea [*Camellia sinensis* (L.) O. Kuntze]. Funct Integr Genom 9:271–275

Roy SC, Chakraborty BN (2009) Cloning and sequencing of chitinase gene specific PCR amplified DNA fragment from tea plant (*Camellia sinensis*) and analyzed the nucleotide sequence using bioinformatics algorithms. Can J Pure Appl Sci 3:795–800

Sahu J, Sarmah R, Dehury B, Sarma K, Sahoo S, Sahu M, Barooah M, Modi MK, Sen P (2012) Mining for SSRs and FDMs from expressed sequence tags of *Camellia sinensis*. Bioinforma 8:260–266

Senthilkumar P, Thirugnanasambantham K, Mandal AKA (2012) Suppressive subtractive hybridization approach revealed differential expression of hypersensitive response and reactive oxygen species production genes in tea (*Camellia sinensis (L.) O. Kuntze*) leaves during *Pestalotiopsis thea* Infection. Appl Biochem Biotech 168:1917–1927

Sharma P, Kumar S (2005) Differential display-mediated identification of three drought-responsive expressed sequence tags in tea [*Camellia sinensis* (L.) O. Kuntze]. J Biosci 30:231–235

Shi Hï¼WangYï¼YangL-Cï¼DZ-T (2012) Analysis of codon bias of the cold regulated transcription factor *ICE1* in tea plant. Acta Hort Sin 39:1341–1352

Shi CY, Yang H, Wei CL, Yu O, Zhang ZZ, Jiang CJ, Sun J, Li YY, Chen Q, Xia T, Wan XC (2011) Deep sequencing of the *Camellia sinensis* transcriptome revealed candidate genes for major metabolic pathways of tea-specific compounds. BMC Genom 12:131–150

Singh K, Kumar S, Ahuja PS (2009d) Differential expression of histone H3 gene in tea (*Camellia sinensis* (L.) O. Kuntze) suggests its role in growing tissue. Mol Biol Rep 36:537–542

Singh K, Kumar S, Rani A, Gulati A, Ahuja PS (2009c) Phenylalanine ammonia-lyase (PAL) and cinnamate 4-hydroxylase (C4H) and catechins (flavan-3-ols) accumulation in tea. Funct Integr Genom 9:125–134

Singh K, Kumar S, Yadav SK, Ahuja PS (2009a) Characterization of dihydroflavonol 4-reductase cDNA in tea [*Camellia sinensis* (L.) O. Kuntze]. Plant Biotech Rep 3:95–101

Singh K, Paul A, Kumar S, Ahuja PS (2009e) Cloning and differential expression of *QM* like protein homologue from tea [*Camellia sinensis* (L.) O. Kuntze]. Mol Biol Rep 36:921–927

Singh K, Rani A, Kumar S, Sood P, Mahajan M, Yadav SK, Singh B, Ahuja PS (2008) An early gene of the flavonoid pathway, flavanone 3-hydroxylase, exhibits a positive relationship with the concentration of catechins in tea (*Camellia sinensis*). Tree Physiol 28:1349–1356

Singh K, Rani A, Paul A, Dutt S, Joshi R, Gulati A, Ahuja PS, Kumar S (2009b) Differential display mediated cloning of anthocyanidin reductase gene from tea (*Camellia sinensis*) and its relationship with the concentration of epicatechins. Tree Physiol 29:837–846

Singh K, Raizada J, Bhardwaj P, Ghawana S, Rani A, Singh H, Kaul K, Kumar S (2004) 26S rRNA-based internal control gene primer pair for reverse transcription-polymerase chain reaction-based quantitative expression studies in diverse plant species. Anal Biochem 335:330–333

Sugiyama T, Sadzuka Y (2003) Theanine and glutamate transporter inhibitors enhance the antitumor efficacy of chemotherapeutic agents. Biochem Biophys Acta 1653:47–59

Sugiyama T, Sadzuka Y (2004) Theanine, a specific glutamate derivative in green tea, reduces the adverse reactions of doxorubicin by changing the glutathione level. Canc Lett 212:177–184

Suzuki T (1972) The participation of *S*-adenosylmethionine in the biosynthesis of caffeine in the tea plants. FEBS Lett 24:18–20

SuXia X, Yuan Z, KaiLin G, Lei W, HongYu Y (2009) Cloning and expression analysis of a flavonol synthase gene from *Camellia sinensis*. Plant Physiol Commun 45:1093–1097

Takeo T (1981) Production of linalool and geraniol by hydrolytic breakdown of bound forms in disrupted tea shoots. Phytochem 120:2145–2147

Takechi A, Matsumoto S (2003) Effect of light on gene expression of catechin biosynthetic expression in tea (*Camellia sinensis*) leaves. Tea Res J 96:27–32

Takeuchi A, Matsumoto S, Hayatsu M (1994a) Chalcone synthase from *Camellia sinensis* isolation of the cDNAs and the organ-specific and sugar-responsive expression of the genes. Plant Cell Physiol 35:1011–1018

Takeuchi A, Matsumoto S, Hayatsu M (1994b) Amplification of β-tubulin cDNA from *Camellia sinensis* by PCR. Bull Natl Res Inst Veg Ornam Plants Tea 7:13–20

Taniguchi F, Tanaka J (2004) Characterization of genes for ammonium assimilation in *Camellia sinensis*. Proc of 2004 Int Con on O-Cha culture and Sci. Shizuaka, Japan, 217–218

Taniguchi F, Fukuoka H, Tanaka J (2012) Expressed sequence tags from organ-specific cDNA libraries of tea (*Camellia sinensis*) and polymorphisms and transferability of EST-SSRs across *Camellia* species. Breed Sci 62:186–195

Thirugnanasambantham K, Prabu G, Palanisamy S, Subhas Chandrabose SR, Mandal AKA (2013) Analysis of dormant bud (Banjhi) specific transcriptome of tea (*Camellia sinensis (L.) O. Kuntze*) from cDNA library revealed dormancy-related genes. Appl Biochem Biotechnol 169:1405–1417

Tomimoto Y, Ikehashi I, Kakeda K, Kowyama Y (1999) A pistal specific PR-1 like protein in Camellia, its expression, sequence and genealogical position. Breed Sci 49:97–104

Venkatesh P, Jaiprakash M, Prasad P, Pillai B, Sadhale PP, Sinkar VP (2006) Flavonoid biosynthesis in tea (*Camellia sinensis*). http://www.teascience.org/pdf/chapter-2–6-2.pdf

Waheed A, Hamid FS, Shah AH, Ahmad H, Khalid A, Abbasi FM, Ahmad N, Aslam S, Sarwar S (2012) Response of different tea (*Camellia sinensis* L.) clones against drought stress. J Master Environ Sci 3:395–410

Wang CX, Li YY, Jiang CJ, Yu YB (2005) Molecular cloning and sequence analysis on cDNA of cystatin gene from tea leaves. J Tea Sci 25:177–182

Wang L, Li X, Zhao Q, Jing S, Chen S, Yuan H (2009) Identification of genes induced in response to low-temperature treatment in tea leaves. Plant Mol Biol Rep 27:257–265

Wang X-C,Yang YJ,Chen L,Ma CL,Yao MZ (2010) Construction and preliminary analysis of the suppression subtractive hybridization cDNA libraries between dormant and sprouting buds of tea plant (*Camellia sinensis*). J Tea Sci 30:129–135

Wang P, Su R, Zheng J, Cheng S, Zhu G (2011a) Molecular cloning and phylogenetic analysis of a novel BURP domain-containing gene from *Camellia sinensis*. Afr J Biotech 10:15470–15476

Wang X, Ma C, Yang C, Yao M, Jin J (2011b) Cloning and expression analysis of auxin-repressed protein gene CsARP1 in tea plant (*Camellia sinensis*). J Nuclear Agric Sci. doi:CNKI:SUN:HNXB.0.2011–05-014

Wang X, Ma C, Yang Y, Jin J, Ma J, Cao H (2012a) cDNA cloning and expression analysis of cyclindependent kinase (CsCDK) gene in tea plant. Acta Hort Sinica 39:333–342

Wang X, Yang Y, Ma C, Jin J, Ma J, Cao H (2011c) Cloning and expression analysis of cyclin gene (*CsCYC1*) of tea plant. Acta Botanica Boreali-Occidentalia Sinica 31:2365–2372

Wang Y, Jiang C, Li Y, Wei C, Deng W (2012b) CsICE1 and CsCBF1:two transcription factors involved in cold responses in *Camellia sinensis*. Plant Cell Rep 31:27–34

Wang YS, Gao LP, Shan Y, Liu YJ, Tian YW, Xia T (2012) Influence of shade on flavonoid biosynthesis in tea (*Camellia sinensis (L.) O.* Kuntze). Scientia Hort 141:7–16

Wang YS, Gao LP, Wang ZR, Liu YJ, Sun M, Zeng W, Yuan H (2012c) Light induced expression of genes involved in phenylpropanoid biosynthetic pathways in callus of tea (*Camellia sinensis* (L.) O.Kuntze). Sci Hort 133:72–83

Wang P, Song P, Li X, Su R, Wang H, Zhu G (2012d) Study on soluble expression of glutamate dehydrogenase from tea plant in *Escherichia coli* using fusion tags. Afr J Biotech 11:6241–6250

Wang X-C, Zhao Q-Y, Ma C-L, Zhang Z-H, Cao H-L, Kong Y-M, Yue C, Hao X-Y, Chen L, Ma J-Q, Jin J-Q, Li X, Yang Y-J (2013) Global transcriptome profiles of *Camellia sinensis* during cold acclimation. BMC Genomics 2013 14:415

Wei CL, Jiang CJ, Tao HZ, Wan XC (2003) Cloning and bioinformatics analysis of sequence signature of violaxanthin de-epoxidase cDNA in tea plant (*Camellia sinensis* (L.) O. Kuntze). J Nanj Agri Univ 26:14–19

Wei CL, Jiang CJ, Tao HZ, Wan XC (2004) Site-directed mutation of violaxanthin de-epoxidase from tea plant (*Camellia sinensis*) *in vitro* and expression of bioactivity assay of the mutants. Chinese J Biochem Mol Biol 20:73–78

Wei K, Wang L, Cheng H, Zhang C, Ma C, Zhang L, Gong W, Wu L (2013) Identification of genes involved in indole-3-butyric acid-induced adventitious root formation in nodal cuttings of *Camellia sinensis* (L.) by suppression subtractive hybridization. Gene 514:91–98

Wu YL, Pan LP, Yu SL, Li HH (2010) Cloning, microbial expression and structure-activity relationship of polyphenol oxidases from *Camellia sinensis*. J Biotech 145:66–72

Wu H, Chen D, Li J, Yu B, Qiao X, Huang H, He Y (2013) *De novo* characterization of leaf transcriptome using 454 sequencing and development of EST-SSR markers in tea (*Camellia sinensis*). Plant Mol Biol Rep 31:524–538

Yadav SK (2009) Computational structural analysis and kinetic studies of a cytosolic glutamine synthetase from *Camellia sinensis* (L.) O. Kuntze. Protein J 28:428–434

Yang H, Xie S, Wang L, Jing S, Zhu X, Li X, Zeng W, Yuan H (2011) Identification of up-regulated genes in tea leaves under mild infestation of green leafhopper. Sci Hort 130:476–448

Yang D, Liu Y, Suna M, Zhao L, Wang Y, Chen X, Wei C, Gao L, Xia T (2012a) Differential gene expression in tea (*Camellia sinensis L.*) calli with different morphologies and catechin contents. J Plant Physiol 169:163–175

Yang YJ, Wang XC, Ma CL (2012b) Cloning and bioinformatics analysis of full-length cDNA of actin gene (*CsActin1*) from tea plant (*Camellia sinensis* (L.) O. Kuntze). Bull Bot Res 32:69–76

Yoshida K, Homma T (2005) Isolation of wound/pathogen inducible cDNA from tea by mRNA differential display. Inter Tea Symp-2005 62–65 (Hanzou, China)

Ye Y, Jiang CJ, Zhu L, Yu M, Wang ZX, Deng WW, Wei CL (2009) Cloning and sequencing of a full length cDNA encoding the RuBPCase small subunit (RbcS) in tea (*Camellia sinensis*). Agric Sci China 8:161–166

Zhao D, Liu ZS, Xi B (2001) Cloning and alignment of polyphenols oxidase cDNA of tea plant. J Tea Sci 21:94–98

Zhao LP, Chen L, Wang XC, Yao MZ (2006a) Quantitative detection of β-glucosidase and β-primeveroside gene expressions in different leaves of tea plant (*Camellia sinensis*) by real-time PCR analysis. J Tea Sci 26:11–16

Zhao LP, Gao QK, Chen L, Wang XC, Yao MZ (2006b) Development and preliminary application of cDNA microarray of tea plant (Camellia sinensis). J Tea Sci 5:3–7

Zhao L, Gao L, Wang H, Chen X, Wang Y, Yang H, Wei C, Wan X, Xia T (2012) The R2R3-MYB, bHLH, WD40, and related transcription factors in flavonoid biosynthesis. Funct Integr Genom. 13:75–98

Zhang YL, Zhao LP, MA CL, Chen L (2007) Molecular identification, bioinformatic analysis and prokaryotic expression of the cyclophilin gene full length cDNA from tea plant (*Camellia sinensis*). J Tea Sci 27:120–126

Zhang YL, Qiao XY, Chen L (2008a) Full-length cDNA cloning and bioinformatic analysis of ACC synthase gene from the tea plant (*Camellia sinensis*). J Tea Sci 28:235–241

Zhang YL, Qiao XY, Chen L (2008b) Molecular cloning and expression analysis of ACC oxidase gene full-length cDNA from the tea plant (*Camellia sinensis*). J Tea Sci 28:459–466

Zhang X, Liu Y, Gao K, Zhao L, Liu L, Wang Y, Sun M, Gao L, Xia T (2012) Characterization of anthocyanidin reductase from Shuchazao green tea. J Sci Food Agric 92:1533–1539

Zhu L, Deng WW, Ye AH, Yu M, Wang ZX, Jiang CJ (2008a) Cloning of two cDNAs encoding a family of ATP sulfurylase from *Camellia sinensis* related to selenium or sulfur metabolism and functional expression in *Escherichia coli*. Plant Physiol Biochem 46:731–738

Zhu L, Jiang CJ, Deng WW, Gao X, Wang RJ, Wan XC (2008b) Cloning and expression of selenocysteine methyltransferase cDNA from *Camellia sinensis*. Acta Physiol Plnt 30:167–174

The manufacturer's authorised representative in the EU is Springer Nature Customer Service Centre GmbH, Europaplatz 3, 69115 Heidelberg, Germany. If you have any concerns regarding our products, please contact ProductSafety@springernature.com

Printed and bound by CPI Group (UK) Ltd, Croydon, CR0 4YY
22/09/2026
02221057-0020